Transport and Transfer Processes in Plants

ACADEMIC PRESS RAPID MANUSCRIPT REPRODUCTION

Transport and Transfer Processes in Plants

Proceedings of a Symposium held under the
auspices of the U.S. — Australia Agreement
for Scientific and Technical Cooperation
Canberra, Australia.   December 1975.

Edited by

I.F. Wardlaw and J.B. Passioura

CSIRO Division of Plant Industry, Canberra,
Australia

1976

Academic Press
New York   San Francisco   London

A SUBSIDIARY OF HARCOURT BRACE JOVANOVICH, PUBLISHERS

ACADEMIC PRESS, INC.
111 Fifth Avenue, New York, New York 10003

United Kingdom Edition published by
ACADEMIC PRESS, INC. (LONDON) LTD.
24/28 Oval Road, London NW1

Library of Congress Cataloging in Publication Data

Main entry under title:

Transport and transfer processes in plants.

    "Proceedings of a symposium held under the auspices of
the U.S.-Australia Agreement for Scientific and Technical
Cooperation, Canberra, Australia, December 1975."
    1.   Plant translocation—Congresses.   I.   Wardlaw,
I.   F.   II.   Passioura, J. B.
QK871.T72   581.1'1   76-27761
ISBN 0-12-734850-6

Contents

LONG DISTANCE TRANSPORT

Participants

CONTRIBUTORS (J — joint author not in attendance)

W. G. Allaway, School of Biological Sciences, University of Sydney, Sydney, N.S.W. 2006, Australia

W. P. Anderson, Research School of Biological Sciences, Australian National University, Canberra, A.C.T. 2600 Australia

R. L. Bieleski, Plant Diseases Division, D.S.I.R., Auckland, New Zealand

S. Bullivant, (J), Department of Cell Biology, University of Auckland, Auckland, New Zealand

A. L. Christy, Agricultural Research Department, Monsanto Agricultural Products Co., St. Louis, Missouri 63166

M. G. Cook, C.S.I.R.O. Division of Plant Industry, Canberra A.C.T. 2601, Australia

W. J. Cram, School of Biological Sciences, University of Sydney, Sydney, N.S.W. 2006, Australia

G. P. Dempsey, (J), Plant Diseases Division, DSIR, Auckland, New Zealand

J. Dunlop, Grasslands Division, DSIR, Palmerston North, New Zealand

L. T. Evans, CSIRO Division of Plant Industry, Canberra A.C.T. 2601, Australia

R. J. Field, (J), Department of Plant Science, Lincoln College, Canterbury, New Zealand

D. B. Fisher, Botany Department, University of Georgia, Athens, Georgia 30602

D. R. Geiger, Department of Biology, University of Dayton, Dayton, Ohio 45409

A. D. M. Glass, Botany and Zoology Department, Massey University, Palmerston North, New Zealand

B. E. S. Gunning, Research School of Biological Sciences, Australian National University, Canberra A.C.T. 2600, Australia

P. W. Hattersley, Research School of Biological Sciences, Australian National University, Canberra A.C.T. 2600, Australia

Katie Helms, CSIRO Division of Plant Industry, Canberra A.C.T. 2601, Australia

J. Hoddinott, (J), Department of Botany, University of British Columbia, Vancouver, B.C., Canada

P. Jarvis, Lincoln College, Canterbury, New Zealand

C. F. Jenner, Waite Agricultural Research Institute, University of Adelaide, Glen Osmond, S.A. 5064, Australia

M. G. K. Jones, Research School of Biological Sciences, Australian National University, Canberra, A.C.T. 2600, Australia

M. R. Kaufmann, Department of Plant Sciences, University of California, Riverside, California 92502

R. W. King, CSIRO Division of Plant Industry, Canberra, A.C.T. 2601, Australia

P. E. Kriedemann, CSIRO Division of Horticultural Research, Merbein, Victoria 3505, Australia

A. Läuchli, Fachbereich Biologie (10), Der Technischen Hochschule, D-6100, Darmstadt, Germany

J. F. Loneragan, School of Environmental and Life Sciences, Murdoch University, Murdoch, W.A. 6153, Australia

B. R. Loveys, (J), CSIRO Division of Horticultural Research, Adelaide, South Australia 5001, Australia

I. A. Newman, Department of Physics, University of Tasmania, Hobart, Tasmania 7001, Australia

T. P. O'Brien, Botany Department, Monash University, Clayton, Victoria 3168, Australia

C. B. Osmond, Research School of Biological Sciences, Australian National University, Canberra A.C.T. 2600, Australia

J. B. Passioura, CSIRO Division of Plant Industry, Canberra A.C.T. 2601, Australia

J. S. Pate, Department of Botany, The University of Western Australia, Nedlands, W.A. 6009, Australia

J. W. Patrick, Department of Biological Sciences, The University of Newcastle, N.S.W. 2308, Australia

M. G. Pitman, School of Biological Sciences, Sydney University, Sydney, N.S.W. 2006, Australia

J. Possingham, (J), CSIRO Division of Horticultural Research, Adelaide, South Australia 5001, Australia

K. Raschke, MSU/ERDA Plant Research Laboratory, Michigan State University, East Lansing, Michigan 48824

A. W. Robards, (J), Department of Biology, University of York Heslington, York Y01 5DD, U.K.

A. D. Robson, (J), Institute of Agriculture, University of Western Australia, Nedlands, W.A. 6009, Australia

M. Satoh, (J), Sericultural Experiment Station, Suginami-Ku, Tokyo, Japan

J. W. Sij, (J), Texas A.&M. University, Agricultural Research and Extension Centre, Beaumont, Texas 77706

K. Snowball, (J), Institute of Agriculture, University of Western Australia, Nedlands, W.A. 6009, Australia

J. K. Sullivan, (J), Physics Department, University of Tasmania, Hobart, Tasmania 7001, Australia

C. A. Swanson, Department of Botany, Ohio State University, Columbus, Ohio 43201

B. Tomkins, (J), Grasslands Division, DSIR Palmerston North, New Zealand

J. H. Troughton, Physics and Engineering Laboratory, DSIR Lower Hutt, New Zealand

N. A. Walker, School of Biological Sciences, University of Sydney, Sydney, N.S.W. 2006, Australia

I. F. Wardlaw, CSIRO Division of Plant Industry, Canberra A.C.T. 2601, Australia

L. Watson, Research School of Biological Sciences, Australian National University, Canberra A.C.T. 2600, Australia

R. E. Williamson, Department of Botany, La Trobe, University, Bundoora, Victoria 3083, Australia

M. H. Zimmermann, Harvard University Forest, Petersham, Massachusetts 01366

SESSION CHAIRMEN (C — indicates contributor)

R. L. Bieleski (C)

D. B. Fisher (C)

B. E. S. Gunning (C)

F. L. Milthorpe, School of Biological Sciences, Macquarie University, North Ryde, N.S.W. 2113, Australia

T. F. Neales, Botany Department, University of Melbourne, Parkville, Victoria 3052, Australia

L. G. Paleg, Waite Agricultural Research Institute, University of Adelaide, Glen Osmond, S.A. 5064, Australia

R. N. Robertson, Research School of Biological Sciences, Australian National University, Canberra, A.C.T. 2600, Australia

C. A. Swanson (C)

OTHER PARTICIPANTS

E. Bachelard, Department of Forestry, Australian National University, Canberra, A.C.T. 2600, Australia

D. Bagnall, CSIRO Division of Plant Industry, Canberra, A.C.T. 2601, Australia

E. W. R. Barlow, School of Biological Sciences, Macquarie University, North Ryde, N.S.W. 2113, Australia

H. Beevers, Thimann Laboratories, University of California, Santa Cruz, California 95064

J. E. Begg, CSIRO Division of Plant Industry, Canberra, A.C.T. 2601, Australia

E. G. Bollard, Plant Diseases Division, DSIR, Auckland, New Zealand

B. T. Brown, CSIRO Division of Plant Industry, Canberra, A.C.T. 2601, Australia

D. J. Carr, Research School of Biological Sciences, Australian National University, Canberra, A.C.T. 2600, Australia

H. Chujo, Osaka University, Japan

I. Cowan, Research School of Biological Sciences, Australian National University, Canberra, A.C.T. 2600, Australia

M. Dalling, School of Agriculture, University of Melbourne, Parkville, Victoria 3052, Australia

A. Davies, Waite Agricultural Research Institute, University of Adelaide, Glen Osmond, S.A. 5069, Australia

G. A. Drake, Research School of Biological Sciences, Australian National University, Canberra, A.C.T. 2600, Australia

R. L. Dunstone, CSIRO Division of Plant Industry, Canberra, A.C.T. 2601, Australia

I. Ferris, N.S.W. Department of Agriculture, N.S.W., Australia

R. M. Gifford, CSIRO Division of Plant Industry, Canberra, A.C.T. 2601, Australia

D. J. Goodchild, CSIRO Division of Plant Industry, Canberra, A.C.T. 2601, Australia

H. G. Jones, Plant Breeding Institute, Trumpington, Cambridge, CB2 2L9, United Kingdom

G. F. Katekar, CSIRO Division of Plant Industry, Canberra, A.C.T. 2601, Australia

A. Lang, CSIRO Division of Plant Industry, Canberra, A.C.T. 2601, Australia

G. H. Lorimer, Research School of Biological Sciences, Australian National University, Canberra, A.C.T. 2600, Australia

P. F. Lumley, Botany Department, Monash University, Clayton, Victoria 3168, Australia

R. C. McDonald, Department of Biology, University of Newcastle, N.S.W. 2308, Australia

I. B. McNulty, University of Utah, Salt Lake City, Utah

C. P. Meyer, Botany Department, University of Melbourne, Parkville, Victoria 3052, Australia

P. A. Morrow, Research School of Biological Sciences, Australian National University, Canberra, A.C.T. 2600, Australia

R. Munns, School of Biological Sciences, Macquarie University, North Ryde, N.S.W. 2113, Australia

D. Parkes, Botany Department, Monash University, Clayton, Victoria 3168, Australia

D. Vince-Prue, University of Reading, U.K.

P. J. Randall, CSIRO Division of Plant Industry, Canberra, A.C.T. 2601, Australia

R. J. Reid, School of Biological Sciences, Sydney University, Sydney, N.S.W. 2006, Australia

G. R. Robinson, Department of Physics, University of Tasmania, Hobart, Tasmania 7001, Australia

N. Schaefer, School of Biological Sciences, Sydney University, Sydney, N.S.W. 2006, Australia

R. F. M. Van Steveninck, Botany Department, University of Queensland, St. Lucia, Queensland 4067, Australia

R. Sward, Botany Department, Monash University, Clayton, Victoria 3168, Australia

N. C. Turner, CSIRO Division of Plant Industry, Canberra, A.C.T. 2601, Australia

Preface

A knowledge of the manner in which organic materials and nutrients are distributed in plants is important for an understanding of the limits to harvestable yield, how environmental factors regulate growth, how nutrients may be used more efficiently, and how diseases spread in plants. To this list can also be added the effectiveness of substances applied to plants such as growth regulators, herbicides, and insecticides.

There has been considerable progress in recent years in a wide range of interrelated disciplines, including salt uptake in cells, water relations, nutrition and translocation, as well as cell to cell transfer processes and plant anatomy. The provisions of the U.S.–Australia Science Agreement made it possible to bring together scientists, working in all of these areas, to a conference and workshop designed to examine how the various forms of both long- and short distance transport operate and how they might interact in the whole plant. By the end of the conference and workshop it was evident that although there has been considerable clarification of many of the facets of transport in plants, many of the basic processes, including the exact nature of the underlying mechanisms, still need to be resolved. However, on a more optimistic note, it is probably correct to say that the techniques are now available to resolve many of the current differences of opinion. The detailed discussion and exchange of ideas that occur at a symposium of this nature are difficult to bring into print. However we are extremely grateful to the chairmen, who at the end of the conference so ably summed up the main ideas from the papers and discussions in each session, and it is these presentations that form the basis of the three short summaries presented at the end of each section. It is hoped that in publishing the proceedings of this meeting, there will be some encouragement for researchers to look at the wider implications of experimental findings relating to specific transport processes.

We are indebted to both the United States Government, through the National Science Foundation, and the Australian Government, through the Department of Science, for their help in arranging this conference and in providing funding for seven participants from the U.S. and twenty from Australia. Also we are indebted to CSIRO for their support, practical help and provision of facilities for the conference. We are particularly grateful to Dr. L. T. Evans, Chief of the Division of Plant Industry, whose initial encouragement and continued support as a member of the organizing committee, played a large part in the success of the proceedings, and also to Professors D. R. Geiger and D. B. Fisher for their part of the organization in America. Finally we would like to express our appreciation of the excellent help we received from many individuals in the Division of Plant Industry. In particular we would like to recognize the efforts of Mrs. V. Ronning in dealing with the correspondence and the reorganization of many of the manuscripts; Mr. T. Buchwald for his patient reading of proofs and the typists Mrs. V. Taylor, Mrs. S. Kelo, Mrs. P. French, Mrs. D. McCann, Mrs. C. Stokes and Mrs. P. Dawson for their team effort in producing the camera-ready copy for this book.

I. F. Wardlaw and
J. B. Passioura

Transport and Transfer Processes in Plants

CHAPTER 1

Transport and Distribution in Plants

L.T. Evans

*CSIRO Division of Plant Industry, Canberra, A.C.T.,
Australia*

INTRODUCTION

The development of a vascular system was a crucial step
in the evolution of plants, and its presence or absence
constitutes a major division of the plant kingdom. The
increase in mechanical strength provided by the tracheary
cells allowed the shoots of plants to be raised into the aerial environment, and thereby made possible the development of
more complex plant canopies and the accumulation of far more
terrestrial biomass. Along with the increased spatial separation of organs went their greater specialization of function,
for absorption, support, assimilation or reproduction. And
with this specialization went an increasing need for efficient
intercommunication between organs and for the integration of
their growth.

The vascular tissue had, therefore, to serve the somewhat
conflicting functions of providing both mechanical strength
and channels for the rapid transport of a wide variety of
substances from one part of the plant to another. The dual
requirements for strength and for relatively open and freely
communicating channels throughout the plant were achieved by
the differentiation of fibres and vessels. In the same way,
the upward flows of water, minerals and substances elaborated
in the roots (such as reduced nitrogen compounds and cytokinins) were separated from the flow of assimilates and other
compounds from the leaves, in the xylem and phloem respectively. Different as these two tissues are, not only in their
predominant direction of transport but also in conduit
diameter, end walls, the role of membranes and other features,
they nevertheless interact and overlap in function, for
example, in the transport of gibberellins, cytokinins and
abscisic acid (cf. King, this volume).

1

Along with the spatial and functional separation of
organs there developed a need for a variety of hormones to
integrate plant growth, to balance roots and shoots, stems
and leaves. The vascular system makes possible their action
away from their site of synthesis, but is also the Trojan
Horse for movement of herbicides and viruses through plants.
The transport mechanisms must cope, therefore, with a wide
variety of substances.

Although the mechanisms of transport may be common to
all plants, the patterns of distribution differ greatly, as
evident from the enormous range of plant form and growth
habit. Shifts in the patterns of assimilate distribution have
not only contributed to the evolution of wild plants but have
played a major role in the domestication and improvement of
crop plants, as may readily be appreciated from comparisons
of modern cultivars of maize, wheat or sunflower with their
wild relatives, or by considering, as Darwin did, the variety
of form among the cabbage-kale-collard-kohlrabi-Brussels
sprout-broccoli-cauliflower complex.

Thus, the topic of this symposium is crucial to an
understanding of both evolution in the plant kingdom and the
improvement of agricultural productivity. Small wonder, then,
that evolutionists of earlier generations, such as F.O. Bower,
paid so much attention to the vascular system. Their modern
counterparts seem less interested: the index of Stebbins'
"Variation and Evolution in Plants" lists *Phleum* but not
phloem.

FORBEARS, OPINIONS AND CALCULATIONS

Exactly 250 years ago Stephen Hales was observing his
Experiment XLVI, in which he grafted a holly to an oak. He
concluded "if there were a free, uniform circulation of the
sap through the oak and ilex why should the leaves of the oak
fall in winter, and not those of the ilex?". It was a perceptive question, even though we cannot accept his conclusion
that this experiment "affords a considerable argument against
circulation" in plants, the idea developed 50 years earlier
by Malpighi following Harvey's demonstration of circulation
in animals.

About 50 years ago H.H.Dixon developed his cohesion
theory for the ascent of sap in trees. Although his proposed
mechanism was for a time discounted by many as too simple –
a distinction it shares with Münch's pressure flow of assimilates – it is now widely accepted. Dixon also did some
calculations on the movement of assimilates into a potato
tuber, from which he estimated their speed at 50 cm h^{-1}. We
now know that this speed is well within the range of those

measured, but Dixon concluded that such fast movement could not possibly occur in the phloem with its small elements and viscous contents.

In fact, the literature on long distance transport is studded with conclusions by our forebears, even the most perceptive of them, that some process is impossible, although subsequently found not to be so. In many cases these wrong conclusions derived from mistaken notions of phloem structure and function, and we are still crippled by uncertainties in many of the assumptions used in contemporary calculations and models of phloem transport. These derive not only from inadequate understanding of phloem properties such as the permeability of the sieve tube membranes but also, as Weatherley and Johnson (1968) point out, because we are unsure of the physical principles governing flow in such small, complex, irregular conduits.

PHLOEM STRUCTURE

Hartig first described sieve tubes in 1837, and presented a comprehensive analysis of their function in his book published 100 years ago. However, there is still no agreement on the answers to the following questions, all relevant to the proposed mechanisms of translocation.

Are the sieve plate pores open or occluded in functioning phloem? - Their possible occlusion, as revealed in many early electron microscope studies, has been a major stumbling block to the acceptance of mass flow. More recent work (eg. Dempsey et al. this volume) suggests that the pores may be open in functioning phloem, but this is not universally agreed (cf. Peel 1974). Allied to this question are two others: *Do transcellular strands traverse the pores?* and *Are the lumina of the sieve elements free enough for rapid mass flow, or occluded by P-protein?* - (cf. Weatherley and Johnson 1968; de Maria and Thaine 1974).

Is P-protein ubiquitous in functional phloem, and is it contractile? - P-protein appears to be abundant in the phloem of some families, such as the Cucurbitaceae, but sparse and mainly parietal in others, such as the Gramineae. In two of these, maize and barley, it appears to be completely absent (Evert et al. 1971; Singh and Srivastava 1972), which poses difficulties for mechanisms based on its presence. Whether or not the P-protein filaments are contractile is an open question (cf. MacRobbie 1971).

Another unresolved question is *Whether the young or the mature sieve elements are functional in translocation?* Most investigators believe it is the mature sieve elements with

3

relatively clear lumina, but Schumacher and Kollmann have asserted that young sieve elements are functional in *Metasequoia* (cf. Kollmann and Dörr 1966).

What role do companion cells play in translocation? - If they are essential, is their function taken over by the albuminous cells in gymnosperms?.

These are not the only debatable questions concerning phloem structure, but they suffice to underline our uncertainties as to how the phloem functions. If translocation is by mass flow, why is the structure of the phloem so different from that of the xylem? Its requirement for a semi-permeable membrane is evident, and its smaller vessel diameter is to be expected in view of rates of flow probably an order of magnitude less than those in the xylem. But why are there sieve plates in the phloem if only pressure flow is involved? Zimmermann (1974) answers on behalf of Münch that they provide sites which can be plugged in the event of injury, while MacRobbie (1971) suggests that they may provide an anchor for P-protein filaments.

PHLOEM MECHANISMS

We have seen that neither anatomical observations nor model calculations can be used with confidence to differentiate between the mechanisms of translocation that have been proposed. Are there more objectively answerable questions which can do so? Figure 1 presents one of several possible sets of such questions.

1. *Do the various solutes move together, in bulk flow?* - Several different substances appear to move together in the phloem, at similar speed and with similar distribution patterns, such as ^{32}P-labelled phosphate and ^{14}C-labelled assimilates in wheat (Marshall and Wardlaw 1973), and the floral stimulus and assimilates in short day plants (King et al. 1968). However, the evidence for togetherness in bulk flow is weakened by the absence of any for associated movement of water, which could be due to technical difficulties (cf. Cataldo et al. 1972).

2. *Is there bidirectional flow within a sieve tube?* - In plants with bicollateral bundles, such as the tomato, movement of ^{14}C-assimilates may be upwards in the internal phloem and downwards in the external phloem (Bonnemain 1965). The very evolution of such an arrangement could be taken to imply that simultaneous bidirectional flow within a single file of sieve elements is not easy in mature phloem. The most convincing evidence for such bidirectional flow is that of Trip and Gorham (1968), but this was in immature leaves.

4

Fig. 1. – Key questions for the mechanism of translocation

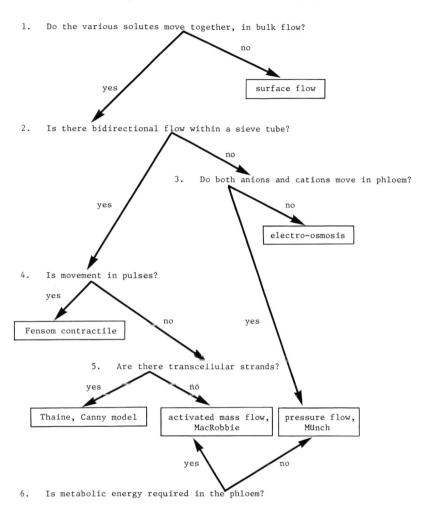

1. Do the various solutes move together, in bulk flow?

    no

    yes

    surface flow

2. Is there bidirectional flow within a sieve tube?

    no

    3. Do both anions and cations move in phloem?

    yes   no

    electro-osmosis

4. Is movement in pulses?

    yes

    Fensom contractile   no   yes

    5. Are there transcellular strands?

    yes   no

    Thaine, Canny model   activated mass flow,   pressure flow,
                           MacRobbie   Münch

    yes   no

6. Is metabolic energy required in the phloem?

3. *Do both anions and cations move in the phloem?* –
There are several questions capable of differentiating between
electro-osmosis and Münch pressure flow, as question 6 does,
but MacRobbie (1971) believes that the mobility of both anions
and cations together in the sieve tubes is argument enough
against electro-osmosis.

4. *Is movement in pulses?* – Fensom (1972) reported
bidirectional movement of pulses of ^{14}C and [42]K in the phloem
of *Heracleum* and interpreted these as due to the action of
contractile proteins. Both the pulses and the bidirectional

movement may, however, have been an artifact of the micro-
cannulation technique used. It is difficult otherwise to
see how the pulses could be so sharp and so far apart (5-10
cm).

5. *Are there transcellular strands?* - The mechanisms
proposed by Thaine and Canny depend on transcellular strands,
whereas that proposed by MacRobbie (1971), involving mass
flow helped along by the contractions of oriented protein
fibrils, does not.

6. *Is metabolic energy required in the phloem?* - This
question is extremely important, since it could differentiate
surface flow and Münch pressure flow from most other proposed
mechanisms, but the answer is uncertain. Partly, this is
due to our inability to draw a sharp distinction between the
metabolic energy needed to maintain the living tissues of
the phloem and that required to power translocation. Estimates of phloem respiration rates are equivocal, and there are
problems of interpretation with the use of respiratory
inhibitors such as cyanide. Anoxia may slow translocation
initially, but it soon recovers (Sij and Swanson 1973).
Chilling has a similar effect (eg. Geiger 1969; Wardlaw 1972;
Giaquinta and Geiger 1973; Lang 1974). Recovery of translocation may occur even when cyclosis remains inhibited
(Swanson and Geiger 1967). On balance the evidence is
suggestive, at the least, that assimilate translocation is
not powered by phloem respiration.

I shall refrain from drawing a conclusion about the most
likely mechanism, in view of the uncertainty of our answers
to each of these questions. Instead, I would ask why there
should not be more than one mechanism operative. Only our
intellectual combativeness demands a unique mechanism.
Thermal diffusion and Münch-type flow are almost inevitable,
as is some transfer by cytoplasmic streaming in at least
young sieve tubes. Fensom (1972) has suggested that several
mechanisms may be involved, and Lush and Evans (1974b) also
present evidence for several components in assimilate translocation.

A better understanding of the mechanism involved might
help us assess the extent to which the transport system in
the phloem limits crop yields. Photosynthetic source or
storage sink are usually examined for the limitations they
impose on yield development, rather than the capacity of the
pathway between them. The relative constancy of the rate of
assimilate transfer through phloem (10-20 pmoles $cm^{-2}s^{-1}$) in
most of the examples listed by Canny led him to suggest that
phloem capacity may limit growth and productivity. However,
far higher rates have been observed, in the flower stalk of

the palm *Arenga* (Tammes 1952), in the leaves of C_4 grasses
(Lush and Evans 1974a), and in the wheat seedling root
(Passioura and Ashford 1974). Severing part of the phloem
may cause only a temporary fall in translocation rate (Wardlaw
and Moncur 1976). Similarly, the translocation of assimilates
from leaves seems capable of increasing in proportion to
photosynthesis (Lush and Evans 1974a; Servaites and Geiger
1974). As photosynthesis increases, so does the speed of
translocation (Troughton et al. 1974; Christy (this volume),
Troughton (this volume)), and likewise as the demand for
assimilate increases (Wardlaw and Moncur 1976).

On present evidence, therefore, phloem capacity seems
unlikely to limit plant growth and development. The close
relation between the area of phloem in the peduncle and the
amount of assimilate imported through it in wheats from all
stages of the evolution of that crop (Evans et al. 1970)
need not suggest a limitation in phloem capacity. Rather,
an optimization phenomenon might be involved. For example,
the advantage to Münch flow of higher sucrose concentrations
in the phloem may be counterbalanced beyond a certain point by
the greater viscosity of the solution.

SHORT DISTANCE TRANSFER

A variety of mechanisms is involved in short distance
transfer. Much cell to cell transfer is probably via plasmodesmatal connections as Tyree proposed. Thermal diffusion
through the plasmodesmata may be adequate, as in the movement of C_4-dicarboxylic acids from mesophyll to bundle sheath
cells in C_4 plants (Hatch and Osmond 1976). However, the
rate of transfer through plasmodesmata may be greatly
augmented by cytoplasmic streaming within the cells, as
Walker (this volume) shows. The movement of sugars from the
sites of photosynthesis to the phloem is probably by diffusion
accelerated by cytoplasmic streaming. Transfer via membranes
may occur where greater selectivity is needed, and transfer
cells may be differentiated at sites of high flux. Some
membrane fluxes may be mediated by reciprocating rearrange-
ments of membrane components, others by active pump mechanisms
which may be driven by respiratory energy, or by light in the
case of the influx of potassium ions into the guard cells of
stomata or the pulvinules of *Albizia* leaves. Loading of
sugars into the phloem is probably against a concentration
gradient, at least in C_3 plants, and requires metabolic
support energy, as Roeckl (1949) first pointed out. Sovonick
et al. (1974) concluded that only 0.3% of the ATP produced is
required for this loading process in sugar beet, which may

nevertheless limit assimilate export when photosynthesis is rapid (Geiger, this volume).

Active loading may be accompanied by sucrose hydrolysis and resynthesis (Porter and May 1955; Ford and Peel 1967), so it is of interest that hydrolysis is not required to get sucrose into the phloem of sugar cane (Hatch and Glasziou 1964). The Kranz anatomy of C_4 plants may render active loading into the phloem from the bundle sheath cells unnecessary (Lush and Evans (1974a).

The unloading of assimilates from the phloem has been little studied and far more work is needed on unloading at the various kinds of sink organs, meristematic, elongating, maintaining or storing soluble or insoluble compounds. Active unloading may not be needed at any of them.

The phloem contents reaching the vascular termini are probably more concentrated than many meristematic tissues. The shoot apices of *Lolium*, for example, have only 17% dry weight, including 5% as soluble compounds. Thermal diffusion of sugars and other substrates through meristems may, in fact, limit their growth. Indeed, their gradients may play an important morphogenetic role. The sucrose gradient from the protophloem termini, for example, may induce the differentiation of new phloem there (Wetmore and Rier 1963), possibly accounting for Esau's old observation that proto phloem sieve tubes differentiate much closer to the apex than do protoxylem elements.

Meristematic tissues could probably use all the contents of the phloem in their growth, as may inflorescences and some fruits. Van Die (1974) has shown that the composition of the phloem exudate from severed *Phoenix* and *Cocos* inflorescences could account for the composition of the fruits, except in respect of calcium, and phloem and xylem imports may account for the composition of developing lupin fruits (Pate, this volume). In organs storing insoluble products, such as the wheat grain, unloading may be a passive process, down a diffusion gradient as Jenner (1974) has shown. Even where soluble compounds are stored, as in the sugar beet and cane stem, unloading from the phloem may be passive even if active loading into the storage compartment is subsequently required.

Thus, in the whole translocation pathway from photosynthetic source to sink, the only energy input required may be for active loading into the phloem, and even that may not be necessary in C_4 plants. If unloading is a passive process, we are left with the problem of how it is controlled, except by variations in local sink activity and membrane permeability.

THE DISTRIBUTION OF TRANSPORTED SUBSTANCES

As mentioned in the introduction, shifts in the patterns of distribution of transported substances have played a major role in both plant evolution and crop domestication. It is important that we understand how they are controlled. As with most things phytovascular, xylem distributions are more readily accounted for than those in the phloem. Although assimilate distribution through the phloem is modified by environmental conditions, it is largely controlled by internal mechanisms. Xylem distributions, on the other hand, are dominated by the water requirements of tissues as determined by their radiation load. The physical principles involved are relatively well known but biological adaptations may, given time, greatly modify water use by plants (cf. Passioura, this volume).

Until we are more sure of the mechanism of phloem translocation and unloading, we are unlikely to understand the principles governing assimilate distribution, so I shall confine myself to three aspects of the problem.

Although vascular systems have a comprehensive network of interconnections, assimilate movement in intact plants is often concentrated in the most direct vascular routes. Zimmermann (1960), for example, found the angle of spread up a tree to be only 44 minutes. Another striking example is the work of Prokofyev et al. (1957) with $^{32}PO_4$ applied to sunflowers, in which movement into the inflorescence was confined to well defined segments and not, in the words of the translation, through the general cauldron of the plant. Presumably, resistance to movement through the interconnecting bundles is significantly greater than that along the most direct routes.

However, the second empirical principle governing the distribution of assimilates, that sinks are supplied by the nearest sources, may imply that there is a significant resistance to movement in the phloem even along the more direct routes. Fruits tend to be supplied by their subtending leaves, shoot tips by the uppermost leaves and roots by the lower ones. It is for this reason that roots may be disadvantaged in the competition for assimilates, since their source leaves are often heavily shaded. This proximity effect is one reason why Canny (1973) argues for phloem translocation by accelerated diffusion. According to his calculations, the higher values for specific mass transfer rate in the phloem should be found only over very short path distances, less than 5-10 cm. In our wheat experiments, however, relatively high rates were obtained with path lengths of more than 50 cm from flag leaf to ear (Evans et al. 1970). Indeed, the length of this major pathway in cereals such as rice and sorghum, in which photosynthesis by the ear itself is slight, usually exceeds 80-90 cm. Trees also provide difficulties for the Canny model, although he does not agree. Thus, although the proximity effect has a major influence on the distribution of assimilates in plants, this does not preclude high rates and speeds of phloem transfer over long distances (Zimmermann 1969).

A third factor possibly influencing the distribution of assimilates is one we may call the Matthew effect, on the basis of the Biblical reference (Matthew 25: 29), "unto everyone that hath shall be given...". In the competition between sinks for assimilate in plants, the larger sinks often appear to have an advantage, i.e. sink size *per se* may be important. We present evidence concerning this size factor later (this volume).

The operation of these principles of distribution is, however, such that considerable flexibility is retained, and the pattern can change rapidly following defoliation, harvest of a sink organ, insect attack or changed climatic conditions, as shown by King et al. (1967) with wheat.

CONCLUSION

It is this very flexibility of the transport system in plants, which may extend not only to the patterns of distribution but also to the mechanism involved, that frustrates so much of our experimentation while being superbly adaptive. If we choose to work on a process of such evolutionary and adaptive significance, we must be prepared to pay the price.

REFERENCES

Bonnemain, J-L. (1965). Sur le transport diurne des produits d'assimilation lors de la floraison chez la Tomate. *Compt. Rend. Acad. Sci. (Paris) 260*, 2054-2057.

Canny, M.J. (1973). Phloem Translocation. Cambridge Univ. Press, London, p.301.

Cataldo, D.A., Christy, A.L., Coulson, C.A. and Ferrier, J.M. (1972). Solution flow in phloem. 1. Theoretical considerations. *Plant Physiol. 49*, 685-689.

de Maria, M.E. and Thaine, R. (1974). Strands of sieve tubes in longitudinal cryostat sections of *Cucurbita pepo* stems. *J. Exp. Bot. 25*, 871-885.

Evans, L.T., Dunstone, R.L., Rawson, H.M. and Williams, R.F. (1970). The phloem of the wheat stem in relation to requirements for assimilate by the ear. *Aust. J. Biol. Sci. 23*, 743-752.

Evert, R.F., Murmanis, L. and Sachs, T.B. (1966). Another
view of the ultrastructure of *Cucurbita* phloem. *Ann.
Bot. 30*, 563–585.

Evert, R.F., Eschrich, W. and Eichhorn, S.E. (1971). Sieve
plate pores in leaf veins of *Hordeum vulgare. Planta
(Berl.) 100*, 262–267.

Fensom, D.S. (1972). A theory of translocation in phloem of
Heracleum by contractile protein microfibrillar material.
Can. J. Bot. 50, 479–497.

Ford, J. and Peel, A.J. (1967). The movement of sugars into
the sieve elements of bark strips of willow. 1. Metabolism during transport. *J. Exp. Bot. 18*, 607–619.

Geiger, D.R. (1969). Chilling and translocation inhibition.
Ohio J. Sci. 64, 356–366.

Giaquinta, R.T. and Geiger, D.R. (1973). Mechanism of
inhibition of translocation by localized chilling.
Plant Physiol. 57, 272–277.

Hatch, M.D. and Glasziou, K.T. (1964). Direct evidence for
translocation of sucrose in sugar cane leaves and stems.
Plant Physiol. 39, 180–184.

Hatch, M.D. and Osmond, C.B. (1976). Compartmentation and
transport in C_4 photosynthesis. In: "Intracellular
Transport and Interaction of Compartments" eds. U. Heber
and C.R. Stocking, Springer-Verlag, Heidelberg. (in
press).

Jenner, C.F. (1974). Factors in the grain regulating the
accumulation of starch. In: "Mechanisms of Regulation
of Plant Growth" eds. R.L. Bieleski, A.R. Ferguson and
M.M.Cresswell, *Bull. Roy. Soc. New Zealand 12*, 901–908.

King, R.W., Evans, L.T. and Wardlaw, I.F. (1968). Translocation of the floral stimulus in *Pharbitis nil* in relation
to that of assimilates. *Z. Pflanzenphysiol. 59*, 377–388.

King, R.W., Wardlaw, I.F. and Evans, L.T. (1967). Effect of
assimilate utilization on photosynthetic rate in wheat.
Planta (Berl.) 77, 261–276.

Kollmann, R. and Dörr, I. (1966). Lokalisierung Funktions-
tüchtiger Siebzellen bei *Juniperus communis* mit Hilfe
von Aphiden. *Z. Pflanzenphysiol. 55*, 131–141.

Lang, A. (1974). The effect of petiolar temperature upon
the translocation rate of ^{137}Cs in the phloem of
Nymphoides peltata. J. Exp. Bot. 25, 71–80.

Lush, W.M. and Evans, L.T. (1974a). Translocation of photosynthetic assimilate from grass leaves as influenced
by environment and species. *Aust. J. Plant. Physiol. 1*,
417–431.

Lush, W.M. and Evans, L.T. (1974b). Longitudinal translocation of ^{14}C-labelled assimilates in leaf blades of

Lolium temulentum. Aust. J. Plant Physiol. 1, 433–443.

MacRobbie, E.A.C. (1971). Phloem translocation. Facts and Mechanisms: a comparative survey. *Biol. Rev. 46*, 429–481.

Marshall, C. and Wardlaw, I.F. (1973). A comparative study of the distribution and speed of movement of ^{14}C assimilates and foliar-applied ^{32}P-labelled phosphate in wheat. *Aust. J. Biol. Sci. 26*, 1–13.

Passioura, J.B. and Ashford, A.E. (1974). Rapid translocation in the phloem of wheat roots. *Aust. J. Plant Physiol. 1*, 521–527.

Peel, A.J. (1974). Transport of Nutrients in Plants. Butterworths, London. p.258.

Porter, H.K. and May, L.F. (1955). Metabolism of radioactive sugars by tobacco leaf discs. *J. Exp. Bot. 6*, 43–63.

Prokofyev, A.A., Zhdanova, L.P. and Sobolev, A.M. (1957). (Certain regularities in the flow of substances from leaves into reproductive organs.) *Fiziol. Rast. 4*, 425–431.

Roeckl, B. (1949). Nachweis eines konzentrationshubs zwischen Palisadenzellen und Siebröhren. *Planta (Berl.) 36*, 530–550.

Servaites, J.C. and Geiger, D.R. (1974). Effects of light intensity and oxygen on photosynthesis and translocation in sugar beet. *Plant Physiol. 54*, 575–578.

Sij, J.W. and Swanson, C.A. (1973). Effect of petiole anoxia on phloem transport in squash. *Plant Physiol. 51*, 368–371.

Singh, A.P. and Srivastava, L.M. (1972). The fine structure of corn phloem. *Can. J. Bot. 50*, 839–846.

Sovonick, S.A., Geiger, D.R. and Fellows, R.J. (1974). Evidence for active phloem loading in the minor veins of sugar beet. *Plant Physiol. 54*, 886–891.

Swanson, C.A. and Geiger, D.R. (1967). Time course of low temperature inhibition of sucrose translocation in sugar beets. *Plant Physiol. 42*, 751–756.

Tammes, P.M.L. (1952). On the rate of translocation of bleeding sap in the fruit stalk of *Arenga. Proc. Sect. Sci. Koninkl. Nederl. Akad. Wet. 55*, 141–143.

Trip, P. and Gorham, P.R. (1968). Bidirectional translocation of sugars in sieve tubes of squash plants. *Plant Physiol. 43*, 877–882.

Troughton, J.H., Chang, F.H. and Currie, B.G. (1974). Estimates of a mean speed of translocation in leaves of *Oryza sativa* L. *Plant Sci. Letters 3*, 49–54.

Van Die, J. (1974). The developing fruits of *Cocos nucifera* and *Phoenix dactylifera* as physiological sinks importing and assimilating the mobile aqueous phase of the sieve tube system. *Acta Bot. Neerl. 23*, 521–540.

Wardlaw, I.F. (1972). Temperature and the translocation of photosynthate through the leaf of *Lolium temulentum*. *Planta (Berl.) 104*, 18–34.

Wardlaw, I.F. and Moncur, L. (1976). Source, sink and hormonal control of translocation in wheat. *Planta (Berl.) 128*, 93–100.

Weatherley, P.E. and Johnson, R.P.C. (1968). The form and function of the sieve tube: a problem in reconciliation. *Intl. Rev. Cytol. 24*, 149–192.

Wetmore, R.H. and Rier, J.P. (1963). Experimental induction of vascular tissues in callus of angiosperms. *Amer. J. Bot. 50*, 418–430.

Zimmermann, M.H. (1960). Longitudinal and tangential movement within the sieve tube system of White Ash (*Fraxinus americana* L.) Beih. *Zn. Schweiz. Forstver. 30*, 289–300.

Zimmermann, M.H. (1969). Translocation velocity and specific mass transfer in sieve tubes of *Fraxinus americana* L. *Planta (Berl.) 84*, 272–278.

Zimmermann, M.H. (1974). Long distance transport. *Plant Physiol. 54*, 472–479.

SHORT DISTANCE TRANSFER

CHAPTER 2

Plasmodesmata and Symplastic Transport

B.E.S. Gunning[1] and A.W. Robards[2]

[1]*Department of Developmental Biology, Research School of Biological Sciences, Australian National University, Canberra City, A.C.T. 2601. Australia*
[2]*Department of Biology, University of York, Heslington, York, YO1 5DD, England.*

STRUCTURE OF PLASMODESMATA

Strasburger (1901) gave us the term "Plasmodesmen" twenty two years after Tangl's original description of "offene Communicationen" between endosperm cells. Strasburger's nomenclature is now established and we more and more realise the truth of his view that plasmodesmata possess the dual properties of being open enough to provide channels for a variety of forms of intercellular communication, while at the same time being closed enough to preserve the genetic identity and morphogenetic potential of the individual cells. As will be seen, it is now to some extent possible to interpret this delicate balance in ultrastructural terms, and we emphasise the concept at the outset because it serves to underline the fundamental fact that plasmodesmata are much more complicated than mere membrane-lined holes in cell walls.

Figure 1 is a generalised diagram that introduces the nomenclature of plasmodesmatal ultrastructure, aspects of which will now be considered in turn.

Cell wall. - There is evidence that in several situations the wall surrounding plasmodesmatal canals is specialised, though the nature of the specialisation can vary. A periplasmodesmatal sleeve resistant to enzyme attack occurs in aleurone (Taiz and Jones 1973); sleeves that are penetrated all along their length by the membranous components grow into the cytoplasm in response to infection with a variety of isometric viruses (no less than 13 examples are cited by Gibbs 1976); differential staining has been observed (Carde 1974; Vian and Rougier 1974; van Went et al. 1975); callose may be present, especially in sieve element-companion cell plasmodesmata (references in Gunning 1976b), but also in others

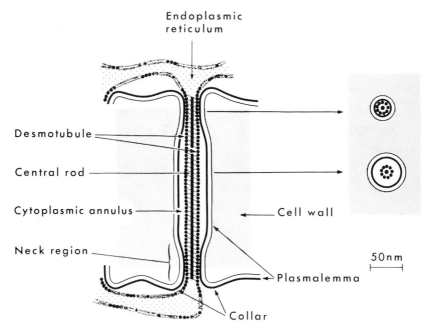

Fig.1.- *Diagram to show the relationships of the plasmodesmatal components discussed in the text.*

(Kollmann and Dörr 1969).

Plasma membrane.- If the diameter of the plasmodesmata in a 1.0 μm thick wall is 60 nm, and if they occur at a frequency of 5 μm^{-2}, there is approximately as much membrane in the plasmodesmatal canals as there is facing the cytoplasm. We do not know if this very considerable fraction of the total is specialised (apart from its position); whether transla-tional diffusion in the plane of the plasmodesmatal plasma membrane is possible; or whether it has special permeability properties. There have been reports of phosphatase and adenosine triphosphatase in plasmodesmata (see Robards 1976). Vian and Rougier (1974) find that in fixed and cryo-sectioned material the plasma membrane in the plasmodesmata is thinner (after negative staining) than that away from the plasmodesmata. Carde (1974) also reports differential staining in conven-tionally-prepared material, and suggests that the plasma membrane within the plasmodesmatal canal is locally depleted in terms of lipids, in accord with the view of Israelachvili and Mitchell (1975) that certain lipids will, depending upon their shapes, tend to diffuse in the plane of the membrane away from regions of marked curvature.

Desmotubule.- Suspicions that the desmotubule might be

some sort of artefact have been allayed by the publication
of a micrograph by Roland (1973) and Vian and Rougier (1974)
showing that it is present in material stated by Roland to
have received no fixation treatment, and that was quick-frozen
and cryo-sectioned without prior dehydration or embedding.
In this, and views of fixed and then cryo-sectioned plasmodesmata, the desmotuble is narrower than is seen in most conventionally prepared specimens (Vian and Rougier 1974), and it is
correspondingly more difficult to discern whether it is a
solid or a hollow cylindrical structure. As shown in earlier
work (reviewed in Robards 1976), many micrographs show desmotubules with a comparatively electron-lucid lumen, but with a
central dot, interpreted as being a transection of a "central
rod". Other micrographs which, on the basis of the ability
to resolve the tripartite image of the trans-sectioned plasma
membrane, would be considered to be accurately aligned .cross
sections of plasmodesmata, show much more solid looking
desmotubules. Whether the two types of image represent alternative states or mere variation in staining cannot yet be
determined.

    The radius of curvature of the desmotubule is much smaller
than that of the plasmodesmatal plasma membrane, and even
greater effects upon migration of mobile lipids can be anticipated (Israelachvili and Mitchell 1975). Jones (1976) uses
this idea to account for the failure to see conventional
membrane substructure in desmotubule walls. He argues that
emigration of lipids would leave a largely proteinaceous tube,
thus agreeing with previous interpretations (see Robards 1976),
and that this should in no way cast doubt upon the postulated
continuity of desmotubule and endoplasmic reticulum membrane.
Continuity of the two systems has often been proposed on the
basis of fixed and embedded material (e.g. Fig. 2) and, more
recently, from cryo-sections (Vian and Rougier 1974). It is

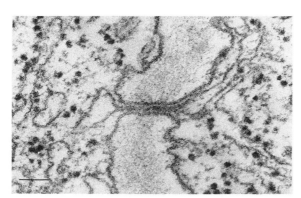

*Fig.2.- Plasmodesma
in* Vicia faba *root
tip, showing con-
tinuity of endo-
plasmic reticulum
and desmotubule.
Scale marker: 0.1μm.*

significant that where plasmodesmatal canals distend, as in central cavities and nodules (below), the desmotubule also distends and becomes rather less granular and more membrane-like in appearance (Wooding 1968; Carde 1974). Branched desmotubules may be seen in branched plasmodesmata (Wooding 1968).

Most, but not all, plasmodesmata have desmotubules, and in most higher plant cells it would seem that they are an integral part of the overall structure. When plasmodesmata arise in cell plates, trapping and subsequent modification of endoplasmic reticulum would seem a convenient mode of formation, but the existence of desmotubules in secondarily formed plasmodesmata (that arise in mature walls) implies a capacity to produce them by other mechanisms (Jones 1976) and, in turn, that they are far from being accidental inclusions.

Cytoplasmic annulus and neck constrictions.- The cytoplasmic annulus, or channel, is the gap between the outer face of the desmotubule and the inner face of the plasma membrane. "Spokes" have been seen traversing it (e.g. Burgess 1971), but as with the "central rod", their genuineness is in doubt. It is very common to find that the annulus is constricted at both orifices, the plasma membrane approaching or even being appressed to the desmotubule so that only a very small gap remains. Once again the familiar image arising from conventionally handled samples has been confirmed by cryo-sectioning (Vian and Rougier 1974). Neck constrictions are not seen in newly formed plasmodesmata in cell plates, but arise later (e.g. Gunning and Steer 1975; Jones 1976). As the nectary trichomes in *Abutilon* flowers approach maturity, so there is a change in the plasmodesmata of the distal wall of the stalk cell, from a state in which all are constricted to one in which about 80% are non-constricted; this being accompanied by a loss of callose from around the plasmodesmata (Hughes and Gunning, unpublished). The potential functional significance of the constrictions is very great (see following discussion), and it is important that other studies be made to see if morphological changes are consistently observed when intensive symplastic transport begins or ceases.

Median nodules, cavities, sinuses; branching.- Single plasmodesmata in mature walls often are distended at the level of the middle lamella. The diameter of the lumen increases, and the desmotubule may become so tortuous in its path that continuity cannot be seen (e.g. O'Brien and Thimann 1967). Krull (1960) described such regions as median nodules ("Mittelknoten"). They may become much more complex if the plasmodesmata become branched, or if anastomoses develop between neighbouring plasmodesmata, and they may even extend for many micrometres in the centre of the wall, with numerous branches leading out to the surrounding cells. Formation of such systems implies that local digestion and removal of wall material can be brought about, not only as plasmodesmata develop median nodules, but also as they develop branches (Jones 1976).

DISTRIBUTION AND FREQUENCY OF PLASMODESMATA

It is easier to state where plasmodesmata do not, or probably do not, occur than to catalogue the very numerous anatomical situations where they have been found.

There are three categories of "absence". One is in simple algae, where there are filamentous organisations (both simple - e.g. *Spirogyra*, and branched - e.g. *Cladophora*) and flat thalli (one cell layer thick, e.g. *Enteromorpha*, and two cell layers thick, e.g. *Ulva*) where they do not occur. Some simple filaments (e.g. *Ulothrix* (but not *U. zonata*)) do, on the other hand, possess plasmodesmata, as do numerous algae at the next higher level of morphological complexity, distinguished as the "heterotrichous" types - at any rate no heterotrichous eukaryotic alga lacking plasmodesmata has yet been found (Marchant 1976).

The second category of absence arises by loss of plasmodesmata during development. This happens in instances of morphogenetic cell death (e.g. xylem maturation) where it is obvious that it must occur, and of greater interest, during development of stomatal complexes, where there is now a number of studies which, while they cannot prove complete absence, do at least indicate a diminution in plasmodesmatal frequency (Vela and Lee 1975; other literature in Carr 1976).

The third category of absence relates to reproductive cells and tissues. Details of the voluminous literature are given in Carr (1976). In summary, there are very few exceptions to the generalisation that plasmodesmata do not interconnect gametophyte with sporophyte, or gametes, spores, zygote, endosperm, with the generation in which they are formed. In that different generations or nuclear phases in the life cycle can be regarded as different individuals, we might say that plasmodesmata do not occur between individuals but only within them. This, however, is not universal, for occasional plasmodesmata connect the host to cells in the chalazal end of the embryo sac in *Capsella* (Schulz and Jensen 1971), and transitory connections form between searching hyphae of the parasite *Cuscuta* and its host (Dörr 1968) and between the parasite *Arceuthobium* and its host (Tainter 1971).

Apart from the above three categories of absence, plasmodesmata are anatomically as well as taxonomically widespread. There is increasing evidence of secondary formation of plasmodesmata, that is, formation in either "division walls" well after the cell plate stage of development, or in "non-division walls" (Jones 1976). They can even appear in walls that normally would never possess them, for example in the outer wall of epidermal cells that have become postgenitally fused back-to-back (Boeke 1971, 1973). The example of stomatal guard cells is a warning, however, against the assumption that all living cells of higher plants are connected to their neighbours, though this can be said with certainty in respect of the very early embryo of *Capsella* (Schulz and Jensen 1968). Primary formation during development of the cell plate is the norm, indeed it would appear that the morphogenetic programmes that are involved are not easily switched off. Thus plasmodesmata even appear in grossly distorted fragments of new cell walls that float freely in the cytoplasm of cells whose division process has been disordered by anti-mitotic agents (Mesquita 1970; Jones 1976). The best, and perhaps the only, example of a cell plate that forms without incorporating long-lived plasmodesmata is found in the pollen mitosis that separates the generative from the vegetative cell: even here, however, transitory connections are sometimes seen (literature in Carr (1976)).

We know little about the controls that the cell seems able to place upon the disposition and frequency of plasmodesmata. One non-random pattern of plasmodesmata that can be seen in the longitudinal walls of root tip cells after a considerable amount of wall stretching is a parallelism between rows of plasmodesmata and transversely oriented microtubules (unpublished work by Jones on *Impatiens* and Hardham on *Azolla*). The origin of primary pit fields, i.e. clusters of plasmodesmata, is also obscure, and might involve selective loss and/or non-random secondary formation of plasmodesmata (Burgess 1971; Jones 1976).

Some recorded frequencies of plasmodesmata are included in Table 3, and will be discussed later. Over whole expanses of wall, values from 1 μm^{-2} to 15 μm^{-2} cover the range from sparse to abundant, though more extreme cases are known. Within pit fields the frequency per unit area can be much higher, e.g. 54-60 μm^{-2} recorded by Olesen (1975) for the bundle sheath outer tangential wall in leaves of the C4 plant *Salsola kali*. Frequency per unit area can alter during development, as can the absolute number per wall or per cell. Juniper and Barlow (1969) have demonstrated conservation of total numbers combined with dilution through cell expansion in maize root tip cells; Schmitz and Srivastava (1974) describe an increase in number and frequency in the equivalent of a sieve plate seen in the brown alga *Laminaria groenlandica*; and as already seen, stomata seem to exemplify diminution in total numbers and frequency per unit area. Other examples of developmental change are given by Jones (1976).

TRANSPORT THROUGH PLASMODESMATA-GENERAL

The ultrastructure of plasmodesmata leads us to consider three possible pathways of transport - the cytoplasmic annulus, the desmotubule, and translational movements in or on the surfaces of the plasmodesmatal membranes. Before doing so, however, it is helpful to examine some general aspects of transport through small pores.

Fick's Law and Poiseuille's Law are assumed to apply, for diffusive and convective transport respectively, but correction factors are necessary. First, the diffusion coefficient in the pore may not be the same as that in free aqueous solution. The colligative properties of cytoplasm were considered by Tyree (1970) to bring about a reduction by a factor of about three, though it is questionable whether the contents of either plasmodesmatal pathway (annulus or desmotubule) can be equated with bulk cytoplasm. Tyree (1970) also recognised the much greater effect that the presence of fixed charges in the pore would have on the mobility of ions, and suggested an effective reduction of some 10-fold.

Another type of correction factor applies regardless of the presence of charges. It is based on the size relationships between the pore and the mobile particles. Clearly, no particle of radius 'a' will be able to enter a pore unless its centre is a distance $2a$ from the pore wall. The effective lumen is reduced by 2a, or, on an area basis where r is the pore radius, by a factor of $(1-a/r)^2$, or, if we are considering entry into an annular pipe with large and small radii r and kr, by a factor of $[1-a/r(2/(1-k))]$. These are the relevant factors for entry by diffusion; those for entry of particles in a bulk flow of solvent are less serious, because the steric hindrance to entry affects only the outermost zone, where the velocity of flow is slight compared with that nearer the axial region of the pipe. Once within the pipe yet another correction factor, the drag coefficient (Paine and Scherr 1975) should be superimposed upon the others, to account for the impedance of particle transport by interaction with the walls. As with the hindrance to entry, the drag coefficient is a function of a/r, and values are available for cylindrical pipes but not for annular pipes (Gunning 1976a). For small molecules (e.g. sucrose, where a = 0.44 nm) in the desmotubule (radius approx. 5 nm), these corrections are comparatively slight, their combined effect for diffusive transport being a reduction to 73% of that expected. However,

with larger a/r values, pronounced ultrafiltration effects
become clear. A protein of molecular weight 60,000-70,000,
with radius about 3 nm, would be transported (by diffusion)
about 50-fold less effectively in the desmobutule, where a/r
= 0.6, than its free diffusion coefficient would suggest.

It is from considerations such as these that we can begin
to interpret Strasburger's (1901) view that plasmodesmata
are large enough to carry small molecules, but small enough
to prevent loss of the genetic and morphogenetic individuality
of the connected cells. The same considerations draw attention
to the potential functional significance of the "neck constrictions" so commonly seen in plasmodesmata. Alteration of the
degree of constriction could be a very sensitive way of
regulating transport through the cytoplasmic annulus. As
already mentioned, we do have one example (*Abutilon* nectaries)
where relaxation of the constriction correlates with the onset
of a rapid symplastic transport process.

Tyree (1970) compared the probable efficiency of the
cytoplasmic annulus with that of the alternative transmembrane
plus wall pathway from cell to cell, and concluded that the
former is superior by a factor of 1-3 orders of magnitude.
This was a very important result, and from then on plasmodesmata were accepted as being quantitatively respectable as
channels of transport. The dimensions are, of course, crucial
(Table 1). The values that Tyree took for the inner and outer
radii do not agree well with the data that are now available,
and his cross sectional areas are 4.5-18 times the <u>average</u>
value for the plasmodesmata in Table 1. The error is much
more serious when the radius is raised to the power of 4, as
in the Poiseuille calculations used to assess hydraulic
conductivities.

Indeed, if the plasmodesmatal dimensions given in the
first two rows of Table 1 are averaged and combined with the
same viscosity, length, and frequency data as used by Tyree,
then the hydraulic conductivity of the plasmodesmatal path-
way becomes $3x10^{-9}$-$3.6x10^{-8}$ m s^{-1} bar^{-1}, i.e. the values lie
within (but at the upper end of) the range found for plant
cell membranes (see note *** in Table 1).

Further, neck constrictions are ignored in Tyree's treatment. It is not known precisely how closely the inner face
of the plasma membrane is appressed to the desmotubule at the
constrictions, but if conventionally-fixed and processed
material is to be trusted, the gap between the two can be no
wider than a few nanometres. It is instructive to use Fick's
Law and Poiseuille's Law (the modified form that applies to
annular pipes (Anderson 1976)) to compare an open annular
pipe with a constricted one. If for simplicity we consider

TABLE 1. Dimensions of Plasmodesmata.

Parameter	Material[*]				Values used by Tyree (1970)
	Azolla young root cortex	*Hordeum* endodermis 4mm from root tip	120mm from root tip	*Abutilon* distal wall of nectary trichome stalk cell	
Diameter at inner face of plasma membrane (nm)	25**	33**	44**	29**	60–120***
Outer diameter of desmotubule (nm)	16	20	20	16	6***
Inner diameter of desmotubule (nm)	7	9	10	10	–
Cross-sectional area of cytoplasmic annulus (nm^2)	290	541	1206	459	2799–11281
Cross-sectional area of desmotubule lumen (nm^2)	38	64	79	79	–
Frequency per μm^2 of wall	–	–	1.05	12.6	1.5–4***
Area fraction: cytoplasmic annulus	–	–	1.26×10^{-3}	5.79×10^{-3}	$3 \times 10^{-3} -10 \times 10^{-3}$
Area fraction: desmotubule	–	–	8.29×10^{-5}	9.95×10^{-4}	–

*   Data from Robards (1976)
**   Measured at middle of plasmodesmata, hence neck constrictions not taken into account
*** Values used by Tyree (1970) in calculating that the hydraulic conductivity of the "plasmodesmatal pathway" ranges from $2.7 \times 10^{-7} -5.4 \times 10^{-5}$ m s^{-1} bar^{-1}, whereas the hydraulic conductivity values that he quoted for the "membrane pathway" range from $2 \times 10^{-11} -6 \times 10^{-8}$ m s^{-1} bar^{-1}.

30 nm long constrictions at either end of a 150 nm long non-constricted zone, and compare the overall permeability and hydraulic conductivity with those for a non-constricted 210 nm long annulus, then the effects of constrictions of a range of severities are as listed in Table 2. It can be seen that the factor by which Tyree found the "plasmodesmatal pathway" to be more effective than the "membrane pathway" (already eroded because plasmodesmata are in fact smaller than in his model) is further diminished, because the constrictions, where present, will reduce the diffusive flux (slightly) and any volume flow that is taking place (very severely). Constrictions will also aggravate the steric hindrance to entry correction factor until it is highly significant, even for small solutes such as sucrose. It is emphasised that we do not known how "tight" the junction between plasma membrane and desmotubule is at the constrictions. For the moment the message must be that in using models to predict the transport capacity or the cytoplasmic annulus, the constrictions must not be neglected.

An extreme interpretation of the ultrastructural evidence is that the cytoplasmic annulus is sealed by constrictions to prevent leakage, and that the desmotubule provides the major transport pathway – one which has the advantage that along it could pass molecules that have been selected by virtue of transmembrane specificity at loading sites on the endoplasmic reticulum. The idea that the symplast reduces to inter-connected endoplasmic reticulum systems has been considered many times, notably by O'Brien and Thimann (1967) and Läuchli (this volume). Bräutigam and Müller (1975) and Robards and Clarkson (1976) emphasise that there is no question of merely replacing one trans-membrane pathway at the plasma membrane with another at the endoplasmic reticulum: it would be difficult to see much biological advantage in this, and the concept of necessity includes the proposal that the endo-plasmic reticulum needs to be loaded only once, at the beginning of the symplastic pathway, and unloaded only once, at the end, perhaps after traversing very many cells. Having said this, it must nevertheless be recognised that the permeability and hydraulic conductivity of the endoplasmic reticulum membranes are unknown quantities, as are the parameters of the intracellular part of the system. Cytoplasmic streaming is almost certain to affect the temporal and spatial continuity of the reticulum, though not nec-essarily to the detriment of transport, for peristaltic effects can be envisaged, also that cisternae might be pinched off and swept along by cyclosis to re-fuse near the cisternae that are connected to the desmotubules. That cytoplasmic

TABLE 2. - Factors by which the presence of constrictions at the terminal 30 nm portions of a 210 nm long plasmodesma reduce the diffusive permeability and hydraulic conductivity of the cytoplasmic annulus, assumed to have inner and outer radii of 10 and 22 nm (as in barley root endodermis (mature, Table 1)), except at the constrictions, where the outer radius is as in column 1.

Outer radius of annulus at constriction (width of annulus in brackets) (nm)	Factors by which constricted annulus is less effective than open annulus			
	For diffusion (Fick's Law)	Correction factor for steric hindrance to entry, for sucrose (radius 0.44 nm)	Combined effects, column 2 and 3*	For volume flow (Poise-uille's Law)**
14 (4)	1.8	0.78	2.3	11.1
13 (3)	2.3	0.71	3.2	26.4
12 (2)	3.2	0.56	5.7	91.3
11 (1)	5.9	0.12	49.5	760

* Drag coefficients are not available for particle movement through annuli, and so are not included. They can be expected to introduce an additional hindrance to transport, of the same order as the factor for steric hindrance to entry.

**Figures exclude effects on solutes such as hindrance to entry and drag.

streaming is an important component of symplastic transport has been shown, both in theory (Anderson 1976) and in practice (Worley 1968; Walker 1976).

Clearly, the desmotubule itself needs to be examined. Although connection to the endoplasmic reticulum seems clear in quite a wide variety of plasmodesmata (see Robards 1976), lumenal continuity through a plasmodesma has not yet been demonstrated satisfactorily. What can be said is that if it is regarded as a hollow cylinder of internal diameter 10 nm, and if plasmodesmatal frequencies are taken into account, then (a) it will cope with the diffusive fluxes across the bundle sheath - mesophyll interface in C_4 plants (Osmond and Smith 1976), (b) in the barley root endodermis (inner tangential wall) it provides a pathway for bulk flow of water that has a hydraulic conductivity (5.15×10^{-8} m s^{-1} bar^{-1}) within the range of values found for plant cell plasma membranes and (c) because of the high frequency and short length of plasmodesmata in the *Abutilon* nectary stalk cell, the hydraulic conductivity of the desmotubular pathway in that situation is greater than that of most plant plasma membranes (see later). Because of its small diameter, high velocities are required if desmotubular transport does occur in the volume-flow systems - up to about 300 $\mu m\ s^{-1}$ in each case. This value, of course, applies only within the desmotubule itself. In the case of *Abutilon* the cytoplasmic annulus is open, so that aside from the desirability of compartmentation of the pre-nectar, one cannot argue strongly that the desmotubular pathway is indeed used. In the C_4 plants, on the other hand, the cytoplasmic annulus is constricted, in many cases at the position of the suberin lamella as well as at the ends of the plasmodesmata, and in the root endodermis constrictions are again evident, so that the argument that the desmotubule is functional is much more compelling.

The case that in addition to transport through the cytoplasmic annulus and the desmotubule there might be symplastic transport based on surface phenomena is purely speculative. As compared with diffusion in three dimensions, diffusion through plasmodesmata (either pathway) involves a reduction in dimensionality, the length usually being much greater than the diameter. Adam and Delbrück (1968) have derived expressions which enable the effects that such a reduction might have to be estimated. They consider a molecule diffusing to a target of specified size along a pathway of specified length, when the molecule is free to diffuse in three dimensions, or is restrained to a two dimensional path by adsorption onto or into a membrane, or when it follows a one dimensional route. The results are surprising, but of potentially very great significance for both intra- and inter-cellular diffusion. For example, the time taken to reach a 1 nm target over a path of 1 μm is some 100 times *less* in two dimensions than in three, provided that the diffusion coefficients are the same in both cases. At the intra-cellular level one can imagine "trapping" of solute molecules by the plasma membrane or endoplasmic reticulum, with thereafter an unexpectedly efficient two dimensional delivery to the plasmodesmatal orifice, and at the inter-cellular level some facilitation of diffusion through the plasmodesma itself would be expected.

TRANSPORT THROUGH PLASMODESMATA - CASE HISTORIES

Unresolved arguments about the pathway of transport through plasmodesmata can be by-passed if we consider transport on a "per plasmodesma" basis. Table 3 (modified from

Gunning and Robards 1976) presents data for solute fluxes and water flows in a variety of situations.

For solute fluxes the transport per plasmodesma has been combined with the frequency of the plasmodesmata to give a flux per unit area of wall, thus allowing comparison with the "membrane pathway", in which the wall between two adjacent cells is ignored, and each plasma membrane is taken to have a permeability of 10^{-7}-10^{-8} mole m^{-2} s^{-1} (1-10 p mole cm^{-2} s^{-1}, considered by MacRobbie (1971) to be average for plant cells). On this basis of comparison the observed flux (for the first 6 rows in the Table) is between 6.5 and 4,300 times larger than the membrane path could be expected to carry. For the roots and the nectary (the last 3 rows of the Table), there is a volume flow that is from 5-1400 times greater than could be carried by a membrane pathway where each plasma membrane has a hydraulic conductivity of 2 x 10^{-8} m s^{-1} bar^{-1} These are unfair bases for comparison in the case of the root endodermis, where suberisation of the cell wall (as in the bundle sheath around leaf veins of many C_4 and C_3 plants) almost certainly reduces the hydraulic conductivity and the solute permeability of the "membrane pathway" below the values used, so that the "superiority" of the plasmodesmatal route is underestimated. It is in this light that the low fluxes of water and ions across the barley and marrow root endodermis should be viewed. Although low, they may still be larger than could be obtained across the suberised wall. If the pre-nectar that traverses the stalk cell of the *Abutilon* nectary is, like the nectar, equivalent to 0.6 M sucrose (Reed *et al*. 1971), then the "superiority factor" for solute transport here is very large, about 1.6 x 10^3-1.6 x 10^4.

In the preceding section the argument that plasmodesmata should be more efficient in cell to cell transport than the membrane pathway was seen to be clouded by the uncertain effects of constrictions and by our present inability to discern which part of the plasmodesma is functional in transport. The data of Table 3 now allow us to argue that for a wide range of situations, fluxes that could not be carried by "average" membranes or walls are in fact sustained. Unless in every case the membranes have above average (in some cases by several orders of magnitude) transport capacities, an alternative and superior pathway must be present, and the plasmodesmata are, of course, the only apparent contenders.

Much the same reasoning has been applied in electrophysiological studies. Measurements of coupling ratios between cells give clear indications that a low-resistance pathway is present (reviewed by Goodwin 1976; Anderson 1976). So too do

Location and Reference	Plasmodesmatal frequency (μm^{-2})	Volume flow Observed: m^3 plasmodesma^{-1} s^{-1}	Volume flow Superiority Factor[a]	Solute flux Observed: mole plasmodesma^{-1} s^{-1}	Solute flux Superiority Factor[b]
Mesophyll C_3 plant[c]	3	–	–	$2-7 \times 10^{-19}$	6.5–210
Bundle sheath C_4 (outer tangential wall)[d]	15 (Salsola)[e]	–	–	5×10^{-19} (Amaranthus) 10×10^{-19} (Zea)	164–1640 296–2960
Bundle sheath, C_3 (inner tangential wall)[f]	7.8	–	–	2.9×10^{-19}	23–230
Sieve element – companion cell[g]	6	–	–	8.2×10^{-19}	48–480
Chara node[h]	4–5	–	–	9.5×10^{-19}	430–4300
Abutilon nectary (distal wall of stalk cell)[i]	12.6	2.1×10^{-20}	230–1400[k]	–	–
Root endodermis (stage III) (inner tangential wall of barley)[j]	1.03	2.4×10^{-20}	3.1–307[k]	3.6×10^{-21} (PO_4)	0.04 0.1
ditto, Cucurbita[j]	6.2	1.5×10^{-21}	30.4–1813[k]	2.2×10^{-20} (K^+)	1.4–14

[a]Factor by which the plasmodesmatal volume flow per unit area of cell junction exceeds the flow that would occur across two successive plasmalemmas of unit area, each with hydraulic conductivity 2×10^{-8} m s^{-1} bar^{-1}. [b]Factor by which the plasmodesmatal flux per unit area of cell junction exceeds the flux that would occur if the solutes passed through two successive plasmalemmas of unit area, each capable of carrying a flux of $10^{-7}-10^{-8}$ mole m^{-2} s^{-1} (1–10 p mole cm^{-2} s^{-1}; see MacRobbie, 1971). [c]Frequency of plasmodesmata estimated by assuming 100 μm^2 of wall junction (Münch, 1930), of which 0.38% is occupied by plasmodesmata (Brinckmann & Lüttge, 1974), each one taken to have radius 20 nm; flux calculated using anatomical and photosynthetic data in Geiger and Cataldo (1969) and Sovonick, Geiger and Fellows (1974). [d]Flux data from Osmond and Smith (1976); [e]Frequency from Olesen (1975); [f]from Kuo, O'Brien and Canny (1974); [g]From Gunning, et al. (1974); [h]From Walker (1976); [i]Gunning and Hughes (unpublished); [j]Data on frequencies and fluxes from Tables 10.2–10.9 of Robards and Clarkson (1976). [k]The two values listed are the desmotubular and cytoplasmic pathways respectively, using plasmodesmatal dimensions as in Table 2 (values for mature *Hordeum* are used for both roots), and viscosities 0.016 poise (for *Abutilon* prenectar) and 0.01 poise (for water transport across the root endodermis). Central rods are ignored in both, as are the constrictions in the endodermal plasmodesmata.

TABLE 3. Comparative Data on Estimated Rates of Transport
Through Plasmodesmata.

---

observations on the intercellular transport of substances that
do not readily traverse the plasma membrane. The work on
Vallisneria by Arisz and co-workers is reviewed by Spanswick
(1976), Goodwin (1976), and Gunning (1976b), and the conclusions from it have now been reinforced by Goodwin's (1976)
detection of intercellular movement of iontophoretically
injected procion dye. Tyree and Tammes (1975) use the rate
of dye movements along *Tradescantia* stamen hairs to deduce
that symplastic transport occurs.

In short, there is excellent evidence for a low-resistance
symplastic pathway in plant tissues. Nevertheless, the *direct*
evidence that plasmodesmata represent this pathway is dis-
turbingly slight, consisting essentially of findings of two
sorts of mobile but recognisable material within plasmodesmata.
Virus particles represent one type (below). The second type
of mobile substance that has been detected in plasmodesmata
is the chloride ion. The procedure that has been used is to
fix the tissue in the presence of a silver salt (a method
long used by light microscopists to stain plasmodesmata, see
Meeuse, 1957). Electron microscopy has then revealed precipitates in plasmodesmata (Ziegler and Lüttge, 1967; Läuchli
(this volume); Van Steveninck, 1976). Campbell and Thomson
(1975) have gone further by identifying the deposits using
electron diffraction, and Lauchli (this volume) and van
Steveninck (1976) have used X-ray microanalysis to prove that
both chlorine and silver are indeed present, and to show that
salt-loaded plants have larger deposits than controls.
Quantitative analysis is, however, fraught with difficulties
(van Steveninck 1976), not only because of technical problems,
but because plasmodesmata can be foci for precipitation of a
variety of unexpected compounds (Gullvag, Skaar and Ophus
1974; Ophus and Gullvag 1974).

There is a number of situations where plasmodesmata may
be pathways of bidirectional transport. For example, Fraser
and Smith (1974) suggest that in a fern gametophyte the
rhizoid supplies the photosynthetic cells with absorbed
material, while the photosynthetic cells supply the rhizoid
with assimilates - all through the plasmodesmata in one wall.
The endodermis of legume root nodules carries a flux of
phloem-derived sugar towards the bacteroid-containing cells
and a return flux of products of nitrogen fixation (Pate
et al. 1969; Gunning et al. 1974). The endodermis of roots
carries a radial flow of water to the stele (Robards and
Clarkson 1976) and a flux of sucrose to the cortex (Dick and

ap Rees, 1975). The dicarboxylic acid pathway of C_4 plants depends upon a two-way flow of metabolites across the bundle sheath/mesophyll interface (Osmond and Smith 1976). The presence of a suberin lamella may direct an outward flow of water against an inward flow of assimilate in the bundle sheath of wheat leaf veins (Kuo, O'Brien and Canny 1974). Retrieval of amino acids and potassium ions from the pre-nectar in *Abutilon* trichomes may have to operate against the outward secretory flow (Gunning 1976b). In all of these cases the oppositely directed fluxes are likely to be symplastic; and while it cannot be proved that it is obligatory for any one plasmodesma to sustain the bidirectional transport, this possibility should be examined.

There is no difficulty where the opposite fluxes are diffusive. Each diffusing species will follow its own gradient and there is no need for separation of the two. Even where diffusion has to operate against an oppositely directed volume flow, the following analysis (for which we thank Dr. N.A. Walker) indicates that the need for compartmentation is not implied by any observations so far available. The solute flux carried in one direction by the volume flow is the average concentration C times the velocity v, while that carried by diffusion in the opposite direction under the influence of a concentration drop Δc is given by the diffusion coefficient D times Δc divided by the length of the plasmodesma L. These opposing fluxes will be equal and opposite in a steady state situation, when $Cv = D\Delta c/L$. Rearranging, $\Delta c/C = vL/D$, from which the concentration drop (relative to the average concentration) needed to counteract a volume flow can be evaluated for any plasmodesma where the velocity of flow can be calculated. In the case of the plasmodesmata of the distal wall of the stalk cell of the *Abutilon* nectary trichome the necessary Δc is only about 0.2% of C (Gunning 1976b); in the most pessimistic view of water transport in barley root endodermis plasmodesmata (Robards and Clarkson 1976, Table 10.5), where the entire water flow is imagined to pass through desmotubules of 10 nm diameter, the limiting Δc for outward diffusion of sugar from the phloem would be 7.6% of C; if we similarly take a pessimistic view of the pathway followed by the transpirational stream in wheat leaves, and average the water loss over all of the desmotubules traversing the bundle sheath inner tangential wall, then the limiting Δc for inward diffusion on assimilate would be 4.8% of C (using data from Kuo et al. 1974). Despite the pessimistic assumptions, none of these Δc values seems impossible, and it may be concluded that bidirectional transport through plasmodesmata is feasible in all situations so far documented.

CONTROL OF SYMPLASTIC TRANSPORT

Two classes of control of symplastic transport pathways can be evisaged; one in which the frequency of plasmodesmata is altered, and the other in which the permeability and conductivity of existing plasmodesmata change. Either of these will influence symplastic transport, except where intracellular transport is the limiting factor (Walker 1976; Anderson 1976).

Increases in plasmodesmatal frequency after the cell plate stage of wall development are known, the best examples being between growing giant cells of roots infected with the nematode *Meloidogyne* (Jones 1976) and in the forming "sieve plates" in the brown alga *Laminaria groenlandica* (Schmitz and Srivastava 1974). A similar phenomenon, single plasmodesmata becoming multiple branched structures, has been described more often. Probably the most important example is the development of the compound plasmodesma that connects sieve element to companion cell (Gunning 1976b).

Just as transfer cells have been said to form where apoplast-symplast exchanges are intensive, so it has long been held that pladmodesmata are most numerous where symplastic fluxes are large. The stalk cells of glands are examples (Lüttge 1971). In *Abutilon* nectary trichomes, high frequencies (about 12 μm^{-2}) are seen from the earliest recognisable stage of development of the distal wall of the stalk cell (Hughes and Gunning, unpublished), so that in this case the plant, in effect, lays down at least the bulk of its symplastic transport system well in advance of the onset of transport. In the giant cells referred to above the opposite pertains and plasmodesmata are secondarily formed in large numbers at a time when presumed transport processes are being augmented. The two contrasting situations could be viewed as, on the one hand, a prediction of a constitutive transport requirement by primary formation of plasmodesmata, and on the other, an adaptive response by secondary formation to an emergency transport requirement - a view which could be tested by experiment.

Loss of plasmodesmata by withdrawal and by wall deposition are relatively common, or else relatively easy to detect, compared with increase in frequency. The case of stomatal guard cells has already been given: its rationalisation in terms of symplastic transport is that control of the delicate and rapid turgor movements of guard cells might be prejudiced if plasmodesmatal pathways of low reflection coefficient connected them to other cells (Kaufman et al 1970). Occlusion of plasmodesmata by adcrusting material in the pigment strand may initiate the changeover from the import

31

phase to the maturation phase of wheat seed development (Zee
and O'Brien 1970). A comparable process may occur in
abscission zones (Carr 1976). Limiting the spread of virus
infections in resistant hosts can in part involve occlusion
of plasmodesmata (Gibbs 1976). Host-parasite plasmodesmata
in *Cuscuta* soon become overlain by wall material (Dörr 1968).
Certain specialised cells such as myrosin cells (Werker and
Vaughan 1974), but most notably a wide spectrum of reproductive cells (Carr 1976), also become isolated, commonly by
deposition of callose. Other categories of loss may be regarded as somewhat accidental, as when cells separate and
intercellular spaces appear; during sliding growth; and when
lignified thickenings or new cell plates come to overlie
plasmodesmata. Isolation from the symplast can have the most
profound morphogenetic consequences (reviewed by Carr (1976)).

Turning to the question of whether the plant or the cell
can alter the permeability of its plasmodesmata, the evidence
is much less direct.

Damaged cells, or their neighbours, seem able to seal
their plasmodesmata, a wound-response that has been observed
at the ultrastructural level by Fulcher and McCully (1971),
electrophysiologically by Pitman et al. (1970) and Jones
et al. (1975), and using physiological evidence by Arisz
(see Holder (1967)) and Geiger et al. (1974). Plasmolysis
stretches (Burgess 1971) and can rupture plasmodesmata,
this presumably being the basis of the plasmolytically induced
inhibition of translocation that has been noted for roots
(Jarvis and House 1970), fern gametophyte filaments (Smith
1972) and mesophyll (Geiger et al. 1974), and perhaps the
basis of plasmolytically induced morphogenetic effects (Carr
1976). It is reported that in *Nitella* nodes (Spanswick and
Costerton 1967), *Chara* nodes (Walker 1976), *Elodea* leaves
(Spanswick 1972) and *Tradescantia* stamen hairs (Tyree and
Tammes 1975), the plasmodesmata are less open than their
dimensions would suggest (by factors of 330, 50-60, and 150
respectively). However, some of the discrepancies can be
accounted for by overestimations of pore dimensions, and by
the presence of desmotubule and neck constrictions (though
not in the Characean nodes). This, together with uncertainty
regarding the values to be given to viscosity and diffusion
coefficients in the pore channels, makes it difficult to
evaluate the observations. Certainly it is doubtful if they
can be general, for instance imposing a hindrance to
transport of two orders of magnitude would cast severe doubt
upon the capacity of the constricted plasmodesmata of
mesophyll-bundle sheath interfaces in C_4 plants to carry the
observed fluxes.

Münch (1930) pointed out that in order to maintain turgor gradients across tissues, there would have to be appreciable resistance to flow through plasmodesmata. Tyree (1970) too emphasised this point, stating that if any abrupt turgor gradient from one cell to another could be demonstrated, this would be strong evidence for a highly gelled state within any plasmodesmata that are present. There is now good evidence of just this situation in the phloem tissue of leaf veins, where the companion cells have a very high osmotic value while the phloem parenchyma cells to which they are symplastic-ally connected (Gunning 1976b) do not (Geiger et al. 1973). The same tissue also provides circumstantial evidence for another form of regulation of plasmodesmatal transport. It is held that companion cells "load" the sieve elements in leaf veins, via the branched plasmodesmata that interconnect the two (Geiger, this volume). But to do so, the companion cell must be able to exercise some selectivity, for plasmodesmata lead from the compansion cell in other directions as well as to the sieve elements (Gunning 1976b). Are the plasmodesmata to the border parenchyma and phloem parenchyma occluded in some way, leaving a path of least resistance to the sieve element?

A reversible gelling of plasmodesmatal contents is one way in which symplastic transport could be regulated, and the neck constrictions offer another. In *Abutilon* nectary trichomes (Gunning and Hughes, unpublished) the plasmodesmata of the distal wall of the stalk cell are nearly all constricted during early phases of development. Electron-lucid regions are seen in the cell wall at the constrictions, and aniline blue staining gives a point fluorescence pattern suggestive of the presence of minute flecks of callose. Shortly before the flower petals elongate beyond the sepals (signalling the beginning of nectar secretion) the plasmodesmatal ultrastruc-ture changes. The constrictions and electron-lucid regions disappear from about 80% of those in the distal wall of the stalk cell. No callose fluorescence can then be detected, provided that precautions such as prolonged fixation in glutaraldehyde are taken to avoid the formation of wound callose. The cytoplasmic annulus is then parallel sided and open, as shown in Fig. 1 of Gunning (1976a). The rate of nectar secretion in this material is measured by observing slices mounted in liquid paraffin (Findlay and Mercer 1971), whereupon some trichomes continue to pump out droplets. The majority do not, however, and by infiltrating slices with aniline blue it can be shown that the walls in most of the quiescent trichomes possess scattered but intense point fluorescence sources. It is difficult to obtain clear

views right down to the stalk cell of trichomes that were
secreting before application of the aniline blue, but in all
cases so far obtained, there has been a lack of callose
fluorescence in those trichomes. The obvious hypothesis is
that wound callose can block symplastic transport of pre-
nectar along the trichomes. Whether the "natural" callose
that is present in young but not in mature trichomes plays a
similar role is less clear. We do not know whether it is
responsible for moulding the neck constrictions. Further, its
removal during maturation is not strictly coupled to the
onset of secretion, for young nectaries that are floated on
very dilute sucrose solutions will remove their "point
fluorescence" callose, but not secrete. Those floated on
stronger sucrose both remove their callose and secrete.

CONCLUSION

In plasmodesmata where the cytoplasmic annulus is open,
there is no obvious reason why it should not be a pathway of
transport. Where it is constricted, the desmotubule/endo-
plasmic reticulum system has to be considered seriously as a
functional pathway. Irrespective of which pathway functions,
however, the dimensions are such that small solutes will be
able to pass while ultrafiltration will become progressively
more serious as the particle size is increased, until the
average protein molecule will be virtually excluded from the
symplastic transport system - a conclusion that is in full
accord with Strasburger's view, as described in the introduction, that genetic information must not pass from cell to
cell.

One type of genetic information, nevertheless, does
pass from cell to cell. A wide variety of virus particles is
mobile within the symplast, and with few exceptions, they
are much too large to enter plasmodesmata (Gibbs 1976). It
would seem that an essential feature of plant viruses is
that they have the ability to modify plasmodesmata. Without
this they could not spread from the wounded cell where they
enter the plant, for, as already pointed out, the plasmodesmata leading to wounded cells become sealed. Without it
they could not spread through healthy tissue, for the plasmodesmatal channels are too small, and viruses do not
move through cell walls. Two of the modifications that have
been seen are enlargement of the plasmodesmatal lumen, and
removal of the desmotubule (see Gibbs 1976). That viruses
have evolved strategies for bringing about such modifications
serves to highlight the idea that the design of plasmodesmata
indeed prevents (under normal conditions) leakage of information from cell to cell, and leads us to two intentionally

provocative and teleological rationalisations of plasmodesmatal ultrastructure with which to conclude.

Both rationalisations are based on a need for cell to cell transport. The first then goes as follows. Cytoplasmic continuity has too many disadvantages to be permitted, except under special circumstances. Hence the transport pathway is protected by enclosing it within a bounding membrane, the protective membrane being that of the endoplasmic reticulum. But the endoplasmic reticulum cannot itself penetrate two plasma membranes and a cell wall, so a cytoplasmic canal is made to accommodate it, and except under special circumstances, the canal is sealed by neck constrictions. The second rationalisation is that cytoplasmic continuity is permitted, provided that very delicate control by constrictions acting as ultrafiltration valves can be imposed. But this is not possible without having an axial strand onto which the constriction valves can seal (because if a cylinder of membrane were to be constricted so finely as to exclude large molecules, there would be an ineffectively small orifice left for transport of small molecules, whereas with the axial strand a narrow annulus with greater total cross-sectional area, and hence greater carrying capacity for small molecules, can be formed).

Faults can be found in both, and no doubt other ideas could be proposed, but these suggestions do meet one perfectly serious and valid need, and that is for some interpretation of the virtual ubiquity of the desmotubule in higher plant plasmodesmata, and the very frequent presence of neck constrictions. Either we must prove these two structural features to be artefacts, which on present evidence is unlikely, or we must attempt to understand their significance.

Note added in proof: The investigation of plasmodesmata in the distal wall of the stalk cell of *Abutilon* nectary trichomes has been continued in the 1976 flowering season, confirming the results described earlier in relation to the changes in plasmodesmatal constrictions and callose sleeves as the trichomes mature. However, by using rapid freezing techniques it has now been shown that the callose around the plasmodesmata of young trichomes arises during fixation with glutaraldehyde. Since mature nectary trichomes do not react in this way, the conclusion that there is a change as the trichomes become secretory remains valid, but doubt is cast on the idea that callose might be responsible for constricting the plasmodesmata and thereby regulating symplastic transport in the young trichomes.

REFERENCES

Adam, G. and Delbrück, M. (1968). Reduction of Dimensionality in Biological Diffusion Processes. In: "Structural Chemistry and Molecular Biology" ed. by A. Rich and N. Davidson, Freeman, San Francisco pp 198–215.

Anderson, W.P. (1976). Physico-Chemical Assessment of Plasmodesmatal Transport. In: "Intercellular Communication in Plants: Studies on Plasmodesmata" ed. by B.E.S. Gunning and A.W. Robards, Springer, Heidelberg pp 107–120.

Boeke, J.H. (1971). Location of the postgenital fusion in the gynoecium of *Capsella bursa-pastoris* (L) Med. *Acta Bot. Neerl. 20*, 570–576.

Boeke, J.H. (1973) The use of light microscopy versus electron microscopy for the location of postgenital fusions in plants. *Proc. K. Ned. Akad. Wet. Ser.* C. *76*, 528–535.

Bräutigam, E. and Müller, E. (1975). Transportprozesse in *Vallisneria* - Blättern und die Wirking von Kinetin und Kolchizin. III. Induktion einer Senke durch Kinetin. *Biochem. Physiol. Pflanzen 167*, 29–39.

Brinckmann, E. and Lüttge, U. (1974). Lichtabhängige Membranpotentialschwankungen und deren interzelluläre Weiterleitung bei panaschierten Photosynthese-Mutanten von *Oenothera*. *Planta (Berl) 119*, 47–57.

Burgess, J. (1971). Observations on structure and differentiation in plasmodesmata. *Protoplasma 73*, 83–95.

Campbell, N. and Thomson, W.W. (1975). Chloride localization in the leaf of *Tamarix*. *Protoplasma 83*, 1–14.

Carde, J.P. (1974). Le tissu de transfert (= cellules de Strasburger) dans les aiguilles du pin maritime (*Pinus pinaster* Ait.). II. Caractères cytochimiques et infrastructuraux de la paroi et des plasmodesmes. *J. Microscopie 20*, 51–72.

Carr, D.J. (1976). Plasmodesmata in Growth and Development. In: "Intercellular Communication in Plants: Studies on Plasmodesmata" ed. by B.E.S. Gunning and A.W. Robards, Springer, Heidelberg pp 243–290.

Dick, P.S. and ap Rees, T. (1975). The pathway of sugar transport in roots of *Pisum sativum*. *J. Exp. Bot. 26*, 305–314.

Dörr, I. (1968). Plasmatische Verbindungen zwischen artfremden Zellen. *Naturwissenschaften 55*, 396.

Findlay, N. and Mercer, F.V. (1971). Nectar production in *Abutilon*. I. Movement of nectar through the cuticle. *Aust. J. Biol. Sci. 24*, 647–656.

Fraser, T.W. and Smith, D.L. (1974). Young gametophytes of

the fern *Polypodium vulgare* L. An ultrastructural study. *Protoplasma 82*, 19–32.

Fulcher, R.G. and McCully, M.E. (1971). Histological studies on the genus *Fucus* V. An autoradiographic and electron microscopic study of the early stages of regeneration. *Can. J. Bot. 49*, 161–165.

Geiger, D.R. and Cataldo, D.A. (1969). Leaf structure and translocation in sugar beet. *Plant Physiol. 44*, 45–54.

Geiger, D.R., Giaquinta, R.T., Sovonick, S.A. and Fellows, R.J. (1973). Solute distribution in sugar beet leaves in relation to phloem loading and translocation. *Plant Physiol. 52*, 585–589.

Geiger, D.R., Sovonick, S.A., Shock, T.L. and Fellows, R.J. (1974). Role of free space in translocation in sugar beet. *Plant Physiol. 54*, 892–898.

Gibbs, A.J. (1976). Viruses and Plasmodesmata. In: "Intercellular Communication in Plants: Studies on Plasmodesmata" ed. by B.E.S. Gunning and A.W. Robards, Springer, Heidelberg pp 149–164.

Goodwin, P.B. (1976). Physiological and Electrophysiological Evidence for Intercellular Communication in Plant Symplasts. In: "Intercellular Communication in Plants: Studies on Plasmodesmata" ed. by B.E.S. Gunning and A.W. Robards, Springer, Heidelberg pp 121–130.

Gullvag, B.M., Skaar, H. and Ophus, E.M. (1974). In ultra-Structural study of lead accumulation within leaves of *Rhytidiadelphus squarrosus* (Hedw.) Warnst. A comparison between experimental and environmental poisoning. *J. Bryol. 8*, 117–122.

Gunning, B.E.S. (1976a). Introduction to Plasmodesmata. In: "Intercellular Communication in Plants: Studies on Plasmodesmata" ed. by B.E.S. Gunning and A.W. Robards, Springer, Heidelberg pp 1–14.

Gunning, B.E.S. (1976b). The Role of Plasmodesmata in Short Distance Transport to and from the Phloem. In: "Intercellular Communication in Plants: Studies on Plasmodesmata" ed. by B.E.S. Gunning and A.W. Robards, Springer, Heidelberg pp 203–228.

Gunning, B.E.S., Pate, J.S., Minchin, F.R. and Marks, I. (1974). Quantitative aspects of transfer cell structure in relation to vein loading in leaves and solute transport in legume nodules. *Symp. Soc. Exp. Biol. 28*, 87–126.

Gunning, B.E.S. and Robards, A.W. (1976). Plasmodesmata: Current Knowledge and Outstanding Problems. In: "Intercellular Communication in Plants: Studies on Plasmodesmata" ed. by B.E.S. Gunning and A.W. Robards, Springer, Heidelberg pp. 297–312.

Gunning, B.E.S. and Steer, M.W. (1975). "Ultrastructure and the biology of plant cells", Edward Arnold, London.

Helder, R.J. (1967). Translocation in *Vallisneria spiralis*. In: "Encyclopedia of Plant Physiology Vol. 13" ed. by W. Ruhland, Springer, Berlin pp 30-43.

Israelachvili, J.N. and Mitchell, D.J. (1975). A model for the packing of lipids in bilayer membranes. *Biochim. biophys. Acta 389*, 13-19.

Jarvis, P. and House, C.R. (1970). Evidence for symplasmic ion transport in maize roots. *J. Exp. Bot. 21*, 83-90.

Jones, M.G.K. (1976). The Origin and Development of Plasmodesmata. In: "Intercellular Communication in Plants: Studies on Plasmodesmata" ed. by B.E.S. Gunning and A.W. Robards. Springer, Heidelberg pp 81-106.

Jones, M.G.K., Novacky, A. and Dropkin, V.H. (1975). Transmembrane potentials of parenchyma cells and nematode-induced transfer cells. *Protoplasma 85*, 15-37.

Juniper, B.E. and Barlow, P.W. (1969). The distribution of plasmodesmata in the root tip of maize. *Planta 89*, 352-360.

Kaufman, P.B., Petering, L.B., Yocum, C.S. and Baic, D. (1970) Ultrastructural studies on stomata development in internodes of *Avena sativa*. *Amer. J. Bot. 57*, 33-49.

Kollmann, R. and Dörr, I. (1969). Strukturelle Grundlagen des zwischenzelligen Stoffaustausches. *Ber. Deutsch Bot. Ges. 82*, 415-425.

Krull, R. (1960). Untersuchungen über den Bau und die Entwicklung der Plasmodesmen im Rindenparenchym von *Viscum album*. *Planta (Berl) 55*, 598-629.

Kuo, J., O'Brien, T.P. and Canny, M.J. (1974). Pit field distribution, plasmodesmatal frequency and assimilate flux in the mestome sheath cells of wheat leaves. *Planta (Berl.) 121*, 97-118.

Lüttge, U. (1971). Structure and function of plant glands. *Ann. Rev. Plant Physiol. 22*, 23-44.

MacRobbie, E.A.C. (1971). Phloem translocation, facts and mechanisms: a comparative survey. *Biol. Rev. 46*, 429-481.

Marchant, H.J. (1976). Plasmodesmata in Algae and Fungi. In: "Intercellular Communication in Plants: Studies on Plasmodesmata" ed. by B.E.S. Gunning and A.W. Robards, Springer, Heidelberg pp 59-80.

Meeuse, A.D.J. (1957). Plasmodesmata (Vegetable kingdom). Protoplasmatologia IIAlc, 1-43.

Mesquita, J.F. (1970). Ultrastructura do meristema radicular de *Allium cepa* L. e suas alterções induzidas por agentes mitoclásicos e radiomiméticos. *Rev. Fac. Cienc. Coimbra*

44, 1-201.

Münch, E. (1930). "Die Stoffbewegung in der Pflanze", Jena, Gustav Fischer.

O'Brien, T.P. and Thimann, K.V. (1967). Observations on the fine structure of the oat coleoptile. II. The paren- chyma cells of the apex. *Protoplasma 63*, 417-442.

Olesen, P. (1975). Plasmodesmata between mesophyll and bundle sheath cells in relation to the exchange of C_4- Acids. *Planta (Berl.) 123*, 199-202.

Ophus, E.M. and Gullvag, B.M. (1974). Localization of lead within leaves of *Rhytidiadelphus squarrosus* (Hedw.) Warnst. by means of transmission electron microscopy and X-ray microanalysis. *Cytobios 10*, 45-58.

Osmond, C.B. and Smith, F.A. (1976). Symplastic Transport of Metabolites during C_4-Photosynthesis. In: "Intercellular Communication in Plants: Studies on Plasmodesmata" ed. by B.E.S. Gunning and A.W. Robards, Springer, Heidelberg pp. 229-242.

Paine, P.L. and Scherr, P. (1975). Drag coefficients for the movement of rigid spheres through liquid-filled cylind- rical pores. *Biophys. J. 15*, 1087-1091.

Pate, J.S., Gunning, B.E.S. and Briarty, L.G. (1969). Ultra- structure and functioning of the transport system of the leguminous root nodule. *Planta (Berl.) 85*, 11-34.

Pitman, M.G., Mertz, S.M., Graves, J.S., Pierce, W.S. and Higinbotham, N. (1970). Electrical potential differences in cells of barley roots and their relation to ion uptake. *Plant Physiol. 17*, 76-80.

Reed, M.L., Findlay, N. and Mercer, F.V. (1971). Nectar production in *Abutilon*. IV. Water and solute relations. *Aust. J. Biol. Sci. 24*, 677-688.

Robards, A.W. (1976). Plasmodesmata in Higher Plants. In: "Intercellular Communication in Plants: Studies on Plasmodesmata" ed. by B.E.S. Gunning and A.W. Robards, Springer, Heidelberg pp 15-58.

Robards, A.W. and Clarkson, D.T. (1976). The Role of Plasmod- esmata in the Transport of Water and Nutrients across Roots. In: "Intercellular Communication in Plants: Studies on Plasmodesmata" ed. by B.E.S. Gunning and A.W. Robards, Springer, Heidelberg pp 181-202.

Roland, J.-C. (1973). The relationship between the plasmalemma and the plant cell wall. *Int. Rev. Cytol 36*, 45-92.

Schmitz, K. and Srivastava, L.M. (1974). Fine structure and development of sieve tubes in *Laminaria groenlandica*, Rosenv. *Cytobiologie 10*, 66-87.

Schulz, P. and Jensen, W.A. (1968). *Capsella* embryogenesis: The early embryo. *J. Ultrastruct. Res. 22*, 376-392.

Schulz, P. and Jensen, W.A. (1971) *Capsella* embryogenesis: The chalazal proliferating tissue. *J. Cell Sci. 8*, 201–227.

Smith, D.L. (1972). Staining and osmotic properties of young gametophytes of *Polypodium vulgare* L. and their bearing on rhizoid formation. *Protoplasma 74*, 465–479.

Sovonick, S.A., Geiger, D.R. and Fellows, R.J. (1974). Evidence for active phloem loading in the minor veins of sugar beet. *Plant Physiol. 54*, 886–891.

Spanswick, R.M. (1972). Electrical coupling between cells of higher plants: A direct demonstration of intercellular communication. *Planta (Berl.) 102*, 215–227.

Spanswick, R.M. (1976). Symplasmic transport in tissues. In: "Encyclopaedia of Plant Physiology", New Series, Vol. 2. Transport in Plants, ed. by U. Lüttge and M. G. Pitman, Springer, Berlin.

Spanswick, R. and Costerton, J. (1967). Plasmodesmata in *Nitella translucens*: Structure and electrical resistance. *J. Cell Sci. 2*, 451–464.

Steveninck, R.F.M. van (1976) Cytochemical Evidence for Ion Transport through Plasmodesmata. In: "Intercellular Communication in Plants: Studies on Plasmodesmata" ed. by B.E.S. Gunning and A.W. Robards, Springer, Heidelberg pp 131–148.

Strasburger, E. (1901). Ueber Plasmaverbindungen pflanzlicher Zellen. *Jb. Wiss. Bot. 36*, 493–610.

Tainter, F.H. (1971). The ultrastructure of *Arceuthobium pusillum. Can. J. Bot. 49*, 1615–1622.

Taiz, L. and Jones, R.L. (1973). Plasmodesmata and an associated cell wall component in barley aleurone tissue. *Amer. J. Bot 60*, 67–75.

Tyree, M.T. (1970). The symplast concept. A general theory of symplastic transport according to the thermodynamics of irreversible processes. *J. Theor. Biol 26*, 181–214.

Tyree, M.T. and Tammes, P.M.L. (1975). Translocation of uranin in the symplasm of staminal hairs of *Tradescantia. Can. J. Bot. 53*, 2038–2046.

Vela, A. and Lee, P.E. (1975). Infection of leaf epidermis by wheat striate mosaic virus. *J. Ultrastruct. Res. 52*, 227–234.

Vian, B. and Rougier, M. (1974). Ultrastructure des plasmodesmes après cryoultramicrotomie. *J. Microscopie 20*, 307–312.

Walker, N.A. (1976). Transport of Solutes through the Plasmodesmata of *Chara* Nodes. In: "Intercellular Communication in Plants: Studies on Plasmodesmata" ed. by B.E.S. Gunning and A.W. Robards, Springer, Heidelberg pp. 165–180.

Went, J.L. van, Aelst, A.C. van and Tammes, P.M.L. (1975) Anatomy of staminal hairs from *Tradescantia* as a background for translocation studies. *Acta. Bot. Neerl.* *24*, 1-6.

Werker, E. and Vaughan, J.G. (1974). Anatomical and ultrastructural changes in aleurone and myrosin cells of *Sinapis alba* during germination. *Planta (Berl.)* *116*, 243-255.

Wooding, F.B.P. (1968). Fine structure of callus phloem in *Pinus pinea*. *Planta (Berl.)* *83*, 99-110.

Worley, J.F. (1968). Rotational streaming in fibre cells and its role in translocation. *Plant Physiol.* *43*, 1648-1655.

Zee, S. -Y. and O'Brien, T.P. (1970). Studies on the ontogeny of the pigment strand in the caryopsis of wheat. *Aust. J. Biol. Sci.* *23*, 1153-1171.

Ziegler, H. and Lüttge, U. (1967). Die Salzdrüsen von *Limonium vulgare*. II Mitt Die Lokalisierung des Chlorids. *Planta (Berl.)* *74*, 1-17.

CHAPTER 3

The Effects of Flow in Solute Transport in Plants

N.A. Walker

School of Biological Sciences, University of Sydney, N.S.W., Australia

INTRODUCTION

TRANSPORT MECHANISMS

The transport of solutes in a plant is accomplished by the following processes:
    (a) membrane transport - covering a distance of 7 nm, and
    (b) transport within an aqueous phase - covering any
        distance.
Here we will only be concerned with transport within one aqueous phase; membrane transport is mentioned for completeness and because both types of transport can be further categorised as involving either or both of diffusion and bulk flow. Their important characteristics from our point of view are:

	SPECIFITY	DISTANCE COVERED
diffusion	depends on D	$\sqrt{tD}$, say 1 mm
flow	quite unspecific	tv, say 1 dm

Although flow may be produced in various strange ways (e.g. in the nerve axon), we can claim that all transport depends on one or other of these mechanisms.

SYMPLAST TRANSPORT

Solute transport within the symplast of the multicellular plant was considered by Tyree (1970); from general considerations and from available data, he concluded:
    (i)   within each cell, diffusion and flow (due to
          cyclosis) will keep the contents well mixed;
    (ii) between cells, the route of transport will be the
          plasmodesmata, where they exist in the usual
          numbers, and
    (iii) the mechanism of movement of solutes through plasmo-
          desmata will be diffusion, since bulk flow would be
          an improvement only if it were very fast or the
          diffusion coefficient were very small.

A corollary of (i) is that the rate of transport of solute through a symplast will be determined by intercellular transport, not by intracellular transport. It would follow, further, that measurement of the rate of intercellular or symplast transport would offer evidence bearing on (iii), since a rate too high for diffusion would require flow.

Measurement of the intercellular transport rate might also be brought to bear on (ii), which is commonsense, but hard to test.

THE RATE LIMITING STEP

Tyree (1970) derived postulate (i) from an argument based on drift velocities of molecules in cytoplasm and in plasmodesmata. Though plausible, this argument neglected a basic physical fact, that (in *Chara* and in many plants) a narrow ribbon of moving cytoplasm crosses the entrances of a large array of plasmodesmata without its solutes being replenished. The argument, given by Walker (1976), can be summarised as follows. If we consider the *Chara* node (Fig. 1), and the route by which solutes can reach the array of plasmodesmata between A and C, we find that only flow of the ribbon of

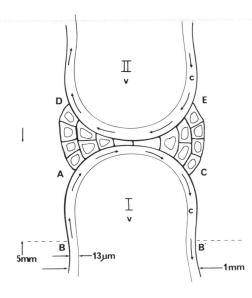

Fig. 1. - Section through node of charophyte plant, showing ends of two internodal cells (I and II). Arrows show streaming directions. c, cytoplasm; v, vacuole; B,B' external barrier, below which the solute is present in radioactive form.

cytoplasm from B to A can make any important contribution. Since the speed of flow is about 100 $\mu m\ s^{-1}$, diffusion of a small solute can make a contribution only over distances of ~ 20 μm (=D/v) in the longitudinal direction. Although for

each solute the case must be examined separately, it does seem
clear that for chloride there will be no significant contribution from the vacuole, during the 10-15 sec that the cytoplasmic ribbon takes to pass from A to C. The temporary isolation of the cytoplasmic ribbon as it passes the node means
that it can equilibrate with the cytoplasm of the node cells,
if the plasmodesmata are permeable enough, or if the flow is
slow enough.

COUNTER-CURRENT FLOW

Fritsch (1965) describes, and observation confirms, the
counter-current flow of the streams of cytoplasm in every
adjacent pair of internodal cells. If cyclosis were rapid
and the plasmodesmata rather impermeable, this could have
little or no functional significance: the stream of cytoplasm
would have essentially the same solute concentrations at C as
at A, as Tyree postulated (see (i) above). For a consideration of the properties of counter-current and of co-current
junctions (e.g. Walker 1976) shows that they only differ
significantly when flow is slow and permeability high. In
this case a counter-current junction can effect essentially
complete transfer of solutes in the stream at A to the stream
at D, while the co-current junction must always transfer less
than half.

Thus it was suggested (Walker and Bostrom 1973) that the
rate of intercellular transport in *Chara* might be determined
by the rate at which solute is brought to the node in the
stream of cytoplasm. The critical condition is PL $\gtrless$ vδ
(Walker 1976); if PL is the greater, the rate of transport is
given by:

$$J_N = \pi v \delta c d \qquad \ldots \ldots (1)$$

while if vδ is the greater, the rate of transport is:

$$J_N = \pi d \, PL \, c$$

where for a hemispherical node $\pi d \, L \doteq 2\pi d^2$, so

$$J_N \doteq 2\pi d^2 \, P \, c \qquad \ldots \ldots (2)$$

Since the interposition of the many small node cells between
the counter-current streams may alter the properties of the
junction, we may write an empirical factor T, the transfer
efficiency, in equation (1):

$$J_n = \pi d v \delta c T \qquad \ldots \ldots (3)$$

Here T will be 1.0 for a perfect counter-current junction,
and less for an imperfect one; when equation (2) obtains,
we could still use (1) by allowing T to have the value

$$T = \frac{2dP}{v\delta} \qquad \cdots \cdots (4)$$

Thus we can express the critical question for any plant system in several equivalent ways:

(a) does PL exceed $v\delta$? Here P will be given by αDw, so the question is equivalent to: does D exceed $v\delta/\alpha wL$? Since D is not known *a priori*, this is not immediately useful.

(b) does J_N depend on v? This is equivalent to asking: is T independent of v?

INFORMATION FROM EXPERIMENTS

CHARA CORALLINA

The experimental system so far used for all work has been that shown in Fig. 2. A pair of *Chara* internodes, trimmed of all laterals, lies in a chamber with three compartments. Of these A contains, during the experimental period, radioactive solute; C contains inactive solution; B may contain moist air, solution, or solid perspex. The cell is sealed in with

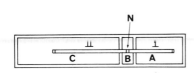

Fig. 2. - Chamber used for intercelluar transport measurements on pairs of adjacent Chara internodes. A, compartment containing radioactive solutes; B, isolating comparment; C, compartment bathing cell II.

silicone grease. At the end of the experimental period, the cells I, II and the intervening node N are separately counted.

Polarity of Transport. - Bostrom's measurements with preparations oriented for acropetal or basipetal transport showed no polarity of chloride transport (Bostrom and Walker 1975, referred to as I). This was not unexpected, but is an important negative, since polar transport would have implied a flow mechanism in the plasmodesmata, the preparations being essentially symmetrical in every other way.

Rate of Transport. - The rate at which radioactive tracer leaves cell I (Fig. 2) is relatively easy to determine, but it will be convertible to a rate of transport only if the specific activity at A can be determined. For a foreign, stable molecule whose specific activity is constant everywhere in the system, the concentration at A is the corresponding unknown. The rate of transport will be given by an equation:

$$J_N = (Y_N + Y_{II})/(\int_o^{\Delta t} s_A \, dt) \qquad \cdots \cdots (5)$$

where no attempt is being made to express J_N as a flux (rate of transport per unit area and time). Ideally we would compare this experimental value of J_N with the value predicted by equation (3): we could then evaluate T, and for good measure see if it is near 1, and independent of v, or low, and dependent on v according to equation (4). Bostrom (I) found formidable difficulties in accomplishing this for *Chara*, both c and s_A being difficult to determine. The value of J_N for chloride was found to be within the interval 6–60 pmol/S. Different experiments (listed e.g. in Hope and Walker 1975, Table 5.5) have given values of c ranging from 10 to 90 mol m^{-3} (mM), so that the value of T remains uncertain, at least for chloride.

Kinetic Model. – Bostrom (I) was however able to show that, for 1800 s $\leq$ Δt $\leq$ 21600 s, the fractions of radioactive solute found in (a) the vacuole of cell I (F_v), (b) the node of cell II (F_t) and (c) the "whole cytoplasm" of cell I (F_s) were all independent of Δt. Thus the kinetic model for intercellular chloride transport is as shown in Fig. 3a, where the halftimes of the three compartments at the top of the diagram are all large (> 10^4 s), and that of the small compartment is of the order of 10^2–10^3 s.

Dependence on v. – Since the dependence of J_N on v could not easily be determined, Bostrom (Bostrom and Walker 1976, referred to as II) investigated the dependence of F_t on v. It turned out that F_t was linearly dependent on v at all speeds up to the normal speed, when v was altered by treatment of cell I with cytochalasin B. It was concluded that F_t, and hence J_N, were proportional to v in untreated cells, and hence that equation (3) applies to chloride in *Chara* with a constant value of T.

Direct Measurement of T. – This may be accomplished by using a foreign solute whose concentrations at A can be approximately determined. We have used 5,5-dimethyloxazolidine-2, 4-dione, suggested by Waddell and Butler for intracellular pH measurement, and applied to *Chara* by Walker and Smith (1975). This weak acid accumulates in the cytoplasm, where its concentration may be determined relatively accurately by classical separation techniques. The major uncertainty is its distribution between flowing and stationary layers of the cytoplasm, and work is in progress to determine this ratio. Assuming that its value is 0.5, current experiments give a median value for T of 0.74 (Walker and Evans, unpublished); these experiments, whose theoretical basis is given in Walker (1976), are based on the kinetic model of Fig. 3b.

DISCUSSION AND CONCLUSIONS

THE *CHARA* NODE

For *Chara* then there is evidence from two quite independ-

a.

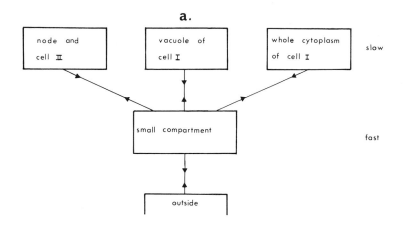

b.

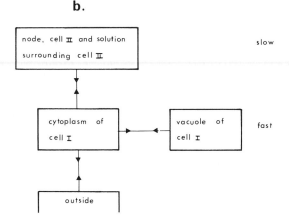

Fig. 3. - Kinetic models for intercellular transport out of cell I in the experimental arrangement of Fig. 2.

ent experiments that Tyree's postulate (i) does not hold. For chloride, T is independent of v, according to the experiments with cytochalasin B: the arguments of Walker (1976) would therefore suggest that T is high (near 1.0) for chloride. For the weak acid DMO, T has been measured as about 0.74 in cells at normal streaming rates, so that it must be nearly or quite independent of v. For both solutes then the rate of intercellular transport is limited by the rate of streaming and not by the permeability of the plasmodesmata.

Tyree's other postulates are however strengthened by the *Chara* results. The rate of intercellular transport of chloride was found to exceed 6 pmol s^{-1}: if this were a membrane flux it would be greater than 6 µmol m^{-2} s^{-1}. The largest chloride fluxes normally found across the plasmalemma are about 60 nmol m^{-2} s^{-1}, and these fluxes are sensitive to darkness and to inhibitors, while intercellular transport is not (Bostrom, unpublished). The plasmodesmata form the only plausible route for such a rate of transport.

The dimensions and frequency of *Chara* plasmodesmata (I) allow the calculation of a maximum diffusion permeability of 40 µm s^{-1}, so that the rate of transport carried by diffusion in the plasmodesmata could be as high as 40 pmol s^{-1} for each mol m^{-3} of chloride in the cytoplasm. Thus a concentration of 1.5 mol m^{-3} would be enough to produce the highest flux required by the experiments, 60 pmol s^{-1}. The actual concentration seems to lie between 10 and 90 mol m^{-3}. We need not assume flow in the plasmodesmata. This might be better tested by the use of a solute with small diffusion coefficient.

COMPARISON WITH *TRADESCANTIA*

Tyree and Tammes (1975), working with the staminal hair, have shown that uranin is transported through a constant number of cells per unit time, not through a constant distance. The inference is that the rate of transport is limited by passage through the intercellular walls, and on this basis they calculate the apparent permeability of the plasmodesmata and the apparent diffusion coefficient of the dye. Their figures show that for the staminal hair PL is about 0.1 of vδ, so that according to the criterion stated above the rate should indeed be limited by the permeability of the plasmodesmata.

REFERENCES

Bostrom, T.E. and Walker, N.A. (1975). - Intercellular transport in plants. I. The rate of transport of chloride and the electric resistance. *J. Exp. Bot.* 26, 767-782.

Bostrom, T.E. and Walker, N.A. (1976). - Intercellular transport in plants. II. Cyclosis and the rate of intercellular transport in *Chara*. *J. Exp. Bot (in press).*

Fritsch, F.E. (1965). - "The structure and reproduction of the algae" Cambridge *Univ. Press*, Cambridge.

Hope, A.B. and Walker, N.A. (1975). - "The physiology of giant algal cells." Cambridge Univ. Press, Cambridge.

Tyree, M.T. (1970). - The symplast concept. A general theory of symplastic transport according to the thermodynamics of irreversible processes. *J. Theor. Biol. 26*, 181-214.

Tyree, M.T. and Tammes, P.M.L. (1975). - Translocation of

uranin in the symplasm of the staminal hairs of *Trades-cantia*. *Can. J. Bot.* (In press).

Walker, N.A. (1976). - Transport of solutes through the plasmo-desmata of *Chara* nodes. In: "Intercellular communication in plants" ed. by B.E.S. Gunning and A.W. Robards, Springer-Verlag, Berlin-Heidelberg.

Walker, N.A. and Bostrom, T.E. (1973). - Intercellular movement of chloride in *Chara*. In: "Ion transport in plants" ed. by W.P. Anderson. Academic Press, London, pp447-458.

Walker, N.A. and Smith, F.A. (1975). - Intracellular pH in *Chara corallina* measured by DMO distribution. *Pl. Sci. Letters 4*, 125-132.

SYMBOLS

c	concentration of solute in cytoplasm
d	diameter of cell
D	diffusion coefficient
F_s	fraction of solute, taken up be cell I, that is found in the "whole cytoplasm" of cell I
F_t	fraction of solute, taken up by cell I, that is transported out of cell I
F_v	fraction of solute, taken up by cell I, that is found in the vacuole of cell I
J_N	rate of transport of solute out of cell I
L	length of array of plasmodesmata (in direction of streaming)
P	permeability of array of plasmodesmata
s_A	specific activity of solute in compartment A
t	time
Δt	solute uptake time
T	transfer efficiency of node (see eqn 3)
v	speed of bulk flow
w	length of plasmodesma (thickness of wall)
Y_N	radioactivity of node
Y_{II}	radioactivity of cell II
δ	thickness of cytoplasm
α	fraction of wall area occupied by plasmodesmata

CHAPTER 4

Cytoplasmic Streaming in Characean Algae

R.E. Williamson
Botany Department, LaTrobe University, Bundoora, Victoria, Australia

Actin has long been known as the major protein of the thin filaments of muscle. The recent demonstrations of actin in the cells of characean algae (Palevitz et al. 1974; Williamson 1974) and angiosperms (Condeelis 1974; Forer and Jackson 1975) have added plants to non-muscular cells from a wide range of animals known to contain the protein (Pollard and Weihing 1974). In spite of its highly conserved structure, it appears able to participate in a wide range of different motile and contractile processes. This apparently results from variation in its pattern of assembly into subcellular structures and in the properties of the myosin and regulatory proteins with which it interacts.

It is of interest to plant physiologists because of its involvement in cytoplasmic streaming and because of the possibility that the P-protein of sieve elements may be actin (Ilker and Currier 1974; but see Palevitz and Hepler 1975a). The giant cells of the characean algae offer unusual experimental opportunities and have made possible the recent formulation of quite detailed hypotheses for the mechanism of force generation. It is the aim of this paper to discuss the two most detailed hypotheses.

THE STRUCTURE OF CHARACEAN CELLS

Characean cells are cylinders with a diameter of about 1 mm and a length which may exceed 15 cm. Within the plasma membrane is a cylinder of stationary cortical cytoplasm containing the cell's complement of microtubules and chloroplasts. This surrounds a cylinder of endoplasm which streams in an unchanging spiral pattern along the length of the cell. Within the endoplasm, almost all the organelles move with the same velocity. However, in the narrow zone at the boundary of the cortex and the endoplasm, both faster and slower moving organelles can be seen (Kamitsubo 1969; Williamson, unpublished).

In characean cells fixed by conventional methods, bundles of microfilaments are the only feature observed likely to be of direct relevance to the mechanism of force generation (Nagai and Rebhun 1966). The orientation of these bundles parallels the direction of streaming and they occur near the boundary of the cortical and endoplasmic layers. In living cells the bundles appear in the light microscope as fibres. Although usually straight (Kamitsubo 1972), they may show abrupt bends (Williamson 1975). No movements of these sub-cortical fibres themselves are observed, although they may become motile if freed from the cortex (Kamitsubo 1972).

A recent light microscope study (Allen 1974) has suggested ed that branches of the sub-cortical fibres hang down into the endoplasm to form a system of endoplasmic fibres. No ultra-structural evidence is currently available to confirm this view.

MECHANISMS OF FORCE GENERATION

Two proposals for the site and method of force generation have recently been made. The first is that endoplasmic fibres generate the force for streaming by the propagation of waves with an amplitude of about 5 μm (Allen 1974). The second (Williamson 1975) invokes only the sub-cortical fibres, this time acting without conformational changes of large amplitude. It is a particular form of earlier, more general proposals for motility invoking "active shearing."

Endoplasmic fibres. - The evidence presented for the occurrence and role of endoplasmic fibres is threefold: the appearance of organelles moving in sinusoidal arrays; the presence of immobile fibres in the endoplasm when streaming is halted by an action potential; the direct visualisation on a ciné film of wave propagation along these fibres during streaming. As only the latter point directly substantiates the model, it is unfortunate that frames from this film were not presented in the paper.

However, it is not the validity of the experimental observations that I wish to examine at this point but rather the validity of the conclusions drawn from the observations. The title of the paper makes the explicit statement that "endoplasmic filaments generate the motive force for rotational streaming in *Nitella*." Can this be justified?

The argument in the paper is that endoplasmic fibres can be seen and the parameters of the waves they propagate measured. These results are then inserted in equations to predict the force they would generate and the velocity of streaming that would result. The match with the observed values is reasonable (0.99 Pa calculated, 0.14-0.36 Pa measured; 27 μm s^{-1} calculated, 88 μm s^{-1} observed).

However, to accept the statement of the title to the paper, it has to be shown that either the experimental observations or the calculations from them preclude the operation of any other mechanism of force generation. This cannot be accepted. The numerous uncertainties in picking values for the terms in the equations are discussed in some detail by Allen. They are such as to make the final values of forces and velocities no more than rough indicators of the quantitative feasibility of the model. They could allow for the endoplasmic fibre system to make a contribution to the motive force ranging from insignificance to being the sole contributor. Similarly, as far as the experimental observations are concerned there is nothing in them to preclude the operation simultaneously of a second system of force generation.

Thus on both theoretical and experimental grounds we are entitled to reject the statement in the title of Allen's paper. Endoplasmic fibres stand as an interesting new suggestion which will doubtless lead to the more detailed studies of their nature and properties which are required.

Sub-cortical fibres. - The sub-cortical fibres are the only established location (Palevitz and Hepler 1975b) of the actin known to occur in characean cells (Palevitz et al. 1974; Williamson 1974). They are part of the stationary cortex in that they are not washed out by perfusion (Williamson 1975) nor dislodged by centrifugation severe enough to displace the endoplasm to the end of the cell (Williamson 1974). No movements of the fibres themselves are observed.

The most direct evidence for the role of the sub-cortical fibres in force generation comes from vacuolar perfusion (Williamson 1975). By opening both ends of the cell and perfusing isotonic solution through the central vacuole, most of the endoplasm can be forced to reverse its direction of movement in response to the forces set up by the flowing solution. Organelles close to the surface of the sub-cortical fibres alone persist in moving forwards. The force for these persistent movements must be generated locally and not transmitted from elsewhere in the endoplasm by viscous properties. The subsequent behaviour of the system lends further strength to the view that force generation occurs around the sub-cortical fibres and suggests something of the possible nature of the process.

When the bulk of the endoplasm has been swept away by perfusion, the forward movements of the organelles near the cortical fibres abruptly cease and the organelles become immobilised along the fibres. The addition of ATP leads to these immobilised organelles resuming movement. These movements are in the original direction of streaming and at velocities up to those found *in vivo*. They are very obviously

53

associated with the sub-cortical fibres and follow closely any
bends in their path.

In addition to inducing motility, ATP also weakens the
linkages between organelles and sub-cortical fibres. In the
absence of ATP, organelles cannot be washed off by rapid perfusion, whereas in the presence of ATP they are carried away
by even fairly gentle perfusion. Inorganic pyrophosphate, a
non-hydrolysable analogue of ATP, makes the organelles easier
to wash out but does not support motility.

These observations were rationalised in terms of the
interaction of actin in the fibres with a myosin-like molecule
in the endoplasm. The myosin would have some ability to link
through its tail to the endoplasmic organelles. No novel
features for the interaction of actin and myosin were envisaged, so that the hypothesis was constructed in terms of the
known interactions of actin and myosin in muscle.

By analogy with muscle, the system in the absence of ATP
would be in rigor. All the myosins would be bonded to the
actin filaments, making the organelles resistant to forces
trying to pull them off the fibres. ATP added to actomyosin
binds to myosin causing it to dissociate from actin. When
the ATP molecule has been hydrolysed by the myosin, the myosin
can reattach. The energy released by hydrolysis is then used
to power movement. Current views of muscle function (e.g.
Reedy et al. 1965) envisage this as a change from perpendicular to oblique in the angle at which the myosin is attached
to the actin. The repetition of this cycle by numerous myosins causes thick and thin filaments to move past each other
in muscle. The similar operation of plant myosin could move
organelles relative to actin filaments in the sub-cortical
fibres (Fig. 1). The formation of arrowheads on *Chara* actin
(Williamson 1974) in itself provides evidence that muscle
myosin at least assumes the tilted configuration on *Chara*
actin needed to generate arrowheads (Moore et al. 1970). The
necessity for myosin to detach in the cycle could account for
the ease with which organelles are washed out by solutions
containing ATP. Similarly pyrophosphate, by its ability to
bind without undergoing subsequent hydrolysis, causes myosin
detachment while blocking reattachment. Organelle detachment
without movement would be expected.

The feasibility of such direct interactions between
organelles and actin is demonstrated by the results of a preliminary electron microscope study of such perfused cells
(Fig. 2). The significance of the presence in some sections
of a more dispersed network of filamentous material (Fig. 2)
is currently unclear.

The proposed mechanism differs from muscle and other

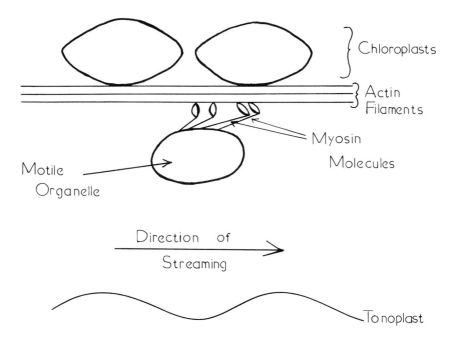

Fig. 1. - A highly diagramatic view of the postulated active shearing mechanism. Myosin is linked through its tail to the organelle. Movement involves by analogy with muscle, a change in the angle made by the myosin head and the actin filament. Two heads show the angle at attachment (perpendicular), while two show the oblique angle after tilting and before detachment. Repetition of the cycle by a number of myosins could generate organelle movements. The arrowheads - composed of the tilted myosin heads - would point to the left, organelles would move to the right.

contractile systems in that no shortening of the system of actin filaments is envisaged. Now shortening is achieved in muscle by having in each sarcomere two sets of actin filaments with opposing polarity, all the arrowheads pointing towards the centre of the sarcomere (Huxley 1963). The two sets of actin filaments are then linked by bipolar myosin aggregates. Now if in this situation the actin were anchored and the myosin free to move, it would do so in the opposite direction to that in which the arrowheads point. These properties have two consequences for an active shearing model in streaming.

Firstly, as all the movements along each sub-cortical fibre are in a single direction, all the actin filaments contributing to these movements should have the same polarity.

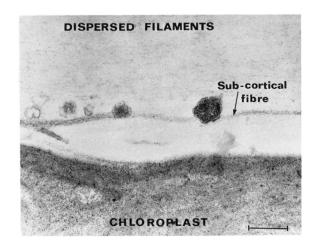

*Fig. 2. - A sub-cortical fibre cut in longitudinal
section. The cell was perfused with an ATP-free
solution and fixed in the immobile state equated
with rigor in muscle. A dispersed network of filamentous material can be seen here and in many other
sections. Its nature and significance are unclear
(Bar = 0.5 μm).*

This has indeed been shown to be the case (Palevitz et al.
1974; Palevitz and Hepler 1975b). Secondly, the arrowheads
should point in the direction opposite to that in which
organelles move. This has also been observed according to
a preliminary report (Kersey 1974).

Thus these two lines of evidence distinguish the sub-
cortical fibres from fibres organised to produce shortening
and both are consistent with an active shear mechanism. It
remains finally to consider some quantitative aspects of
feasibility and whether the sub-cortical fibres could be the
sole source of the motive force for streaming.

The forces required to be generated in association with
an actin filament of given length in a characean cell appear
modest compared with the forces generated in muscle. For
muscle, a force of about 10^{-10}N per actin filament is indicated for frog sartorius muscle (Lowy et al. 1964), For
Nitella, the corresponding figure is some 3 to 4 orders of
magnitude smaller. This figure is derived from measurements
of motive force and number of filaments; the calculation was
carried out to the level of the *Nitella* fibre by MacRobbie
(1971), and can be easily extended to the level of the individual filament.

While the forces involved may in comparative terms be small, there is an intuitive doubt that force could be generated in such a narrow zone and still be able to move the entire endoplasmic mass. The theoretical treatment of Donaldson (1972) suggests that this doubt cannot be sustained and that the viscous properties of the endoplasm are such as to enable it to respond as a unit to force applied at the sub-cortical zone.

CONCLUSIONS

Both undulations of endoplasmic fibres and active shearing by sub-cortical fibres have some degree of experimental support and both appear quantitatively feasible. There is no justification at this stage for attempts to present either as the sole mechanism for force production. Further experiments may lead to the establishment or rejection of both theories. If both can be substantiated, it then remains to decide their relative contributions to force generation.

ACKNOWLEDGEMENTS

My thanks are due to Ms Helen Forward for her valuable assistance with the electron microscopy.

REFERENCES

Allen, N.S. (1974). Endoplasmic filaments generate the motive force for rotational streaming in *Nitella*. *J. Cell. Biol. 63*, 270–287.

Condeelis, J.S. (1974). The identification of F-actin in the pollen tube and protoplast of *Amaryllis belladonna*. *Exp. Cell Res. 88*, 435–439.

Donaldson, I.G. (1972). The estimation of the motive force for protoplasmic streaming in *Nitella*. *Protoplasma 74*, 329–344.

Forer, A., and Jackson, W.T. (1975). Actin in the higher plant *Haemanthus katherinae* Baker. *Cytobiologie 10*, 217–226.

Huxley, H.E. (1963). Electron microscope studies on the structure of natural and synthetic protein filaments from striated muscle. *J. Mol. Biol. 7*, 281–308.

Ilker, R., and Currier, H.B. (1974). HMM complexing filaments in the phloem of *Vicia faba* and *Xylosma congestum*. *Planta (Berl.) 120*, 311–316.

Kamitsubo, E. (1969). Motile protoplasmic fibrils in cells of *Characeae*. *J. Cell Biol. 43*, 166a.

Kamitsubo, E. (1972). Motile protoplasmic fibrils in cells of the *Characeae*. *Protoplasma 74*, 53–70.

Kersey, Y.M. (1974). Correlation of polarity of actin filaments with protoplasmic streaming in characean cells. *J. Cell Biol. 63*, 165a.

Lowy, J., Millman, B.M., and Hanson, J. (1964). Structure and function in smooth tonic muscles of lamellibranch molluscs. *Proc. Roy. Soc., Lond. B. Biol. Sci. 160*, 525–536.

MacRobbie, E.A.C. (1971). Phloem translocation. Facts and mechanisms: a comparative survey. *Biol. Rev. (Camb.) 46*, 429–481.

Moore, P.B., Huxley, H.E., and deRosier, D.J. (1970). Three dimensional reconstruction of F-actin, thin filaments and decorated thin filaments. *J. Mol. Biol. 50*, 279–295.

Nagai, R., and Rebhun, L.I. (1966). Cytoplasmic microfilaments in streaming *Nitella* cells. *J. Ultrastruct. Res. 14*, 571–589.

Palevitz, B.A., Ash, J.F., and Hepler, P.K. (1974). Actin in the green alga *Nitella*. *Proc. Nat. Acad. Sci. U.S.A. 71*, 363–366.

Palevitz, B.A., and Hepler, P.K. (1975a). Is P-protein actin-like? – not yet. *Planta (Berl.) 125*, 261–272.

Palevitz, B.A., and Hepler, P.K. (1975b). Identification of actin *in situ* at the ectoplasm – endoplasm interface of *Nitella*. *J. Cell Biol. 65*, 29–38.

Pollard, T.D., and Weihing, R.R. (1974). Actin and myosin and cell movement. *C.R.C. Crit. Rev. Biochem. 2*, 1–65.

Reedy, M.K., Holmes, K.C., and Tregear, R.T. (1965). Induced changes in orientation of the cross-bridges of glycerinated insect flight muscle. *Nature (Lond.) 207*, 1276–1280.

Williamson, R.E. (1974). Actin in the alga *Chara corralina*. *Nature (Lond.) 241*, 801–802.

Williamson, R.E. (1975). Cytoplasmic streaming in *Chara*: a cell model activated by ATP and inhibited by cytochalasin B. *J. Cell Sci. 17*, 655–668.

CHAPTER 5

Transfer Cells

T.P. O'Brien

Botany Department, Monash University, Clayton, Victoria
Australia

Transfer cells can be defined in functional terms as
cells engaged in transport but in this account, attention is
focussed on cells that display a distinctive morphological
feature, namely the occurrence of wall ingrowths which sharply
enhance the area (3-10 fold) of the cell membrane-apoplast
boundary.

OCCURRENCE AND FUNCTION

My first encounter with transfer cells was in 1963 when
I mistook them for differentiating tracheary elements in the
coleoptilar node of wheat. They were seen, and again unrecog
nized, by Gunning and myself, in pea leaves in 1966 but the
Belfast group soon afterwards recognized their common occurr
ence in minor veins of leaves and drew attention to their
probable function (Gunning and Pate 1969, Pate and Gunning
1969, 1972). Such cells have since been located in a wide
variety of situations in which there is good reason to believe
that there is an intensive flux of solutes. Meanwhile, we had
been studying the anatomy of the vascular system that feeds
the developing grain of wheat. Alerted by preprints of the
work in Belfast, we recognized that the vascular parenchyma
cells at the base of the ovary, sterile glumes, and lemma and
palea of wheat showed transfer cell characters (see Figs. 1
and 2). Such modified cells were present in these reproductive tissues only in the nodal regions where fusions were
occurring and where bundles lacked a mestome sheath (see Zee
and O'Brien 1971 for details). A check of vegetative material showed that the same was true for all nodes (O'Brien and
Zee 1971). Our discoveries in wheat were shown soon afterwards by the Belfast group to be but a special case of a
widespread phenomenon when they reported the widespread association of nodes and vascular transfer cells (Gunning et al.

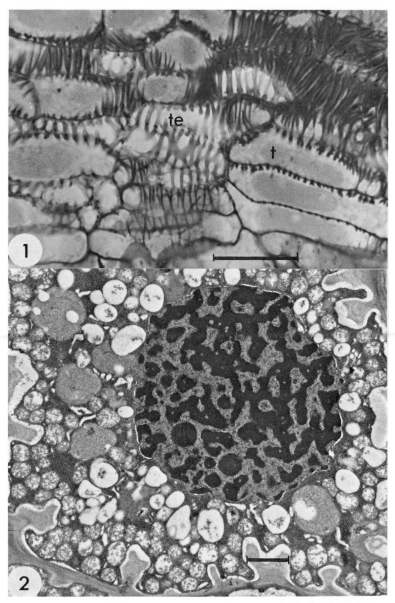

Fig. 1. - Xylem transfer cells (t) in longitudinal section at the junction of the sterile glumes in a spikelet of wheat, **Triticum aestivum** *L. cv. Heron. Note the development of the complex flanges of inwardly directed wall thickenings in these cells. te, tracheary element. Scale is 30 µm.*

Fig. 2. - Electron micrograph of a phloem transfer cell at the base of the attachment of the ovary in Heron wheat. Note the smaller wall ingrowths. Scale is 2 μm.

1970).

In the wheat plant, the distribution of transfer cells was very specific. In the lower region of the node, departing leaf-traces assume an elliptical outline in cross section and contain a horse-shoe shaped mass of tracheary elements embedded in a profusion of xylem parenchyma transfer cells. Slightly higher in the node, these bundles are connected to the larger bundles of the internode above by phloem bridges and in these strands the companion cells develop wall ingrowths. Small bundles, consisting of single vessels and sieve tubes, interconnect all bundles in the more central tissues of the node and the parenchyma cells that accompany these tiny strands are transfer cells. Wherever the transfer cell is associated with a tracheary element, the wall ingrowths take the form of flanges and often occur only on the side that faces the tracheary element.

It is plain that the sites where these cells occur in wheat are potential sites of nutrient exchange and in particular, cells so placed could be transferring solutes from the transpiration stream before it leaves the stem and moves into the leaf. Their distribution in this system agrees with the suggestion of the Delfast group that such cells form in structural bottlenecks that are sites of high intensity solute fluxes, the wall ingrowths increasing the area of membrane/apoplast boundary sufficiently to meet the demands imposed. Save only that such wall ingrowths are permeable to colloidal lanthanum solutions, are unlignified and contain no stainable levels of β 1-3 polysaccharides, little is known of their composition, structure, manner of deposition, or the factors that evoke their formation.

Further evidence that such cells are associated with a constraint upon solute fluxes comes from their presence in the aleurone layer of *Echinochloa utilis*. In wheat, nutrients enter the developing grain across the long furrow that runs along the chalazal side of the ovule. However, in *Echinochloa* the comparable loading zone is very much smaller, and the aleurone cells in this region, and in this region only, are developed as transfer cells (see Figs. 3 and 4).

It is unfortunate that there is so little hard core data that transfer cells actually carry out the very plausible functions assigned to them. Following a suggestion that they might be accumulating nitrate and secreting glutamine, we attempted to show histochemically that vascular transfer cells

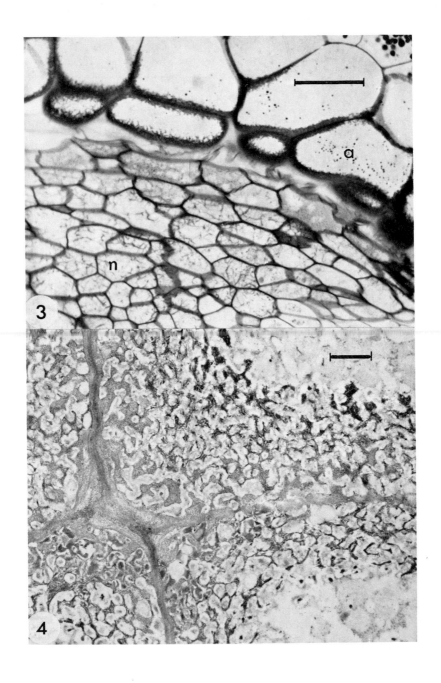

Fig. 3. - Light micrograph of aleurone transfer cells (a) at the base of the grain in millet (Echinochloa utilis Ohwi and Yabuno) n, nucellar projection. At this magnification the fine filiform wall ingrowths can only just be resolved in favourable sections. Scale is 50 μm.

Fig. 4. - Electron micrograph showing the thickenings in the aleurone transfer cells in greater detail. Scale is 1.5 μm.

---

in wheat nodes had high levels of glutamate transaminase activity, using methods known to work in animal cells. These tests simply failed to work at all. Clearly, we must be able to test for such activity if we are to establish the functions of these modified cells.

REFERENCES

Gunning, B.E.S. and Pate, J.S. (1969). Transfer cells. Plant cells with wall ingrowths specialized in relation to short distance transport of solutes - their occurrence, structure and development. *Protoplasma 68*, 107-133.

Gunning, B.E.S., Pate, J.S. and Green, L.W. (1970). Transfer cells in the vascular system of stems: taxonomy, association with nodes, and structure. *Protoplasma 71*, 141-171.

O'Brien, T.P. and Zee, S-Y. (1971). Vascular transfer cells in the vegetative nodes of wheat. *Aust. J. Biol. Sci. 24*, 207-217.

Pate, J.S. and Gunning, B.E.S. (1969). Vascular transfer cells in Angiosperm leaves - a taxonomic and morphological survey. *Protoplasma 68*, 135-156.

Pate, J.S. and Gunning, B.E.S. (1972). Transfer cells. *Ann. Rev. Plant Physiol. 23*, 173-196.

Zee, S-Y. and O'Brien, T.P. (1971). Vascular transfer cells in the wheat spikelet. *Aust. J. Biol. Sci. 24*, 35-49.

CHAPTER 6

Movement of Solutes from Host to Parasite in Nematode Infected Roots

M.G.K. Jones

Research School of Biological Sciences, Australian National University, Canberra, A.C.T. 2601, Australia

INTRODUCTION

Certain phytoparasitic nematodes have evolved the ability to control the morphogenesis of a limited number of cells of susceptible host plants. The nematodes feed from the transformed cells, lose the ability to move on, and are completely dependent on the integrity of the transformed cells. A rapprochement between host and parasite occurs: if the transformed cells are destroyed, the nematode will die.

This type of host-parasite relationship occurs with nematodes of the family Heteroderidae (*Heterodera, Meloidogyne, Meloidodera*) and to a lesser extent the family Tylenchidae (*Nacobbus, Rotylenchulus*).

ULTRASTRUCTURAL OBSERVATIONS OF TRANSFORMED CELLS

I have examined the cells transformed by *Meloidogyne incognita, arenaria* and *javanica; Heterodera rostochiensis, glycines* and *tabacum,* and *Nacobbus serendipiticus* and *aberrans* at both the light and electron microscope levels, and *Rotylenchulus reniformis* by light microscopy. In addition, some of these associations have been examined ultrastructurally by others (Bird 1961, Huang and Maggenti 1969a,b, Paulson and Webster 1970, Gipson et al. 1971, Rebois et al. 1975) and changes induced by *M. hapla* (Paulson and Webster 1970) and *H. carotae* (Ambrogioni and Porcinai 1972) have also been studied by electron microscopy.

*Root-knot nematode (*Meloidogyne *spp).* - Root-knot nematodes induce the formation of between 2 and 12 discrete giant cells (Figs. 2A and 3A). In the literature there are frequent reports that during early stages of giant cell formation wall breakdown occurs, allowing adjacent protoplasts to fuse

(see review by Bird 1974). Ultrastructural observations do not
support this view (Huang and Maggenti 1969a,b, Paulson and
Webster 1970, Jones and Northcote 1972b, Jones unpublished).
At later stages, small cells on the periphery of giant cells
may occasionally be incorporated after wall fracture (Bird
1972, Jones and Northcote 1972b), but this is incidental to
the basic mechanism of formation, which is repeated nuclear
mitosis without cytokinesis. Mitotic stimulation is also
evident outside the giant cells as a gall forms around them.
Under favourable circumstances, giant cells may reach up to
600 μm long and several hundred μm in diameter. The central
cell vacuole is replaced by many small vacuoles, mitochondria
are numerous, nuclei are enlarged with amoeboid profiles, and
the general impression is that the cells have an extremely
active metabolism. A conspicuous feature which develops between 3 and 6 days after induction is the presence of wall ingrowths typical of transfer cells. These occur on giant cell
walls adjacent to xylem elements, sieve elements and also in
localized regions on walls between neighbouring giant cells
(Figs. 2B, C, 3A). The wall ingrowths in giant cells are morphologically similar to those found in normal transfer cells
in the same plant; where there are two ingrowth morphologies,
e.g. in nodal xylem transfer cells and leaf minor vein phloem
transfer cells in *Helianthemum*, the ingrowths in giant cells,
both next to xylem and phloem elements are of the xylem transfer cell type (Jones and Gunning 1976). From scanning electron
micrographs of such wall ingrowths (Figs. 2B, C), an estimate
of the increase in surface area of wall over wall without
ingrowths is about tenfold.

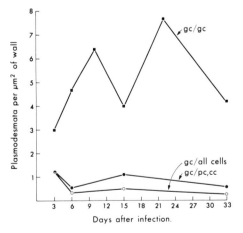

*Fig. 1. - The frequency
of plasmodesmata in giant
cell walls. gc/gc -
walls separating adjacent
giant cells. gc/all
cells - walls between
giant cells and all cells
outside them. gc/pc,cc -
walls between giant cells
and parenchyma or com-
panion cells, i.e. ex-
cluding vascular elements*

As the giant cells expand, a striking asymmetry in the
distribution of plasmodesmata is apparent. Between adjacent

giant cells large pit fields with numerous plasmodesmata are
seen (Fig. 2C), whereas pit fields are not obvious between
giant cells and normal cells (Fig. 2B). At first, as a result of this scanning electron microscope study, it was
thought that all plasmodesmata were lost between giant and
normal cells (Jones et al. 1975), but subsequent work has
shown that although plasmodesmatal frequency is reduced such
connections do occur. The frequencies of plasmodesmata are

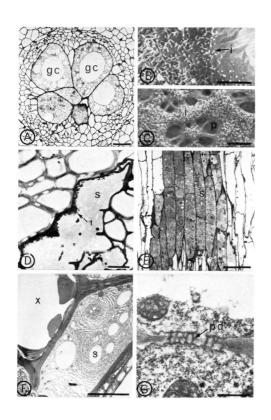

*Fig. 2. - Micrographs
of cells transformed
by nematodes. A. Trans-
verse section of giant
cells (gc) induced in*
Impatiens balsamina
by M. incognita, *six
days after induction.
Bar - 50 μm. B. Scan-
ning electron micro-
graph of giant cell
wall next to normal
cells. Note wall in-
growths on left (i)
and smooth wall on
right. The cytoplasm
has been digested
away. Bar 10 μm.
C. Processing as B.
Wall between two
giant cells. Note the
large pit fields (p),
and wall ingrowths
(i). Bar - 10 μm.
D. Light micrograph
of a syncytium (s)
induced in potato by*
H. rostochiensis,
*transverse section.
Wall ingrowths (i)
are present next to xylem. Bar - 10 μm.   E. Light micrograph
of a syncytium (s) induced in soybean by* **R. reniformis.** *Longitudinal section through curved sheet of transformed cells.
Bar - 50 μm.   F. Electron micrograph of syncytium (s) induced
by* **N. aberrans,** *showing that no ingrowths form next to xylem
elements (x), although walls are evenly thickened.   Bar -
10 μm.   G. Junction of syncytium (s) induced by* **N. aberrans**
*and a normal cell, showing numerous plasmodesmata (pd). Bar -
0.5 μm.*

shown in Fig. 1. With increasing age the frequency of plasmodesmata in the walls separating adjacent giant cells appears
to increase as the walls expand, and this suggests that
secondary formation of plasmodesmata occurs. As giant cells
age, more xylem and sieve elements differentiate outside them
and these lack plasmodesmatal connections with giant cells.
Although the plasmodesmatal frequency between giant cells and
neighbouring cells is low, membrane potential experiments
suggest that there is some electrical coupling between the
cells (Jones et al. 1975).

Cyst-nematode (Heterodera *spp.*). Cyst-nematodes induce
the formation of syncytia by wall breakdown and subsequent
coalescence of cell contents. There is no mitotic stimulation, but considerable expansion of incorporated cells occurs,
and syncytia extend for 2-3 mm up and down the root in contact
with xylem elements (Figs. 2D and 3B). Cytoplasmic changes
are similar to those for root-knot giant cells. Ingrowths
develop where syncytial walls abut xylem elements: they are
rarely observed next to sieve elements, since these appear to
be damaged by the developing syncytia.

In particular, cells on the cortical side of the syncytia
are crushed by this expansion, and the outer syncytial wall
becomes thickened and staining indicates it becomes impreg-
nated with lignin or polyphenols. Plasmodesmata are overlaid
with wall polysaccharides, and the only sites where functional
plasmodesmata may still occur are at the ends of the syncytia
where further incorporation is occurring (Jones and Northcote
1972a, Jones 1972).

Reniform nematode (Rotylenchulus). The reniform nematode
causes the formation of a syncytium in a very specific cell
layer, the pericycle, although the endodermis may also be involved at the initial site. Cytoplasmic changes are similar
to those described above; wall breakdown is evident for at
least 5-6 cells away from the head of the nematode: these
cells also expand. A curved sheet of cells progressing both
radially and longitudinally from the initial site is transformed (Figs. 3D, 2E), but cells further away from the nematode, although filled with cytoplasm, are uninucleate and not
connected to their neighbours through wall gaps. At protoxylem
poles the sheet of cells contacts xylem elements, and sieve
elements are also contacted at protophloem poles. Wall ingrowths do not form at either site (Rebois et al. 1975, Jones
and Dropkin 1975). Plasmodesmata are frequent between syncytial cells and non-transformed cells (Rebois et al. 1975),
and the syncytium clearly has a relatively large surface to
volume ratio.

Nacobbus. - Syncytia induced by *Nacobbus* form through wall dissolution, and are accompanied by mitotic stimulation, so that a spindle- or crescent-shaped mass of interconnected cells is surrounded by a gall (Fig. 3C). The appearance of the cytoplasm of syncytia is similar to the preceding examples although whorls of endoplasmic reticulum are more common. Syncytial cells can abut both xylem and sieve elements,

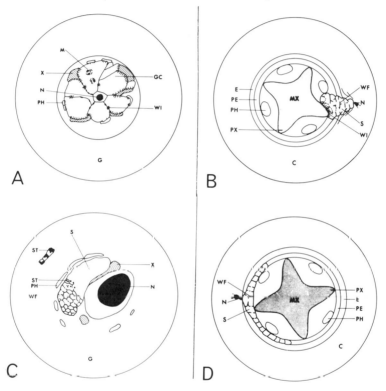

Fig. 3 - Diagrammatic representations (transverse sections) of cells transformed by endoparasitic nematodes. Abbreviations - C cortex, E endodermis, G gall, GC giant cell, M mitosis, MX metaxylem, N nematode, PE pericycle, PH phloem, PX protoxylem, S syncytium, ST starch, WI wall ingrowths, WF wall fragments, X xylem.
    A. Giant cells induced by **Meloidogyne** *spp. B. Syncytium induced by* **Heterodera** *spp.   C. Syncytium induced by* **Nacobbus** *spp.   D. Syncytium induced by* **Rotylenchulus.**

which differentiate outside the transformed cells. A proliferation of phloem elements, produced by cambial cells just outside the syncytium, is noteworthy. No wall ingrowths are found (Fig. 2F). Plasmodesmata are numerous between syncytial cells and gall parenchyma cells, and pit fields are rather obvious because elsewhere the cell walls are extensively thickened (Fig. 2G). A particularly striking feature is that numerous large starch grains are present in the gall parenchyma cells and in the syncytial cells (Schuster et al. 1965).

DISCUSSION

It is probable that the number of carrier sites per unit area of plasmalemma for a given solute has a limiting value. One way of increasing solute transport at that location is to increase the surface area of the plasmalemma. If we accept the postulate that the function of wall ingrowths is to provide a stable increase in surface area at such a location, then the wall ingrowths mark the regions in cells where the greatest solute flow across the plasmalemma is occurring, and their extent of development for the same solute will indicate the relative size of the fluxes (Jones and Northcote 1972b; Jones and Gunning 1976). However, if different carriers are saturated at different locations, e.g. an amino acid carrier next to the xylem and a sugar carrier next to the phloem, since the maximum number of carriers per unit area of plasmalemma may be different, the comparison of ingrowth development and solute flux will not hold.

When taken together, the paucity of plasmodesmatal connections between root-knot giant cells and neighbouring parenchyma cells and the extensive ingrowths developed in giant cells adjacent to vascular elements, suggest that the bulk of the solutes enter giant cells from the apoplast and cross the plasmalemma where the ingrowths occur. A similar mode of solute entry is suggested for cyst-nematode syncytia. In contrast, the numerous plasmodesmatal connections between syncytia and untransformed cells induced by the reniform nematode and *Nacobbus*, and the lack of formation of wall ingrowths despite similar overall solute fluxes to the nutrient sink (nematode), suggest that there are no preferential sites where the bulk of the solutes enter the syncytia from the apoplast, and that solutes are more likely to enter these syncytia symplastically via the plasmodesmata.

REFERENCES

Ambrogioni, L. and Porcinai, G.M. (1972). Studio ultrastutturistico delle cellule giganti prodotte da *Heterodera carotae* Jones, 1950 (Nematoda:Heteroderidae) in radici di carota. *Redia 53*, 437-448.

Bird, A.F. (1961). The ultrastructure and histochemistry of a nematode-induced giant cell. *J. Biophys. Biochem.Cytol. 11*, 701-715.

Bird, A.F. (1972). Cell wall breakdown during the formation of syncytia induced in plants by root-knot nematodes. *Int. J. Parasitol.* 2, 431-432.

Bird, A.F. (1974). Plant response to root-knot nematode. *Ann. Rev. Phytopathol.* 12, 69-85.

Gipson, I., Kim, D.S. and Riggs, R.D. (1971). An ultra-structural study of syncytium development in soybean roots infected with *Heterodera glycines*. *Phytopathology* 61, 347-353.

Huang, C.S. and Maggenti, A.R. (1969a). Mitotic aberrations and nuclear changes of developing giant cells in *Vicia faba* caused by root-knot nematode, *Meloidogyne javanica*. *Phytopathology* 59, 447-455.

Huang, C.S. and Maggenti, A.R. (1969b). Wall modifications in developing giant cells of *Vicia faba* and *Cucumis sativus* induced by root-knot nematode, *Meloidogyne javanica*. *Phytopathology* 59, 931-937.

Jones, M.G.K. (1972). Cellular alterations induced in plant tissues by nematode pathogens. Thesis, University of Cambridge.

Jones, M.G.K. and Dropkin, V.H. (1975). Cellular alterations induced in soybean roots by three endoparasitic nematodes. *Physiol. Plant Pathol.* 5, 119-124.

Jones, M.G.K. and Gunning, B.E.S. (1976). Transfer cells and nematode induced giant cells in *Helianthemum*. *Protoplasma* (in press).

Jones, M.G.K. and Northcote, D.H. (1972a). Nematode-induced syncytium - a multinucleate transfer cell. *J. Cell Sci.* 10, 789-809.

Jones, M.G.K. and Northcote, D.H. (1972b). Multinucleate transfer cells induced in coleus roots by the root-knot nematode, *Meloidogyne arenaria*. *Protoplasma* 75, 381-395.

Jones, M.G.K., Novacky, A. and Dropkin, V.H. (1975). Trans-membrane potentials of parenchyma cells and nematode-induced transfer cells. *Protoplasma* 85, 15-37.

Paulson, R.E. and Webster, J.M. (1970). Giant cell formation in tomato roots caused by *Meloidogyne incognita* and *Meloidogyne hapla* (Nematoda) infection. A light and electron microscope study. *Can. J. Bot.* 48, 271-276.

Rebois, R.V., Madden, P.A. and Eldridge, B.J. (1975). Some ultrastructural changes induced in resistant and suscept-ible soybean roots following infection by *Rotylenchulus reniformis*. *J. Nematol.* 7, 122-139.

Schuster, M.L., Sandstedt, R. and Estes, L.W. (1965). Host-parasite relations of *Nacobbus batatiformis* and the sugar beet and other hosts. *J.Amer.Soc.Sugar Beet Technol.* 13, 523-537.

CHAPTER 7

Wheat Grains and Spinach Leaves as Accumulators of Starch

C.F. Jenner

Waite Agricultural Research Institute, The University of Adelaide, Glen Osmond, South Australia 5064. Australia

INTRODUCTION

No attempt is made here to review broadly all aspects of the storage of starch in plant tissues. Instead, attention is confined to what is known of the physiological and biochemical mechanisms regulating the accumulation of starch. Two contrasting situations are considered: the synthesis of starch in chloroplasts of spinach leaves, and the accumulation of starch in the endosperm of developing wheat grains. Although in these two diverse tissues there are marked differences in the temporal pattern and environmental responsiveness of the production of starch, control mechanisms at the biochemical level operating in both may closely resemble each other.

BASIC REQUIREMENTS

Plastids. - Starch is normally only deposited in the form of discrete granules within plastids. In the chloroplast one or more granules lie in the stroma, usually in close association with grana (Kirk and Tilney-Bassett 1967) while in the amyloplast one to several granules often completely fill the internal volume of the plastid. As there is no doubt that polysaccharide synthesis takes place within the plastid, it follows that the appropriate enzymic system must operate inside the plastid.

Enzymes. - The main outlines of the biochemical pathway, discovered almost 20 years ago by Leloir and his associates (see article by Akazawa 1965), and involving nucleoside diphosphate sugars is now well established. Turner's (1969) scheme for wheat incorporates much of what is generally accepted. Sucrose, translocated from photosynthetic organs, is the source of carbohydrate. Both hexose moieties are converted into starch, and at approximately the same rates

(Jenner 1973) but hydrolysis is not the first step in the conversion (Jenner 1974b). Instead, uridine diphosphate glucose (UDPG) is produced by the sucrose synthetase reaction operating in reverse. The fructosyl moiety so released is phosphorylated, converted to glucose-6-phosphate and finally to glucose-1-phosphate (G-1-P) which is the precursor for adenosine diphosphate glucose (ADPG). Turner (1969) proposes that the glucosyl moiety of UDPG enters the pool of G-1-P by way of the UDPG-pyrophosphorylase reaction. The step in the production of ADPG from G-1-P, catalysed by ADPG-pyrophosphory-lase, is held to be the major point in the regulation of the rate of starch synthesis at the biochemical level, and the mechanism is dealt with more fully later. Thus in this scheme UDPG is envisaged as being involved in the breakdown of sucrose, while ADPG is the principal precursor for starch (see also Akazawa 1965).

In the spinach leaf the source of carbon for starch synthesis derives directly from the carbon-reduction cycle of photosynthesis (Gibbs 1971). Otherwise the sequence from the hexose phosphates to ADPG and finally to starch is similar to the one described above for wheat endosperm.

Substrate. – In general terms, starch accumulates where-ever and whenever carbohydrate is in plentiful supply. As the supply of carbohydrate available for starch synthesis is determined by the production of carbohydrate as well as by its utilization in other, competing, processes, alterations in either production or utilization can have marked effects on the amounts of starch laid down. As might be excepted, more starch appears in leaves illuminated with bright light than with dim light, and more in continuous light than in alternating light and dark. When growth of the plant is restricted by mineral deficiency or low temperatures (Porter 1962) more starch accumulates than in conditions favouring growth.

CONCURRENT SYNTHESIS AND DEGRADATION

Whether they are formed in chloroplasts or in amylo-plasts, starch granules grow in size by apposition (Porter 1962). There is however a striking difference between spin-ach leaves and wheat endosperms in the temporal pattern of the deposition of starch. In the former, most of the starch deposited during the day disappears at night, while in the latter there has been no suggestion of any breakdown of starch, at least during the early period of the grain's development. Indeed, starch did not break down in detached ears of wheat cultured for four days on water (Jenner 1968).

In leaves of tobacco, starch is being degraded in the light even when the intensity of light is high enough to support net accumulation (Chan and Bird 1960). However, ADPG-starch transglucosylase catalyses the almost complete conversion of ADPG to starch, and the free energy change of the reaction greatly favours synthesis (Akazawa 1965). Little or no breakdown of starch can therefore be expected by reversal of the synthetic pathway. Although there is no agreement on which enzymes catalyse the dissolution of leaf starch (compare, for example, Akazawa 1965; Manners 1974; de Fekete and Vieweg 1974), degradation by phosphorylase seems the most plausible hypothesis on the basis of present evidence. Thus conjoint action of transglucosylase and phosphorylase can account for the transient nature of the starch deposited in chloroplasts.

REGULATION

Physiological and biochemical mechanisms regulating the metabolism of starch operate within the spatial framework provided by the structure of the accumulating systems. Although something is known about physiological aspects of

PASSAGE OF SUCROSE INTO WHEAT GRAIN

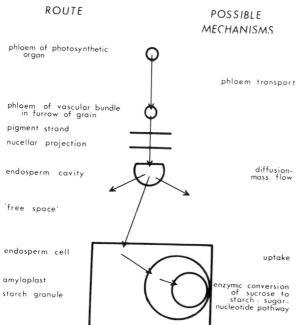

ROUTE

phloem of photosynthetic organ

phloem of vascular bundle in furrow of grain

pigment strand

nucellar projection

endosperm cavity

'free space'

endosperm cell

amyloplast

starch granule

POSSIBLE MECHANISMS

phloem transport

diffusion-mass flow

uptake

enzymic conversion of sucrose to starch : sugar-nucleotide pathway

Fig.1. - Hypothetical scheme illustrating the pathway of movement of sucrose into wheat edosperm. From the vascular bundle in the furrow sucrose flows down a gradient of concentration to the cells of the endosperm. Possible sites of resistance to the flow of sucrose are (1) the tissues of the pigment strand and nucellar projection and (2) uptake into the cells of the endosperm.

the accumulation of starch in wheat endosperm, the biochemistry of regulation has been much more extensively investigated in the chloroplast. An attempt is made here to integrate what is known in both systems with a view to suggesting
a general mechanism.

Wheat Endosperm. - All sucrose entering the endosperm is
transported in the phloem running along the length of the
furrow of the wheat grain (Frazier and Appalanaidu 1965). The
sugar is presumed to move (Fig. 1, see also Jenner 1974a)
across the tissues of the nucellar projection and pigment
strand and into the endosperm cavity (Bradbury et al. 1956).
From the cavity, at the axis of the grain, the sugar moves in
the free space between the cells radially outwards across the
endosperm where it is absorbed by the cells.

From estimates of the concentration of sucrose in different parts of the grain (Jenner 1974a), it appears as though
sucrose travels down a gradient of concentration from its
point of entry into the grain to its destination within the
cells of the endosperm. Culturing dissected endosperm for a
few hours on solutions of ^{14}C-sucrose has provided information
on the kinetics of the uptake of sucrose by the cells, and
the synthesis of starch (Jenner 1974a). At equilibrium the
amounts of sucrose in the free space and in the cells are
directly related to the external concentration. This linear
relationship holds within the (external) range 3 to 100 mg/ml
which spans the normal metabolic level of sucrose in the
endosperm (viz. about 25 mg/ml in the endosperm cavity).
Within the range 3 to 50 mg/ml the rate of incorporation of
radioactivity into starch is also a linear function of the
external concentration. It is concluded that as the
direction and flux of sucrose through the grain depends on the
concentration gradient, physico-chemical processes such as
diffusion and mass flow might transport the sugar. Certainly,
on the basis of present evidence, there seems no need to
invoke any special active transport process.

The direct relationship between the concentration of
sucrose in the cells of the endosperm and the rate of starch
synthesis, not only in isolated endosperm but in detached
ears as well (Jenner 1970) suggests that synthesis of starch
is regulated by the concentration of sucrose. Thus, regulation of the movement of sucrose into and through the grain is
part of the overall mechanism controlling the deposition of
starch.

There are two sections of the pathway from the phloem
into the endosperm (Fig. 1) where the gradient of concentration is relatively steep: from the phloem to the endosperm
cavity, and between the free space and the internal volume

of the cells. The gradient between the endosperm cavity and
the free space seems much shallower than in the other two
sections. Steep gradients are considered in this account to
indicate sections of the pathway where the movement of sugar
meets resistance. It has been proposed therefore (Jenner
1974a) that the grain can be regarded as a series of conductors of sucrose, each having the properties of resistance;
that the two sections of the pathway outlined above are
conductors of relatively high resistance, and that the overall
flux of sucrose from the phloem of the furrow to the sites of
starch synthesis is determined by the resistances of the conductors operating on the gradient of concentration (as well as
sucrose, this mechanism may also regulate amino acid inflow,
(Davies and Jenner unpublished)). At the physiological level
then, the cardinal variable factor in the grain regulating
the rate of accumulation of starch is the concentration of
sucrose in the vascular bundle. In turn, the concentration
of sucrose at this site is influenced by the supply of sucrose
to the grain. So the question arising next is what regulates
the supply of sucrose from the photosynthetic tissues to the
vascular bundle of the furrow?

Culturing detached ears on solutions of sucrose has
provided some information on this point. Sucrose accumulates
in all the tissues of the cultured ear except the grains, in
direct proportion to the concentration supplied to the cut
peduncle (Jenner 1970). In the grains themselves, and in all
of the parts investigated including the tissues of the furrow
(Jenner 1976) the level of sucrose is responsive to changes
in the external concentration up to about 25 mg/ml, but at
higher external concentrations the amount in the grain is
constant. Evidently, the flow of sucrose into the grain is
restricted and the restriction operates in a section of the
pathway leading to the furrow. Precisely where the restriction operates it is not possible to say with certainty. It
may be that the whole of the transporting system within the
ear is involved. Indeed, it has been suggested (Jenner 1976)
that the restriction is imposed by the kinetic properties of
loading and transport in the phloem, and that saturation of
one or other of these processes is manifested as a restricted
inflow of sucrose into the grain. Whatever the nature of the
restriction, it is clear that it determines the upper limit
to the flow of sucrose to the grain, and so limits the
accumulation of starch and the growth of the grain itself
(Jenner and Rathjen 1972).

Spinach Chloroplasts.- The cardinal point in the regulation of starch synthesis is the enzyme, ADPG pyrophosphorylase,
catalyzing the production of ADPG from G-1-P, and the

following account of the mechanism (Fig. 2) has been proposed by Preiss and his colleagues (Preiss and Kosuge 1970). The pyrophosphorylase is activated by intermediates derived from the carbon-reduction cycle, especially by 3-phosphoglycerate (3-PGA). Inorganic phosphate (P_i) inhibits the production of ADPG, but this inhibition is overcome by 3-PGA. In the light the level of triose-phosphates (triose-P) is high and that of P_i is low, so it is envisaged that ADPG is produced and starch is synthesised. In darkness the concentrations of triose-P fall, that of P_i rises, the activity of the enzyme is reduced to very low levels, and synthesis of starch ceases as ADPG is depleted.

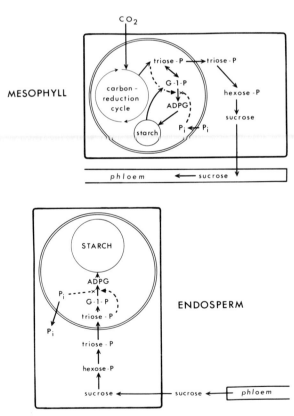

Fig.2. - Schemes depicting a proposed common mechanism regulating the accumulation of starch in the mesophyll of spinach leaves and cells of wheat endosperm. Chloroplast and amyloplast are drawn with a double outline. Arrows with continuous lines indicate the direction of flow of carbon, and do not necessarily imply single steps or reactions. Dotted lines depict regulatory mechanisms: solid arrow-heads indicate activation and crosses inhibition. The scheme for the mesophyll is adapted from Sheu-Hwa, Lewis and Walker (1975). Abbreviations are defined in the text; the activator of ADPG pyrophosphorylase is probably 3-PGA.

A scheme accounting for the simultaneous breakdown of starch, and for its disappearance at night, is conceptually easy to derive from the known properties of phosphorylase

(Akazawa 1965). This enzyme catalyzes reversibly the transfer of the glycosyl moiety from G-1-P to starch. At night, when the level of P_i in the chloroplast is higher than during the day, the ratio of the amounts of P_i to G-1-P would be more favourable for the degradation of starch than it would during the day. Even during the day, the ratio probably favours degradation rather than synthesis by this enzyme.

There is now substantial evidence (reviewed by Walker 1974) that the envelope of the chloroplast is relatively impermeable to hexoses and hexose phosphates, and that the flux of sucrose through the envelope is too low to account for the observed efflux of carbon. Instead of sugars and sugar phosphates, Heber and his associates (Heber and Krause 1971) propose that carbon is transported from the chloroplast as triose-P, 3-PGA and possibly other intermediates too and that export of these compounds is linked with influx of P_i into the chloroplast. Sheu-Hwa, Lewis and Walker (1975) suggest a mechanism regulating the efflux of carbon compounds (on which Fig. 2-Mesophyll is based) and point to the interesting possibility that the regulation of the whole scheme of carbon-flow within and from the chloroplast is controlled by a balance between P_i and triose-P. Thus low levels of P_i inside the chloroplast during the day favour ADPG production and starch synthesis, and influx of P_i is balanced by efflux of triose-P. In darkness, higher levels of P_i accelerate the breakdown of starch providing the source of carbon leaving the chloroplast during the night. Sucrose, it seems, is mostly synthesised outside the chloroplast (Bird et al. 1974).

A Common System of Regulation.- No comparable work on the passage of carbon into and out of amyloplasts has been reported. However, by adopting the principal of parsimony, and in the absence of any evidence to the contrary, the scheme outlined above for chloroplasts could operate in amyloplasts in much the same way (Fig.2 - Endosperm). There is a need to invoke one major difference between the two systems to account for the differing patterns of accumulation of starch: in the amyloplast the system degrading starch either lacks activity, or it is absent altogether.

Although there is now a great deal of information on the chemical composition of chloroplasts, there is a complete lack of similar information for amyloplasts (Kirk and Tilney-Bassett 1967). However chloroplasts and amyloplasts have a common origin, and amyloplasts develop into chloroplasts under certain circumstances (see Mühlethaler 1971). It is assumed therefore, that the permeability of the amyloplast envelope is like that of the chloroplast envelope and that a similar transporting mechanism for triose-P operates in amyloplasts. This concept is not entirely lacking experimental support. More ^{14}C-starch is produced from a source of ^{14}C-sucrose in wheat endosperm cultured in phosphate buffer than in solutions lacking P_i (Table 1). Although 17% more ^{14}C-sucrose is taken up in the presence of P_i than in its absence, 56% more radioactive starch accumulates, a greater increment than can be accounted for on the basis of enhanced uptake of ^{14}C-sucrose alone. On the other hand the proportionate increase in radioactivity due to P_i in other insoluble radioactive material in the endosperm (22%) can be ascribed almost wholly to increased uptake of ^{14}C-sucrose. Evidently, phosphate has a specific effect on ^{14}C-starch synthesis, and one way to explain its specificity is to suggest that starch is accumulating within a compartment into which P_i influences the inflow of carbon (i.e., the amyloplast) whereas the other insoluble material (mostly protein)

TABLE 1.   Effect of phosphate on synthesis of starch in wheat endosperm[1]

Endosperm isolated from wheat grain was cultured at 25°C for 24 hr in solution of ^{14}C-sucrose (25 mg/ml) alone or combined with KH_2PO_4 at 50 mMolar and pH 4.5. The tissue was rinsed for 1.5 hr in ice cold water to remove soluble material from the free space, and then boiled in 80% ethanol. Sucrose and inorganic phosphate were assayed on portions of the soluble fraction, and starch and other material in the residue insoluble in ethanol. Inorganic phosphate was assayed by Bartlett's (1959) method; all other procedures and methods are found in Jenner (1974a). Figures in parentheses are values for phosphate relative to no phosphate.

Fraction of the endosperm	KH_2PO_4 (mMolar)		
	0	50	
Inorganic phosphate[2] (μmoles per grain)	0.20	0.80	(4.00)
[14]C-sucrose (mg per grain)	0.18	0.21	(1.17)
[14]C-starch produced (mg ^{14}C-sucrose equivalents per grain)	0.55	0.86	(1.56)
Other insoluble radioactive material (mg ^{14}C-sucrose equivalents per grain)	0.36	0.44	(1.22)

[1]unpublished data – Ruiter and Jenner.

[2]amount of inorganic phosphate in the cells of freshly dissected endosperm = 0.29 μmoles per grain.

is synthesised in the cytoplasm.

Other similarities between chloroplasts and amyloplasts have already been pointed out (e.g. dependence of the rate of starch synthesis on the supply of substrate) but what is seen as the chief distinguishing feature is of equal importance. What engenders the production (or activation) of the degradative system within the plastid? One requirement for the conversion of amyloplasts (or proplastids) into chloroplasts is light. Starch is commonly observed in proplastids, and it is known to disappear as the plastid develops into a functional chloroplast. Price, Mitrakos and Klein (1964) have shown that the dissolution of starch in the leaves of corn seedlings is mediated by the phytochrome system. Perhaps, as Kirk and Tilney-Bassett (1967) suggest, phytochrome may activate or induce the formation of starch-degrading enzymes. The possibility that degradative systems can be induced within amyloplasts in wheat endosperm remains so far unexplored, but in the endosperm of another species (*Oxalis dispar*; Sunderland and Wells 1968) amyloplasts can be transformed into chloroplasts.

Lastly, not only do the schemes in Fig.2 depict the outlines of the regulation of carbohydrate metabolism in the source and in the destination of these compounds, but at least one factor determining the polarity and rate of flow through the system (see also Fig.1) is explicitly suggested as well: differences between chloroplasts and amyloplasts in the degradation of starch. Whether this factor alone is adequate to account for the observed rate of flow, or whether there are other differences between the ends of the system (transfer into and out of the phloem for example) are matters for further investigation.

ACKNOWLEDGMENTS

The author's work was supported by grants from the Commonwealth Wheat Industry Research Council and the Australian Research Grants Committee.

REFERENCES

Akazawa, T. (1965). Starch, inulin and other reserve polysaccharides. In: "Plant Biochemistry" ed. by J. Bonner and J.E. Varner, Academic Press, New York pp. 258-297.

Bartlett, G.R. (1959). Phosphorus assay in column chromatography. *J. Biol. Chem.* *234*, 466-8.

Bird, I.F., Cornelius, M.J., Keys, A.J. and Whittingham, C.P. (1974). Intracellular site of sucrose synthesis in leaves. *Phytochem.* *13*, 59-64.

Bradbury, D., Cull, I.M. and MacMasters, M.M (1956). Structure of the mature wheat kernel. I. Gross anatomy and relationship of parts. *Cereal Chem. 33*, 329-42.

Chan, T.T. and Bird, I.F. (1960). Starch dissolution in tobacco leaves in the light. *J. Exp. Bot. 11*, 335-40.

De Fekete, M.A.R. and Vieweg, G.H. (1974). Starch metabolism: Synthesis versus degradation pathways. In: "Plant Carbohydrate Biochemistry" ed. by J.B. Pridham, Academic Press, London pp. 127-44.

Frazier, J.C. and Appalanaidu, B. (1965). The wheat grain during development with reference to nature, location, and role of its translocatory tissues. *Amer. J. Bot. 52*, 193-8.

Gibbs, M. (1971). Carbohydrate metabolism by chloroplasts. In: "Structure and Function of Chloroplasts" ed. by M. Gibbs, Springer-Verlag, Berlin pp. 169-214.

Heber, U. and Krause, G.H. (1971). Transfer of carbon, phosphate energy and reducing equivalents across the chloroplast envelope. In: "Photosynthesis and Photorespiration" ed. by M.D. Hatch, C.B. Osmond and R.O. Slatyer, Wiley Interscience, New York pp. 218-25.

Jenner, C.F. (1968). Synthesis of starch in detached ears of wheat. *Aust. ᵉ. Biol. Sci. 21*, 597-608.

Jenner, C.F. (1970). Relationship between levels of soluble carbohydrates and starch synthesis in detached ears of wheat. *Aust. J. Biol. Sci. 23*, 991-1003.

Jenner, C.F. (1973). The uptake of sucrose and its conversion to starch in detached ears of wheat. *J. Exp. Bot. 24*, 295-306.

Jenner, C.F. (1974a). Factors in the grain regulating the accumulation of starch. In: "Mechanisms of Regulation of Plant Growth" ed. by R.L. Bieleski, A.R. Ferguson and M.M. Cresswell, *Royal Soc. N.Z., Wellington. Bull. 12*, 901-8.

Jenner, C.F. (1974b). An investigation of the association between the hydrolysis of sucrose and its absorption by grains of wheat. *Aust. J. Plant. Physiol. 1*, 319-29.

Jenner, C.F. (1976). Physiological investigations on restrictions to transport of sucrose in ears of wheat. *Aust. J. Plant Physiol. 3*, (In press).

Jenner, C.F. and Rathjen, A.J. (1972). Limitations to the accumulation of starch in the developing wheat grain. *Ann. Bot. (N.S.) 36*, 743-54.

Kirk, J.T.O. and Tilney-Bassett, R.A.E. (1967). "The Plastids", W.H. Freeman, London.

Manners, D.J. (1974). The structure and metabolism of starch. *Essays Biochem. 10*, 37-71.

Mühlethaler, K. (1971). The ultrastructure of plastids. In: "The Structure and Function of Chloroplasts" ed. by M. Gibbs, Springer-Verlag, Berlin pp. 7-34.

Porter, H.K. (1962). Synthesis of polysaccharides in higher plants. *Ann. Rev. Plant Physiol. 13*, 303-28.

Preiss, J. and Kosuge, T. (1970). Regulation of enzyme activity in photosynthetic systems. *Ann. Rev. Plant Physiol. 21*, 433-66.

Price, L., Mitrakos, K. and Klein, W.H. (1964). Photomorphogenesis and carbohydrate changes in etiolated leaf tissue. *Quart. Rev. Biol. 39*, 11-18.

Sheu-Hwa, C., Lewis, D.H. and Walker, D.A. (1975). Stimulation of photosynthetic starch formation by sequestration of cytoplasmic orthophosphate. *New Phytol. 74*, 383-92.

Sunderland, N. and Wells, B. (1968). Plastid structure and development in green callus tissues of *Oxalis dispar*. *Ann. Bot. (N.S.) 32*, 327-46.

Turner, J.F. (1969). Starch synthesis and changes in uridine diphosphate glucose pyrophosphorylase and adenosine diphosphate glucose pyrophosphorylase in the developing wheat grain. *Aust. J. Biol. Sci. 22*, 1321-7.

Walker, D.A. (1974). Chloroplast and cell- The movement of certain key substances, etc. across the chloroplast envelope. In: "Plant Biochemistry" ed. by D.H. Northcote, Butterworths, London Biochemistry Series One, Volume 11 pp. 1-49.

CHAPTER 8

Nutrient Uptake by Roots and Transport to the Xylem: Uptake Processes

M.G. Pitman

*School of Biological Sciences, University
of Sydney, N.S.W. Australia*

INTRODUCTION

This paper is intended to introduce the following papers by Dr. Läuchli on transport in the symplasm and by Dr. Cram on regulation of uptake, as well as discussing the absorption of nutrients by plants.

Uptake by the plant is a complex process because it is affected by external factors such as soil or nutrient availability, by the flux of water through the system and by the organisation within the root. This complexity is summarised in Figure 1. The root operates as a system for absorption of ions from the soil and for supply of ions to the shoot, but it is convenient to separate the overall activity into components, which have particular characteristics. For example, the absorption is a property primarily associated with the outer cell membranes, and there is much information about this process from different kinds of study. Within the root ions can be accumulated to high concentration in cell vacuoles or pass in the symplasm into the stele and eventually to the xylem. This process of transport across the root and release to the xylem is discussed further by Dr. Läuchli. In a transpiring plant, concentrations of ions in the xylem sap can be low (about 1.2 mol m^{-3} total) but when evaporation is low or when the shoot is cut off the ions secreted to the xylem set up a standing gradient in osmotic potential resulting in water flow which appears as guttation from the leaves or exudation from the cut end of the root.

We have reasons to think that the release of ions to the xylem producing this exudation is separate from ion absorption, based on the differential response of these processes to inhibitors.

A   B

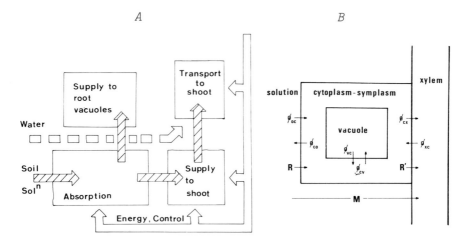

Fig. 1A. - Inter-relationship between uptake processes. Ions absorbed from soil or solution can be transported to root vacuoles or into the stele, where they can pass in the xylem to the shoot. Water flow through the root may interact with uptake and supply to the shoot. Translocation from the shoot supplies energy (sugars) and possibly regulators of uptake.

Fig. 1B. - The processes in Fig. 1B can be set out in a formal system where ϕ represents fluxes into and out of different compartments. R is net flux into the root and M is water-flow-dependent uptake.

For example, the amino acid analogue p-fluorophenylalanine (FPA) inhibits release of ions from the stele but does not inhibit ion uptake or oxygen uptake (Schaefer et al. 1975). The phytohormone abscisic acid has the same effect under certain conditions (Cram and Pitman 1972).

Supply of ions to cells in the root and supply of ions to the shoot appears to be regulated. This aspect will be dealt with in more detail by Dr. Cram, and in a separate paper by Dr. Glass but two examples will serve to show what is meant by this claim. Firstly, the total concentration of the univalent cations K^+ and Na^+ in the cell vacuoles of certain roots grown in nutrient solution can be very nearly independent of the external total concentration, over a wide range of concentration. Secondly, the average concentration of K^+ in the shoot can be nearly constant and independent of relative growth rate, though the rate of transport from root to shoot increases proportional to relative growth rate (Pitman 1972).

The transport processes can be thought of as in Fig.1. The absorption process appears to be located at the plasmalemmas of the root cortical cells and has properties like transport at the plasmalemma of other plant cells. There is a complication in roots though where the cortical cells effective in absorption may be those at the surface, especially when external concentration is low (see later discussion).

UPTAKE PROCESSES

Uptake of most ions to roots can be rapidly and severely decreased by inhibitors of energy metabolism, implying that entry to the cell is due to a carrier system driven by metabolism (e.g. ATP) or that entry is in response to an electrochemical potential gradient set up by active transport of another ion or a proton. In some cases the presence and properties of active transport systems can be inferred from measurements of uptake, concentrations in the cell and cell potential differences.

Arguments for active transport have been discussed in more detail elsewhere (e.g. Pitman 1976) and are summarised here. There seems to be a need to locate a carrier for K^+ (and Na^+?) at the outer membrane of the cell that can act at very low external concentrations, and is equivalent to Mechanism I described from kinetic analysis by Epstein (e.g. 1966). Evidence for this carrier comes from comparison of observed rates of uptake with estimated rates of diffusion of ions across the cell membrane (Pitman 1969). When external K^+ is less than 1 mol m^{-3}, the observed rates are at least 30 times larger than the calculated diffusive fluxes. Similar calculations show a need for active anion transport (Cl^-, NO_3^-) as again observed rates of uptake are much larger than estimated diffusive fluxes; when external Cl^- is 25 mol m^{-3} the estimated diffusive influx is only about 2% of the uptake. Similar arguments can be applied to phosphate and sulphate uptakes.

The Mechanism II component of Epstein's analysis for K^+ uptake, dominant above about 2-5 mol m^{-3}, may well be due to electrogenic coupling with H^+ efflux. Firstly, the calculated diffusive K^+ flux is adequate to meet observed rates of uptake. Secondly, the amount of H^+ efflux from low-salt barley roots during salt accumulation increases with external KCl concentration, and this H^+ efflux is against its electrochemical gradient. In addition, measurements of ion uptake, cell potential and H^+ efflux in roots treated with fusicoccin show that the increased K^+ and Na^+ uptakes accompanying the stimulated H^+ efflux are coupled electrogenically with the H^+ efflux (Pitman et al. 1975).

The OH^- production accompanying the H^+ efflux is dissipated in some conditions by production of organic acids which are accumulated in the vacuoles. However, the OH^- may also be important as a component of anion uptake, as this could operate as an OH^-/anion antiport system.

There is some evidence that an Na^+ efflux is required at the plasmalemma. Many plants show selectivity for K^+ relative to Na^+ and the electro-chemical activity in the cells is often much less than in the external solution. Selectivity for K appears to be due to higher ratios of K^+/Na^+ in the cytoplasm so that low fluxes of Na^+ to the vacuoles are due to low concentrations of Na^+ in the cytoplasm, not to low permeability of the tonoplast to Na^+ (Pitman and Saddler 1967, Jeschke 1973). These results argue for an Na^+ efflux, that may be related to the H^+ efflux as proposed for *Neurospora* (Slayman and Slayman 1968). In addition to Na^+ efflux, there is also clear evidence from electrical measurements that the diffusive permeability of the plasmalemma is much higher to K^+ than to Na^+.

While Na^+ can substitute for K^+ in the vacuoles it cannot substitute for K^+ as an activator of various enzymes (e.g. pyruvic kinase). However it is well known that addition of Na^+ to plants growing in soils where K^+ is limiting can lead to stimulation of growth (see, for example, *Chloris gayana*, Smith 1974). The simplest explanation is that the mechanisms achieving high K^+/Na^+ in the cytoplasm allow the small amounts of K^+ in the plant to be concentrated in the cytoplasm.

Information about other processes is less well developed. Uptake of phosphate seems to be by a different process since phosphate uptake is not inhibited by extremely high concentrations of Cl^- (Carter and Lathwell 1967). Phosphate uptake differs too in having a very low K_m; 1 to 5 mmol m^{-3} produce maximum uptake of phosphate.

Sulphate also seems to have a low K_m and 4 mmol m^{-3} has been been suggested to be adequate for growth of wheat (Reisenauer 1968). It is, of course, metabolised in the cell and its uptake can be severely inhibited by DNP, KCN and sodium azide.

Nitrate uptake in some ways resembles Cl^- uptake since KNO_3 can be taken up to the same extent as KCl by low salt roots, but this similarity may be because uptake of both ions is limited by the capacity of the cell vacuoles. NO_3^- is taken up preferentially to Cl^- from mixtures of KCl and KNO_3 and has been suggested to be transported by a separate system. The preferential uptake may be partly due to metabolism of NO_3 in the cytoplasm. The NO_3^--transport system has been suggested to be inducible (Jackson et al. 1973) as is the NO_3^--reductase in the roots.

Uptake of Fe has been extensively studied for its importance in plant nutrition. An important feature of its uptake is that Fe^{3+} is reduced to Fe^{2+} at the surface of the cell and the root cell absorbs Fe^{2+} (Ambler et al. 1970; Chaney et al. 1972). Iron uptake is related to H^+ transport in some systems and Fe-deficient plants have been shown to lower the pH of the solution around the roots (Raju et al. 1972). This H^+ release could be an alternative to reduction of Fe^{3+}, and the roots might then be analogous to yeast (Conway 1951).

Uptake of divalent cations (including Fe^{2+}) is little understood. There seems to be separate "divalent" as opposed to "univalent" carriers and often there is competitive inhibition between pairs of divalent cations such as Ca^{2+} and Mg^{2+}; Mn^{2+} and Ca^{2+}. Roots differ in their rates of uptake of Ca^{2+} and it is well known that excised roots of maize and wheat, for example, take up Ca^{2+} at a much more rapid rate than do roots of barley, *Phaseolus*, soybean and others, even though this latter group of plants absorbs Ca^{2+} when intact at a rapid rate. Leggett and Gilbert (1969) suggested for soybean that the low rate of uptake to excised roots was because Ca^{2+} was excluded from the cell vacuoles, and only small amounts saturated the small volume of the symplasm. They pointed out that Moore et al. (1965) showed for barley that there was Ca^{2+} in the xylem exudate at high concentrations even though little Ca^{2+} was absorbed in the root. Uptakes of Mn^{2+} and Zn^{2+} have been shown to be affected by external H^+ concentration (Robson and Loneragan 1970; Chaudhry and Loneragan, 1972).

These suggestions give the impression of the cell membrane being the site for a spectrum of transport sites. Of these sites only a few have any identity and even then they are not widely accepted.

Hodges and colleagues (e.g. Leonard and Hodges 1973) have described the isolation of a K^+-stimulated ATPase from oat roots that has the same pattern of kinetics as the uptake process found by Epstein (see earlier). The ATPase is associated with a membrane fraction that has been inferred to be the plasmalemma, though some workers have not been satisfied with the evidence, based largely on staining reactions. It is interesting to note that the K^+-stimulated ATPases isolated from oats and wheat have different requirements for Mg^{2+} and Ca^{2+} (Kähr and Kylin 1974), and that the ATPase from oats can be inhibited by Ca^{2+} (Balke et al. 1974).

There have been various reports too of anion-stimulated ATPases associated with membrane fractions (e.g. Rungie and

Wiskich 1973, Balke et al. 1974, Leigh et al. 1974). It is interesting that the Cl-stimulated ATPase was effective at a higher pH range than the K-stimulated ATPase (Balke et al. 1974) and that the Cl-stimulated ATPase was on a different membrane. It has been suggested in all these reports that it may be located at the tonoplast.

Returning to the problem of K^+ and Na^+ selectivity, R.A. Wildes (unpublished) has recently been comparing the relative stimulation produced by K^+ and Na^+ using membrane fractions isolated from low-salt and salt saturated roots. Both preparations show Na^+ as about 60-80% effective as K^+ in the range of Mechanism II, so that the activity of the ATPase does not explain the difference in selectivity observed between low-salt and salt saturated roots.

UPTAKE BY WHOLE PLANTS

Rates of uptake to whole plants can be estimated from analysis of the plant content at two different harvests, together with the measurement of plant weight, (for comparison with excised roots it is useful if fresh as well as dry weight is known). Such measurements are of course average rates over the period of the harvest. Details of calculation of uptake are given by Milthorpe and Moorby (1974).

Much good data is available for calculation of rates of uptake to intact plants from studies with flowing solution culture. Pioneers in this field have been C. Asher, J. Loneragan and P. Ozanne and their colleagues in Western Australia. Figure 2 shows some rates of uptake for several nutrients collected from their results and from experiments by Clement et al. (1974) and Lycklama (1963) for N uptake. For the same relative growth rate the rate of uptake rises to a maximum and then is independent of external concentration. The "kinetics" for the whole plant thus differs from the "dual" kinetics of uptake to excised barley roots, where influx continues to rise with external concentration in the range 1 to 50 mol m^{-3}.

This difference is also shown in Fig.3, which compares rates of uptake from experiments using excised roots with rates of uptake to whole plants. There is a further difference in the data of Fig.3 as at low concentrations uptake to the whole plant was much lower than uptake to excised roots. At low concentrations growth of the whole plants was impaired and this presumably led to reduced nutrient uptake. The measurements with excised roots were over short periods (20-60 min) where uptake was not affected by growth and only by the external concentration.

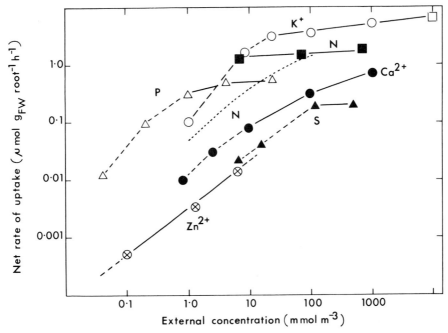

Fig. 2. - Rates of net uptake of various elements calculated from change in external concentration or internal content. Dashed lines show where growth limitation was observed. Note this is a log/log plot to increase the range of data given. (○) K^+, from Asher and Ozanne, 1967, assuming R_W = 0.13 d^{-1}, mean of 14 spp; (□) from Pitman 1972 for same R_W for barley; (△) P from Asher and Loneragan 1967, Loneragan and Asher 1967, mean of 8 spp; (■) N from Clement et al. 1974, dotted line N from Lycklama 1963, both for Lolium perenne; (●) Ca^{2+} from Loneragan and Snowball 1969a,b, mean of 30 spp; (▲) S from Bouma, 1967a,b,c, Trifolium subterraneum; (⊗) Zn^{2+} from Carroll and Loneragan 1968,1969, mean of 8 spp (from Pitman 1976).

Both these differences emphasise that the equations relating uptake to external concentration should include terms for availability of energy or control of the process. For example, the Michaelis-Menten formulation might be better expressed as

$$J_P = J_{max} \text{ (energy, control)} . \frac{C_o}{C_o + K_m}$$

In this way it is recognised that the carrier is not always operating at the same efficiency but its rate can be affected by "metabolic status" of the roots.

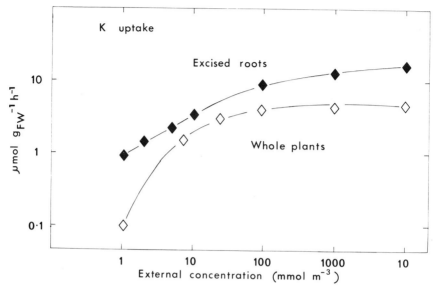

Fig. 3. - Rates of K⁺ uptake from excised roots (◆) compared with rates of uptake to whole plants (◇).

Figure 2 also shows the high efficiency of phosphate uptake compared with K^+ or N uptakes. In this example the apparent K_m is about 0.5 mmol m^{-3}, though this is only a description of the shape of the curve. I will return to this aspect of uptake by the root later.

DIFFUSION LIMITATIONS AND UPTAKE

This topic is a complex one especially when dealing with roots in the soil and I propose only to look at a particular aspect as it affects physiological type experiments. A root in solution will depend on circulation of the solution and on diffusion through the non-stirred layer near the surface of the root for supply of ions. The effect of this layer depends on its thickness, on the rate of uptake, and of course on external concentration. For example, Polle and Jenny (1971) found stirring affected uptake of Rb to barley roots only when the concentration was below 10 mmol m^{-3}, and suggested the effective thickness of the layer was between 30-70 μm for moderately stirred solutions. House (1974) gives an extensive treatment of unstirred layers in which he suggests

the layer has an equivalent thickness of about 1 to 500 µm
depending on stirring speed, but takes 100 µm as an "unstirred
layer".

The effect of this layer on diffusion of ions to the
root can be estimated by assuming that the relation between
rate of transport (J) and surface concentration (C_S) is

$$J = J_\infty \frac{C_s}{C_s + K_m}$$

The rate of diffusion through the unstirred layer is

$$J = \frac{D}{\delta} (C - C_s)$$

where δ is an equivalent thickness that can take into account
the curvature of a cylindrical layer around the root (for
example the effective thickness of a layer 100 µm thick
around a root of external diameter 500 µ would be 117 µm)
C_s can be eliminated from these equations to give an expression for J. Figure 4 shows calculated values of J assuming
δ = 100 µm and 10 µm; clearly if δ is as large as 100 µm
there could be appreciable reduction in uptake due to diffusion. An effect of diffusion then could be to increase the

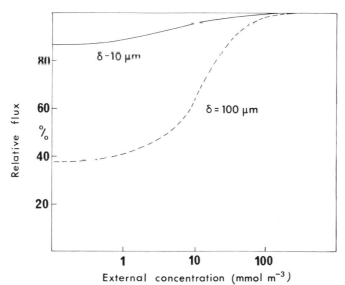

*Fig. 4. - Calculated flux to the root assuming
different thickness of unstirred layer (δ) and
expressed relative to rate of uptake expected from
$J = J_\infty C_0/(C_0 + K_m)$ where $J_\infty = 1.6 \times 10^{-7}$ mol m^{-2} s^{-1}
and $K_m = 10^{-2}$ mol m^{-3}.*

estimated value of K_m for the uptake site from its "real" value. At very low concentrations J will tend to

$$J = \frac{D}{\delta} \cdot C_o$$

when limited by diffusion, or to

$$J = \frac{J_\infty}{K_m} \cdot C_o$$

if limited by the activity of the exchange sites when J_∞ = 9 μmol $g_{FW}^{-1} h^{-1}$ (approx 3×10^{-7} mol m^{-2} s^{-1} at the surface), and $K_m = 12$ mmol m^{-3} (data from Epstein 1972) then

$$\frac{J_\infty}{K_m} = \frac{3 \times 10^{-7}}{1.2 \times 10^{-3}} = 2.5 \times 10^{-5}.$$

It is clearly critical whether δ is 10 μm ($D/\delta = 10^{-4}$) or nearer 100 μm ($D/\delta = 10^{-5}$) in interpreting K_m. (The same limitations should apply to K_m for a suspension of vesicles carrying ATPase.)

A complication to this simple approach to uptake is that the boundary layer of the root appears to be a thin mucigel with fixed negative charges that will contain more exchangeable cation than anion. The mobility of both cations and anions in such a gel may be almost as large as in water, but as Teorell has shown in classical studies of charged membranes, net diffusion of cations and anions will depend on there being equality of charge transfer across the boundary and so in a negatively charged layer, cation diffusion may be limited by slower net transfer of anions. Jenny (e.g. 1966) has discussed diffusion in the mucigel in contact with solution or clay particles, but I do not propose to explore this model further here.

Diffusion in the soil can be expected to be slower for two other reasons. One is that ions may be adsorbed on the solid phase, reducing concentration in the soil solution and another is that the amount of soil solution, and its presentation to the root may increase the effective diffusion path to the root. Barley (1970) suggested that the diffusive coefficients for ions in the soil were about 10^{-9}-10^{-10}; 10^{-10}; and 10^{-10}-10^{-11} m^2 s^{-1} for NO_3^-; K^+; and phosphate respectively, while the apparent diffusion coefficient, taking into account reaction with the solid phase was about 10^{-9}-10^{-10}; 10^{-10}; and 10^{-11}-10^{-15} m^2 s^{-1} respectively.

Barley also suggested that the "dual kinetics" for uptake was replacelable for whole plants in the soil by a simple kinetic in which K_m had the values 0.1, 0.1 and 1.0 mol m^{-3} for NO_3^-, K^+ and phosphate. These values are 10 times

greater than K_m from excised roots studied for NO_3^- and K^+
and some 1000 times greater than K_m for phosphate. By
analogy with "unstirred layers" in solution studies, the soil
has the effect of restricting supply to the root and the plant
hardly achieves the efficiency of its absorption sites, as
measured in other studies with excised roots, though clearly
the plant achieved adequate rates of phosphate uptake from
soils. The value of the low K_m to the plant may be more in
facilitating removal of phosphate from adsorption on the
solid phase of the soil, than in overcoming boundary layer
diffusion.

Diffusion also affects the interpretation of rates of
uptake to roots when C_0 is low because the concentration
within the root will fall below C_0. The effect will be that
at very low concentrations ions are absorbed predominantly
by cells at the surface of the root, whereas at higher
concentrations cells within the cortex will receive ions
by diffusion in the free space (cell walls). Free space
depletion in an absorbing root when external concentration
is low is one reason for thinking that transport across the
root occurs in the symplasm. Vakhmistrov (1967) and Bange
(1973) have suggested that uptake to roots may in fact be
limited to the epidermal cells and that the cortical cells
have little role in absorption. This extreme view does not
agree with measurements of ion fluxes made using half-
cortices of maize roots (Cram 1973).

WATER MOVEMENT AND UPTAKE

The supply of ions to the root can be seen intuitively
to be affected by any flow of water to the root surface due
to transpiration. The combined effects can be expressed
by a differential equation including bulk flow (proportional
to C) and diffusion (proportional to dC/dx). [See Barley
(1970) and Milthorpe and Moorby (1974)]. Opinions differ on
the extent to which water flow in the soil affects uptake,
but the effect of transpiration seems to be not more than
about 10-20% increase (Barley 1970).

Transpiration and uptake have been studied extensively
with more emphasis on the other side of the plant/soil
barrier. For certain neutral molecules uptake seems to be
proportional to their concentration and to water flow. Many
organic molecules, silica and also borate come in this class.

At the other extreme are ions such as NO_3^- and K^+ where
uptake is often quite independent of water flow across the
root. Uptake in this case seems to be limited by the movement in the symplasm and release to the xylem.

95

There are other examples where uptake shows combinations of a component dependent on, and another independent of water flow and external concentration. Uptake of Ca^{2+} and Mg^{2+} comes in this category. As Dr. Läuchli will discuss, these ions are transported across the root in regions where the endodermis is now entirely sealed by suberin, but there is evidence too that the process depends on energy metabolism so uptake is not simply bulk flow of solution.

Between these extremes there are many examples where K^+ for example, is taken up proportionately to water flow. In these cases it is tempting to think that K^+ is moving in a kind of "passive" pathway different from the symplasm and release to the xylem. Many of these examples though refer to conditions where plants have been transferred from one solution to another and there are consequent adjustments of net transport. The moral is, perhaps, that the physiologist has to develop models that can handle the behaviour of the plant in transient situations as well as in the steady state that is more amenable to experimentation.

REFERENCES

Ambler, J.E., Brown, J.C. and Gauch, H.G. (1970). Effect of zinc on translocation of iron in soybean plants. *Plant Physiol. 46*, 320-323.

Asher, C.J. and Loneragan, J.F. (1967). Response of plants to phosphate concentration in solution culture: I. Growth and phosphorus content. *Soil Sci. 103*, 225-233.

Asher, C.J. and Ozanne, P.G. (1967). Growth and potassium content of plants in solution cultures maintained at constant potassium concentrations. *Soil Sci. 103*, 155-161.

Balke, N.E., Sze, Heven, Leonard, R.T. and Hodges, T.K. (1974) Cation sensitivity of the plasma membrane ATPase of oat roots. In; "Membrane Transport in Plants" ed. by U. Zimmermann and J. Dainty, Springer-Verlag, Berlin Heidelberg New York pp.301-306.

Bange, G.G.J. (1973). Diffusion and absorption of ions in plant tissue III. The role of the root cortex cells in ion absorption. *Acta Bot. Neerl. 22*, 529-542.

Barley, K.P. (1970). The configuration of the root system in relation to nutrient uptake. *Advan. Agron. 22*, 159-201.

Bouma, D. (1967a). Growth changes of subterranean clover during recovery from phosphorus and sulphur stresses. *Aust. J. Biol. Sci. 20*, 51-66.

Bouma, D. (1967b). Nutrient uptake and distribution in sub-terranean clover during recovery from nutritional stresses. I. Experiments with phosphorus. *Aust. J. Biol. Sci. 20,* 601–612.

Bouma, D. (1967c). Nutrient uptake and distribution in sub-terranean clover during recovery from nutritional stresses. II. Experiments with sulphur. *Aust. J. Biol. Sci. 20,* 613–622.

Carroll, M.D. and Loneragan, J.F. (1968). Response of plant species to concentrations of zinc in solution. I. Growth and zinc content of plants. *Aust. J. Agric. Res. 19,* 859–868.

Carroll, M.D. and Loneragan, J.F. (1969). Response of plant species to concentrations of zinc in solution. II. Rates of zinc absorption and their relation to growth. *Aust. J. Agric. Res. 20,* 457–463.

Carter, O.G. and Lathwell, D.J. (1967). Effect of chloride on phosphorus uptake by corn roots. *Agron. J. 59,* 250–253.

Chaney, R.L., Brown, J.C. and Tiffin, L.O. (1972). Obligatory reduction of Fe chelates in Fe uptake by soybeans. *Plant Physiol. 50,* 208–213.

Chaudhry, F.M. and Loneragan, J.F. (1972). Zinc absorption by wheat seedlings: II. Inhibition by hydrogen ions and by micronutrient cations. *Soil Sci. Soc. Am. Proc. 36,* 327–331.

Clement, C.R., Hopper, M.J., Canaway, R.J. and Jones, L.H.P. (1974). A system for measuring the uptake of ions by plants from flowing solutions of controlled composition. *J. Exp. Bot. 25,* 81–99.

Conway, E.J. (1951). The biological performance of osmotic work. A redox pump. *Science 113,* 270–273.

Cram, W.J. (1973). Chloride fluxes in cells of the isolated root cortex of *Zea mays*. *Aust. J. Biol. Sci. 26,* 757–779.

Cram, W.J. and Pitman, M.G. (1972). The action of abscisic acid on ion uptake and water flow in plant roots. *Aust. J. Biol. Sci. 25,* 1125–1132.

Epstein, E. (1966). Dual pattern of ion absorption by plant cells and by plants. *Nature 212,* 1324–1327.

Epstein, E. (1972). Mineral nutrition of plants: principles and perspectives. New York, London, Sydney, Toronto: John Wiley and Sons.

House, C.R. (1974). Water transport in cells and tissues. London: Edward Arnold.

Jackson, W.A., Flesher, D. and Hageman, R.N. (1973). Nitrate uptake by dark–grown corn seedlings. Some characteristics of apparent induction. *Plant Physiol. 51,* 120–127.

Jenny, H. (1966). Pathways of ions from soil into roots

according to diffusion models. *Plant Soil 25*, 255–285.

Jeschke, W.D. (1973). K$^+$-stimulated Na$^+$ efflux and selective transport in barley roots. In: "Ion Transport in Plants" ed. by W.P. Anderson, Academic Press, London, New York pp. 285–296.

Kähr, M. and Kylin, A. (1974). Effects of divalent cations and oligomycin on membrane ATPase from roots of wheat and oat in relation to salt status and cultivation. In: "Membrane Transport in Plants" ed. by U. Zimmermann and J. Dainty, Springer-Verlag, Berlin, Heidelberg, New York pp. 321–325.

Leggett, J.E. and Gilbert, W.A. (1969). Magnesium uptake by soybeans. *Plant Physiol. 44*, 1182–1186.

Leigh, R.A., Wyn Jones, R.G. and Williamson, F.A. (1974). Ion fluxes and ion-stimulated ATPase activities. In: "Membrane Transport in Plants" ed. by U. Zimmermann and J. Dainty, Springer-Verlag, Berlin, Heidelberg, New York pp. 307–316.

Leonard, R.T. and Hodges, T.K. (1973). Characterization of plasma membrane-associated adenosine triphosphatase activity in oat roots. *Plant Physiol. 52*, 6–12.

Loneragan, J.F. and Asher, C.F. (1967). Response of plants to phosphate concentration in solution culture. II. Rate of phosphate absorption and its relation to growth. *Soil Sci. 103*, 311–318.

Loneragan, J.F. and Snowball, K. (1969a). Calcium requirements of plants. *Aust. J. Agric. Res. 20*, 465–467.

Loneragan, J.F. and Snowball, K. (1969b). Rate of calcium absorption of plant roots and its relation to growth. *Aust. J. Agric. Res. 20*, 479–490.

Lycklama, J.C. (1963). The absorption of ammonium and nitrate by perennial rye-grass. *Acta Botan. Neerl. 12*, 361–423.

Milthorpe, F.L. and Moorby, J. (1974). An introduction to crop physiology. Cambridge, Cambridge University Press.

Moore, D.P., Mason, B.J. and Maas, E.V. (1965). Accumulation of calcium in exudate of individual barley roots. *Plant Physiol. 40*, 641–644.

Pitman, M.G. (1969). Simulation of Cl$^-$ uptake by low-salt barley roots as a test of models of salt uptake. *Plant Physiol. 44*, 1417–1427.

Pitman, M.G. (1972). Uptake and transport of ions in barley seedlings. III. Correlation of potassium transport to the shoot with plant growth. *Aust. J. Biol. Sci. 25*, 905–919.

Pitman, M.G. (1976). Ion uptake by plant roots. In: Encyclopedia of Plant Physiology, New Series Vol. 2 Plant Cells and Tissues ed. by U. Lüttge and M.G. Pitman,

Springer-Verlag, Berlin, Heidelberg, New York pp. 95-128.

Pitman, M.G. and Saddler, H.D.W. (1967). Active sodium and potassium transport in cells of barley roots. *Proc. Natl. Acad. Sci. (U.S.) 57*, 44-49.

Pitman, M.G., Schaefer, N. and Wildes, R.A. (1975). Relation between permeability to potassium and sodium ions and fusicoccin-stimulated hydrogen-ion efflux in barley roots. *Planta (Berl.) 126*, 61-73.

Polle, E.O. and Jenny, H. (1971). Boundary layer effects in ion absorption by roots and storage organs of plants. *Physiol. Plant. 25*, 219-224.

Raju, K.V. von, Marschner, H. and Römheld, V. (1972). Effect of iron nutritional status on ion uptake, substrate pH and production and release of organic acids and riboflavin by sunflower plants. *Z. Pflanzenernähr. Düng. Bodenk. 132*, 177-190.

Reisenauer, H.M. (1968). Growth and nutrient uptake by wheat from dilute solution cultures. *Agron. Abst.* p. 108.

Robson, A.D. and Loneragan, J.F. (1970). Sensitivity of annual *Medicago* species to manganese toxicity as affected by calcium and pH. *Aust. J. Agric. Res. 21*, 223-232.

Rungie, J.M. and Wiskich, J.T. (1973). Salt-stimulated adenosine tri-phosphatase from smooth microsomes of turnip. *Plant Physiol. 51*, 1064-1068.

Schaefer, N., Wildes, R.A. and Pitman, M.G. (1975). Inhibition by p-fluorophenylalanine of protein synthesis and of ion transport across the roots in barley seedlings. *Aust. J. Plant Physiol. 2*, 61-74.

Slayman, C.L. and Slayman, C.W. (1968). Net uptake of potassium in *Neurospora*. Exchange for sodium and hydrogen ions. *J. Gen. Physiol. 52*, 424-443.

Smith, F.W. (1974). The effect of sodium on potassium nutrition and ionic relations in Rhodes grass. *Aust. J. Agric. Res. 25*, 407-414.

Vakhmistrov, D.B. (1967). On function of apparent free space in plant roots. A study of the absorbing power of epidermal and cortical cells in barley roots. *Soviet Plant Physiol. 14*, 103-107.

CHAPTER 9

Symplasmic Transport and Ion Release to the Xylem

A. Läuchli

Fachbereich Biologie (10), Der Technischen Hochschule, D-6100 Darmstadt, Germany

PATHWAYS OF ION TRANSPORT IN ROOTS

The basic anatomy of primary roots is well understood. Epidermis and cortex are separated from the stele containing the vascular tissues, by the endodermis. A short distance behind the root apex, the endodermal cells are in their primary state of development with the Casparian strip forming a band in the radial cell walls. The Casparian strip is produced by incrustation of the wall with suberin and lignin and decreases drastically the permeability of the radial walls of the endodermis. Thus, the presence of the Casparian strip has important implications for ion transport from the cortex to the stele. Although ions appear to move readily through the cortex in the apoplasmic pathway of the cell walls, the Casparian strip blocks effectively the apoplasmic transport through the endodermis into the stele (Läuchli 1976a). Hence, there are only two possibilities for ions to pass through an endodermal cell: (i) by uptake through the plasmalemma at the outer tangential wall, transport through the cytoplasm, and release through the plasmalemma at the inner tangential wall into the apoplast of the stele, or (ii) by symplasmic transport through plasmodesmata in the outer and inner tangential walls, respectively (Clarkson 1974; Läuchli 1976a). The symplasmic pathway extends from the epidermis all the way up to the vessels (Läuchli 1972; Kurkova et al. 1974; Läuchli et al. 1974a; Robards and Clarkson 1976). At the xylem parenchyma cells, ions are released from the symplasm to the xylem vessels. The transport pathways in roots are shown schematically in Fig. 1.

In the laboratory of Clarkson and Robards, extensive studies have been made on the transport of ions to the xylem in relation to endodermal development. The results of these studies have considerable bearing on the pathways

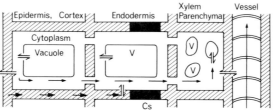

Fig. 1. Pathways of ion transport in roots: a simplified model. Cs = Casparian strip. Modified after Lüttge (1974).

involved in ion transport through the root. The development of a secondary endodermis, where a suberin lamella is deposited over the entire inner wall surface, is detectable in barley roots about 8 cm from the root tip (Clarkson 1974). Movement of Ca^{2+} into the stele occurs predominantly in the region of the root where the endodermis is in its primary state (Robards et al. 1973; Harrison-Murray and Clarkson 1973). Since Ca^{2+} apparently is not transported in the symplasm (Spanswick 1976) and apoplasmic Ca^{2+} transport is hindered by the Casparian strip, this ion may only enter the stele by uptake at the outer and release at the inner tangential plasmalemma of the endodermis. On the other hand, K^+ and $H_2PO_4^-$ move readily into the stele of even the oldest parts of the root where the endodermis is in its tertiary state of development (Clarkson et al. 1971; Harrison-Murray and Clarkson 1973). Since numerous intact plasmodesmata were found in the inner tangential wall of the tertiary endodermis of barley roots, transport of K^+ and $H_2PO_4^-$ to the stele of these roots was most probably via the symplasm. Ziegler et al. (1963), however, were not able to detect plasmodesmata in the tertiary endodermis of *Iris* roots, and symplasmic transport could only be demonstrated through the passage cells.

The demonstration of symplasmic transport of ions through roots rests on the assumption that the Casparian strip is essentially impermeable to ions. That this assumption is in fact correct, has been shown by several methods (Läuchli 1976a). The apoplasmic pathway can be "labelled" with electron-dense metal ions such as UO_2^{2+} or La^{3+}. Electron microscopic observation of the distribution of these metal ions in roots of barley and corn demonstrated that the Casparian strip is an effective barrier to apoplasmic transport into the stele, at least for the metal ions used (Robards and Robb 1972, 1974; Nagahashi et al. 1974). Another useful method is the electron microscopic observation of the pathway of Cl^- transport after precipitation of Cl^- with organic Ag^+ salts (Läuchli 1975a). Using the latter method, the suberin

lamella in a tertiary endodermal cell of the root of
Puccinellia peisonis was found to block apoplasmic transport
of Cl⁻ to the stele (Läuchli 1976a, Fig. 1.14).

SYMPLASMIC TRANSPORT FROM CORTEX TO STELE

Lüttge (1973) and Clarkson (1974) have reviewed the
largely indirect evidence that the symplasm is an important
pathway for ion transport in roots. The main points are as
follows: (i) Ions can be released to the xylem under conditions where their movement in the apoplasmic pathway is most
unlikely (e.g. Jarvis and House 1970). (ii) Cells in roots
are coupled electrically through junctions of low resistance
which are most probably formed by the plasmodesmata (Spanswick
1976). (iii) The endoplasmic reticulum appears to be continuous through the plasmodesmata, thus representing the cytological compartment for the intercellular symplasmic transport
(Gunning and Robards, this volume).

Is there any direct evidence of ion localization in the
symplasm of the root from which symplasmic transport could be
inferred? The AgCl-precipitation technique for intracellular
localization of Cl⁻ mentioned above was applied to a study of
intercellular Cl⁻ transport in barley roots (Läuchli et al.
1974b; Stelzer et al. 1975; Läuchli 1976b). The precipitates were found in plasmodesmata of cells of the cortex,
endodermis, pericycle, and xylem parenchyma. Furthermore,
they were localized particularly in the endoplasmic reticulum
of root cells. It was therefore concluded that ion transport
through the root proceeds in the symplasm and that the endoplasmic reticulum is involved in symplasmic ion transport. Of
particular significance is the fact that precipitates were
detectable in the endoplasmic reticulum of xylem parenchyma
cells after an absorption period of only 20 minutes from 5 mM
Cl⁻ (Läuchli et al. 1974b). This emphasizes the efficiency
of the symplasmic pathway for ion transport to the xylem of
roots. Van Iren and Van der Spiegel (1975) have also obtained
evidence for ion localization in the endoplasmic reticulum of
barley roots. They used an *in vivo* precipitation technique in
which the roots were exposed to Tl⁺ followed by a treatment
with I⁻, leading to precipitation of TlI in the roots. The
distribution of precipitates was then observed electron micro-
scopically. The administration of Tl⁺ and I⁻ as analogues of
nutrient ions, however, is questionable (Läuchli 1976b). In
summary, there is direct evidence of ion localization in the
symplasm of barley roots. This conclusion is corroborated by
the demonstration of Cl⁻ in the endoplasmic reticulum of cells
of *Nitella* (Van Steveninck et al. 1974) which may be related
to the fast component of Cl⁻ transport to the vacuole of these
cells (MacRobbie 1973).

ION RELEASE TO THE XYLEM: LOCATION AND POSSIBLE MECHANISMS

Transport through the root involving the symplasm as an integral component is envisioned as a three-steps-process, that is loading the symplasmic pathway in the cortex - passive movement through the symplasm - unloading the symplasmic pathway in the xylem parenchyma and release to the xylem vessels. We shall now turn to a discussion of the process of unloading the symplasmic pathway coupled to the release of ions to the xylem vessels.

It has been proposed that the unloading process is located at the plasmalemma of the xylem parenchyma cells (Läuchli et al. 1971; Pitman 1972; Anderson 1972; Epstein 1972). What are the structural and functional characteristics of the xylem parenchyma cells in roots? First of all, it is well established that many plasmodesmata connect the xylem parenchyma cells with the endodermis and with one another (Kurkova et al. 1974; Läuchli et al. 1974a). Thus, the symplasmic pathway extends to the xylem parenchyma surrounding the vessels. An ultrastructural study of the xylem parenchyma cells in barley roots showed that they are rich in cytoplasm with extensive rough endoplasmic reticulum and well-developed mitochondria (Läuchli et al. 1974a). Hence, their structural features make them suitable to play an active role in the stele.

X-ray microanalysis of the distribution of K^+ in roots of *Zea mays* revealed that the xylem parenchyma cells had the highest K^+ contents of all the root cells (Läuchli et al. 1971). When the vacuolar K^+ activities were measured by means of K^+ sensitive microelectrodes, however, no gradients in K^+ activity were found in roots of several species (Dunlop and Bowling 1971; Bowling 1972; Dunlop 1973). Yet the technique of advancing a microelectrode from the epidermis through the root tissue to the xylem parenchyma can give rise to false measurements (Anderson and Higinbotham 1975). Furthermore, X-ray microanalyses of deep-frozen, hydrated specimens from barley roots confirmed the accumulation of K^+ in xylem parenchyma cells (Läuchli 1975b; Läuchli et al. 1975). In the latter study, a new technique of specimen preparation was used by which leaching or redistribution of soluble ions is virtually eliminated during preparation and analysis. Root segments about 3 mm long were cut with a razor blade, placed on a copper holder and quench-frozen with liquid N_2. The top parts of the frozen root segments were fractured away under liquid N_2 to provide an uncontaminated root surface for analysis. The holder with the specimens was transferred under liquid N_2 to the cryostage of a scanning electron microscope fitted with an energy dispersive X-ray detector. The cryostage was cooled with liquid N_2 throughout the observation and analysis of the specimens. Note that the specimens were not exposed to any aqueous solutions or organic fluids that could have given rise to artificial redistribution of soluble ions. One may conclude that xylem parenchyma cells are the sites of ion accumulation, probably acting as sinks for ions that move in the symplasmic pathway to the stele. Part of the ions arriving at the xylem parenchyma cells by symplasmic transport may be accumulated in the vacuoles of these cells (sinks) from where they could be readily accessible for release to the xylem vessels.

The evidence thus far reviewed suggests an ion secretory process operating at the xylem parenchyma cells. Ions secreted by such a process could be most easily released into the vessels through the pits of the vessel walls. Except for the pits the walls of the vessels are highly lignified which pro- bably hinders movement of ions (Läuchli 1976a). This secret- ory process can only operate into mature vessels lacking cyto- plasm. There is the difficulty in maize roots, however, that xylem vessels develop slowly and gradually along the length of the root (Anderson and House 1967; Burley et al. 1970; Higinbotham et al. 1973). This led Higinbotham et al. (1973) to suggest that living, immature vessels could generate ion transport to the shoot. The electron microscopic evidence presented by these authors does indeed show the presence of cytoplasm in vessels several cm from the tip, but it is not clear whether this cytoplasm is already degenerating or still active. In addition, the situation in barley is quite differ- ent. Metaxylem vessels in barley roots mature within 3-4 mm (Läuchli et al. 1974a) or about 1 cm from the apex (Robards and Clarkson 1976). Furthermore, ion transport through the root is not stopped by removal of that part of the apical root containing vessels with living cytoplasm (Läuchli et al. 1974a). In consequence, the xylem parenchyma cells are a more suitable site for ion secretion than the immature vessels, at least in barley.

The secretory process delivering ions into the vessels is sensitive to metabolic inhibitors, to inhibitors of protein synthesis and to certain hormones such as abscisic acid and cytokinins. The uncoupler CCCP inhibits both uptake of ions by the root and secretion into the xylem (Läuchli and Epstein 1971; Pitman 1972). When the effect of the protein synthesis inhibitor cycloheximide was tested (Läuchli et al. 1973; Lüttge et al. 1974), ion transport to the xylem of barley roots was greatly reduced but uptake by the root was initially not affected. It was furthermore shown that cycloheximide in- hibited protein synthesis but did not uncouple respiration; the ATP level in the root was slightly enhanced. Thus, ion

105

transport through the root appears directly dependent on concurrent protein synthesis. It is difficult, however, to deduce the site of inhibition by cycloheximide. This inhibitor
could affect protein synthesis at the endoplasmic reticulum
and, hence, could interfere with symplasmic transport. Or
else, cycloheximide could actually block the ion release from
the symplasm to the vessels. The phytohormones abscisic acid
and cytokinins also inhibited ion transport through the root
without reducing uptake (Cram and Pitman 1972; Pitman et al.
1974a). The effect of abscisic acid appeared to be located in
the stele and there was no effect on symplasmic transport
(Pitman et al. 1974b). One may therefore argue that ion release to the vessels is a process different in kind from uptake
by the root and possibly under hormonal regulation.

Cycloheximide interferes with the process of protein
synthesis *per se*. On the other hand, the amino acid analogue
p-fluorophenylalanine is incorporated into proteins in the
place of phenylalanine causing the cell to produce ineffective
proteins. This inhibitor therefore acts only on the final
product of protein synthesis. In testing the effect of p-
fluorophenylalanine at a concentration of 2 mM, Schaefer et al.
(1975) showed that ion transport through the root was inhibited drastically but uptake by the root was not affected for
several hours. In comparison, L-phenylalanine applied at the
same concentration as its analogue caused only a small reduction in transport. As was discussed previously with regard to
cycloheximide, p-fluorophenylalanine could inhibit symplasmic
transport of ions or their release to the xylem. Schaefer et
al. (1975) therefore attempted to measure the effect of the
amino acid analogue on transport from the cortex to the stele,
that is on symplasmic transport. The roots were pre-treated
with the analogue, exposed to ^{36}Cl, and then separated into
cortex and stele. No decrease in the tracer content of the
stele was detectable. Hence, it was concluded that symplasmic
transport is not affected but the site of the analogue's
action is within the stele, probably between xylem parenchyma
and vessels. This is in line with our assumption of the ion
secretory process being located at the plasmalemma of the
xylem parenchyma cells. Schaefer et al. (1975) furthermore
concluded that ion secretion to the vessels involves some
specific protein(s) having a short half-life. Such a protein
could represent a component of a transport system or it could
act as a permease with a relatively rapid turnover. In this
context, it may be pertinent to note that in our laboratory
Sluiter (1975) has cytochemically demonstrated ATPase activity
at the plasmalemma of xylem parenchyma cells in barley roots.
This ATPase was inhibited by diethylstilbestrol, an inhibitor

of the isolated plasma membrane ATPase as first found by Balke and Hodges (1975). It is not known if this ATPase is linked to the transport protein proposed by Schaefer et al. (1975).

A possible model of ion transport through the root is represented in Fig. 2. Ions can move readily in the apoplast of the epidermis and cortex up to the Casparian strip. The loading of the symplasmic pathway occurs predominantly in the cortex. The symplasmic pathway extends into the xylem parenchyma cells and is composed of the endoplasmic reticulum and the plasmodesmata. The xylem parenchyma cells and in particular their vacuoles act as sinks for ions from the symplasmic pathway. The final release of ions into the vessels occurs mainly through the pits and possibly involves a specific protein that operates at the plasmalemma of the xylem parenchyma cells.

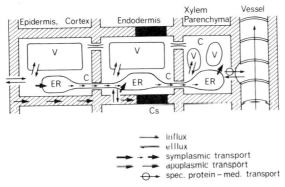

Fig.2. - Model of ion transport through the root. C = cytoplasm, V = vacuole, ER = endoplasmic reticulum, Cs = Casparian strip.

—→ influx
←— efflux
→ - → symplasmic transport
→ apoplasmic transport
⊖→ spec. protein – med. transport

STRUCTURE AND POSSIBLE FUNCTION OF XYLEM PARENCHYMA CELLS IN THE BASAL ROOT

Jacoby (1964) first demonstrated in physiological experiments that bean roots accumulate Na^+ in their upper, basal regions. Similarly, the moderately salt tolerant soybean mutant "Lee" accumulates Na^+ and Cl^- in the roots (Läuchli, unpublished results). The bean plant is thus rated a Na^+ excluder (that is exclusion of Na^+ from the leaves) and the "Lee" mutant of soybean is a salt excluding plant. This physiological property appears to be correlated with the occurrence of wall ingrowths in xylem parenchyma cells of the older region of the roots. Läuchli et al. (1974b) found wall ingrowths in the soybean mutant "Lee", predominantly in the area of the pits; salt stress did not alter the basic structures of the xylem parenchyma cells. This emphasizes the function of these cells as transfer cells according to the terminology of Pate and Gunning (1972). Comparative studies of the salt sensitive soybean mutant "Jackson", however, produced evidence that the xylem parenchyma cells in the basal root were damaged heavily by salt stress and only formed wall ingrowths in the absence, or in the presence of very little salt (Läuchli et al. 1974b). Furthermore, the transfer cells in the upper root of "Lee" were shown by X-ray microanalysis of deep-frozen, hydrated specimens to accumulate Na^+ under salt stress, but K^+ when no Na^+ was present in the medium (Läuchli, un-published results). Hence, their probable function is the reabsorption of Na^+ from the xylem sap and its exclusion from the leaves. The xylem parenchyma cells in the basal parts of roots of two bean species are also differentiated as transfer cells with well-developed wall ingrowths bordering the pits of the vessels (Kramer and Läuchli, in preparation). They may have a function comparable to that in soybean roots.

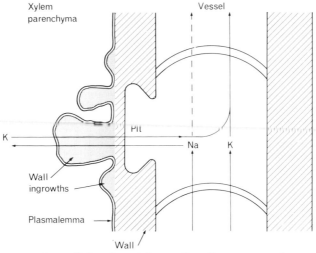

Fig.3. - Possible function of xylem parenchyma cells containing wall ingrowths in the basal root

Fig. 3 summarizes the possible function of xylem par-enchyma cells in the basal root of plant species where these cells are differentiated as transfer cells. In the absence of Na^+ these cells accumulate K^+. When Na^+ is transported upwards in the xylem sap, however, Na^+ is reabsorbed in ex-change for K^+, possibly by a Na^+/K^+ exchange mechanism opera-ting at the plasmalemma in the region of the wall ingrowths. Thus, these cells appear to have an important function in the exclusion of Na^+ from the leaves of salt excluders. It is pertinent to note that the xylem parenchyma cells in the basal root of barley, which is considered a salt includer, are not differentiated as transfer cells (Läuchli et al.

1974b).

ACKNOWLEDGEMENTS

This review was prepared while the author was a Visiting Fellow at the Research School of Biological Sciences, the Australian National University, during sabbatical leave from the Technische Hochschule Darmstadt, Germany. The author's leave was also supported by a travel grant from the Deutsche Forschungsgemeinschaft.

REFERENCES

Anderson, W.P. (1972). Ion transport in the cells of higher plant tissues. *Ann. Rev. Plant Physiol. 23*, 51-72.

Anderson, W.P. and Higinbotham, N. (1975). A cautionary note on plant root electrophysiology. *J. Exp. Bot. 26*, 533-535.

Anderson, W.P. and House, C.R. (1967). A correlation between structure and function in the root of *Zea mays. J. Exp. Bot. 18*, 544-555.

Balke, N.E. and Hodges, T.K. (1975). Inhibition of the plasma membrane ATPase by phenolics. *Plant Physiol. Suppl. 56*, 221.

Bowling, D.J.F. (1972). Measurement of profiles of potassium activity and electrical potential in the intact root. *Planta 108*, 147-151.

Burley, J.W.A., Nwoke, F.I.O., Leister, G.L. and Popham, R.A. (1970). The relationship of xylem maturation to the absorption and translocation of P^{32}. *Amer. J. Bot. 57*, 504-511.

Clarkson, D.T. (1974). Ion Transport and Cell Structure in Plants. McGraw-Hill, London.

Clarkson, D.T., Robards, A.W. and Sanderson, J. (1971). The tertiary endodermis in barley roots: fine structure in relation to radial transport of ions and water. *Planta 96*, 292-305.

Cram, W.J. and Pitman, M.G. (1972). The action of abscisic acid on ion uptake and water flow in plant roots. *Aust. J. Biol. Sci. 25*, 1125-1132.

Dunlop, J. (1973). The transport of potassium to the xylem exudate of ryegrass. I. Membrane potentials and vacuolar potassium activities in seminal roots. *J. Exp. Bot. 24*, 995-1002.

Dunlop, J. and Bowling, D.J.F. (1971). The movement of ions to the xylem exudate of maize roots. I. Profiles of membrane potential and vacuolar potassium activity across the root. *J. Exp. Bot. 22*, 434-444.

Epstein, E. (1972). Mineral nutrition of plants: principles and perspectives. Wiley, New York-London-Sydney-Toronto.

Harrison-Murray, R.S. and Clarkson, D.T. (1973). Relationships between structural development and the absorption
of ions by the root system of *Cucurbita pepo*. *Planta 114*,
1-16.

Higinbotham, N., Davis, R.F., Mertz, S.M. and Shumway, L.K.
(1973). Some evidence that radial transport in maize
roots is into living vessels. In: "Ion Transport in
Plants" ed. W.P. Anderson, Academic Press, London,
pp. 493-506.

Iren, F. van and Spiegel, A. van der (1975). Subcellular
localization of inorganic ions in plant cells by *in vivo*
precipitation. *Science 187*, 1210-1211.

Jacoby, B. (1964). Function of bean roots and stems in sodium
retention. *Plant Physiol. 39*, 445-449.

Jarvis, P. and House, C.R. (1970). Evidence for symplasmic
ion transport in maize roots. *J. Exp. Bot. 21*, 83-90.

Kurkova, J.B., Vakhmistrov, D.B. and Solovev, V.A. (1974).
Ultrastructure of some cells in the barley root as re-
lated to the transport of substances. In: "Structure
and Function of Primary Root Tissues" ed. by J. Kolek,
Veda Publ. House of the Slovak Academy of Sciences,
Bratislava, pp. 75-86.

Läuchli, A. (1972). Translocation of inorganic solutes. *Ann.
Rev. Plant Physiol. 23*, 197-210.

Läuchli, A. (1975a). Precipitation technique for diffusible
substances. *J. Microscopie Biol. Cell. 22*, 239-246.

Läuchli, A. (1975b). X-ray microanalysis in botany. *J.
Microscopie Biol. Cell. 22*, 433-440.

Läuchli, A. (1976a). Apoplasmic transport in tissues. In:
"Encyclopedia of Plant Physiology, New Series, Vol. 2,
Transport in Plants 2, Part B" ed by U. Lüttge and
M.G. Pitman, Springer, Berlin-Heidelberg-New York. pp.
3-34.

Läuchli, A. (1976b). Subcellular localization of inorganic
ions in plant cells. *Science 191*, 492.

Läuchli, A. and Epstein, E. (1971). Lateral transport of
ions into the xylem of corn roots. I. Kinetics and
energetics. *Plant Physiol. 48*, 111-117.

Läuchli, A., Kramer, D. and Gullasch, J. (1975). Ion distribution in barley roots measured by electron microprobe
analysis of deep-frozen specimens. *Plant Physiol.
Suppl. 56*, 223.

Läuchli, A., Kramer, D., Pitman, M.G. and Lüttge, U. (1974a).
Ultrastructure of xylem parenchyma cells of barley roots
in relation to ion transport to the xylem. *Planta 119*,
85-99.

Läuchli, A., Kramer, D. and Stelzer, R. (1974b). Ultrastructure and ion localization in xylem parenchyma cells
of roots. In: "Membrane Transport in Plants" ed. by
U. Zimmermann and J. Dainty, Springer, Berlin-Heidelberg-
New York, pp. 363-371.

Läuchli, A., Lüttge, U. and Pitman, M.G. (1973). Ion uptake
and transport through barley seedlings: differential
effect of cycloheximide. *Z. Naturforsch. 28c*, 431-434.

Läuchli, A., Spurr, A.R. and Epstein, E. (1971). Lateral
transport of ions into the xylem of corn roots. II.
Evaluation of a stelar pump. *Plant Physiol. 48*, 118-124.

Lüttge, U. (1973). Stofftransport der Pflanzen. Springer,
Berlin-Heidelberg-New York.

Lüttge, U. (1974). Co-operation of organs in intact higher
plants: a review. In: "Membrane Transport in Plants"
ed. by U. Zimmermann and J. Dainty, Springer, Berlin-
Heidelberg-New York, pp. 353-362.

Lüttge, U., Läuchli, A., Ball, E. and Pitman, M.G. (1974).
Cycloheximide: a specific inhibitor of protein synthesis
and intercellular ion transport in plant roots.
Experientia 30, 470-471.

MacRobbie, E.A.C. (1973). Vacuolar ion transport in *Nitella*.
In: "Ion Transport in Plants" ed. by W.P. Anderson,
Academic Press, London, pp. 431-446.

Nagahashi, G., Thomson, W.W. and Leonard, R.T. (1974). The
Casparian strip as a barrier to the movement of lanthanum
in corn roots. *Science 183*, 670-671.

Pate, J.S. and Gunning, B.E.S. (1972). Transfer cells. *Ann.
Rev. Plant Physiol. 23*, 173-196.

Pitman, M.G. (1972). Uptake and transport of ions in barley
seedlings. II. Evidence for two active stages in transport to the shoot. *Aust. J. Biol. Sci. 25*, 243-257.

Pitman, M.G., Lüttge, U., Läuchli, A. and Ball, E. (1974a).
Action of abscisic acid on ion transport as affected by
root temperature and nutrient status. *J. Exp. Bot. 25*,
147-155.

Pitman, M.G., Schaefer, N. and Wildes, R.A. (1974b). Effect
of abscisic acid on fluxes of ions in barley roots. In:
"Membrane Transport in Plants" ed. by U. Zimmermann and
J. Dainty, Springer, Berlin-Heidelberg-New York, pp.
391-396.

Robards, A.W. and Clarkson, D.T. (1976). The role of plasmodesmata in the transport of water and nutrients across
roots. In: "Intercellular Communication in Plants:
Studies on Plasmodesmata" ed. by B.E.S. Gunning and
A.W. Robards, Springer, Berlin-Heidelberg-New York,
pp. 181-201.

Robards, A.W., Jackson, S.M., Clarkson, D.T. and Sanderson, J. (1973). The structure of barley roots in relation to the transport of ions into the stele. *Protoplasma 77*, 291-311.

Robards, A.W. and Robb, M.E. (1972). Uptake and binding of uranyl ions by barley roots. *Science 178*, 980-982.

Robards, A.W. and Robb, M.E. (1974). The entry of ions and molecules into roots: an investigation using electron-opaque tracers. *Planta 120*, 1-12.

Schaefer, N., Wildes, R.A. and Pitman, M.G. (1975). Inhibition by p-fluorophenylalanine of protein synthesis and of ion transport across the roots in barley seedlings. *Aust. J. Plant Physiol. 2*, 61-73.

Sluiter, E. (1975). Cytochemische Lokalisation von ATPasen in Gerstenwurzeln und ihre mögliche Bedeutung beim Ionentransport. Diplomarbeit Technische Hochschule Darmstadt.

Spanswick, R.M. (1976). Symplasmic transport in tissues. In: Encyclopedia of Plant Physiology, New Series Vol. 2, Transport in Plants 2, Part B" ed. by U. Lüttge and M.G. Pitman, Springer, Berlin-Heidelberg-New York, (in press).

Stelzer, R., Läuchli, A. and Kramer, D. (1975). Pathways of intercellular chloride transport in roots of intact barley seedlings. *Cytobiologie 10*, 449-457.

Steveninck, R.F.M. van, Steveninck, M.E. van, Hall, T.A. and Peters, P.D. (1974). X-ray microanalysis and distribution of halides in *Nitella translucens*. In: "Electron Microscopy 1974, Vol.II" ed. by J.V. Sanders and D.J. Goodchild, The Australian Academy of Science, Canberra, pp. 602-603.

Ziegler, H., Weigl, J. and Lüttge, U. (1963). Mikroautoradiographischer Nachweis der Wanderung von $^{35}SO_4^{--}$ durch die Tertiärendodermis der *Iris*-Wurzel. *Protoplasma 56*, 362-370.

CHAPTER 10

The Regulation of Nutrient Uptake by Cells and Roots

W.J. Cram

School of Biological Sciences, University of Sydney, N.S.W. Australia

INTRODUCTION

While the main limitation on plant growth is often the supply of nutrients in the soil, it is also true that at the later stages of growth and development many plants do not continue to take up the nutrients that are available (Williams 1955; Halse *et al.* 1969). Even early during growth a well-nourished seedling can be found to take up nutrients more slowly than a starved ("low salt") one. Although the interpretation of these observations is uncertain, they do suggest that the plant has it in its power to dampen down the mechanisms taking nutrients into itself, when it is in some sense internally sufficient.

NEGATIVE FEEDBACK REGULATION

As a basis for analysing and interpreting observations related to this, and for framing hypotheses, one can consider the characteristics of types of control systems which have as their basic regulatory component the feedback of information about the state of the system to the process governing that state.

THE GENERALISED NEGATIVE FEEDBACK LOOP

The elementary negative feedback loop consists of a process (e.g. active transport), its output (e.g. the concentration of the substance transported), and a signal from the output to the process. The signal is in the direction that an increase in the output leads to a decrease in the rate of the process. In such a system there is obviously a tendency for the output concentration to be restrained from varying. Such elementary concepts of negative feedback will be familiar to everybody. A quantitative and fuller description can be found in Toates (1975), and some elementary applications to the transport of substances in plant cells in Cram (1976).

THE STRUCTURE OF BIOLOGICAL CONTROL SYSTEMS

A biological control system never works in isolation, but to the extent that it can be isolated conceptually or experimentally it consists most frequently of the output state; a passive response to changes in that state which tends to restore it to its thermodynamically equilibrium value (Le Chatelier's principle); and one, or, more frequently, two, potentially opposing regulated processes which act to restore the output from above or below towards its "desired" value. When the system maintains the output constant it is said to be acting as a homeostat. Computationally the simplest arrangement to ensure constancy of output is one where the primary negative feedback signal (the instantaneous value of the output) is subtracted from an inbuilt reference or set point (the "desired" value of the output), and the difference (which signals how far the output is from its desired value) is transduced into a controlling signal to which the controlled process responds.

Each of these elements has a physical counterpart in the control system maintaining body temperature in warm-blooded animals (Hardy 1966), but the negative feedback system found in end product inhibition of an enzyme sequence does *not* appear to have a physically distinct set point. Since the uptake of a metabolite is the first step in its utilization, any regulation of nutrient uptake may be more analogous to end product inhibition than to the thermoregulation type of control system. The next section therefore compares some properties of a simple feedback system with and without a distinct set point.

COMPARISON OF TWO SIMPLIFIED TRANSPORT CONTROL SYSTEMS WITH AND WITHOUT A SET POINT

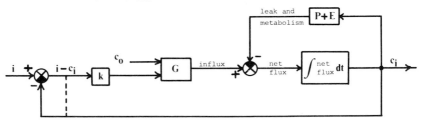

Fig. 1. - Block diagram of a simple hypothetical system for accumulating a metabolised nutrient. The rate of accumulation is equal to the regulated influx minus the rate of leakage ($P\ c_i$) and the rate of metabolism ($E\ c_i$). Regulation of influx is via negative feedback with a set point (Case I in the text), or via negative feedback without a set point (dashed connection, Case 2 in the text).

Fig. 1 shows a simple system consisting of a regulated influx of a metabolite and its passive disappearance from the cell through metabolism and leakage. All relationships are shown as linear for simplicity. Influx is represented as $G\, C_o$ x (the actuating feedback signal). The leak is represented as $P\, C_i$ and metabolic utilization as $E\, c_i$.

Case 1. Negative feedback with a set point.- The different-ial equation describing the rate of change of internal con-centration of the metabolite is

$$\frac{dc_i}{dt} = k(i - c_i)\, G\, c_o - (P+E)\, c_i$$

Solving for c_i with the boundary condition $c_i = c_z$ at $t = 0$ gives

$$c_i = \frac{k\, i\, G\, c_o}{(P+E) + k\, G\, c_o} + \left[c_z - \frac{k\, i\, G\, c_o}{(P+E) + K\, G\, c_o} \right] e^{-\{(P+E) + k\, G\, c_o\}t}$$

$$..1$$

and as $t \to \infty$

$$c_i = \frac{k\, i\, G\, c_o}{(P+E) + k\, G\, c_o} \qquad ..2$$

Equation 2 shows that when c_o is relatively high ($k\, G\, c_o >$ (P+E)) the internal concentration of the metabolite (c_i) approaches the set point, i; and anything which tends to alter the internal concentration, influx, passive leak, or metabolic utilization is compensated for by the system so that the internal concentration remains nearly constant.

Case 2. Negative feedback without a set point. - In this case the influx is treated as being reduced directly in proportion to the internal concentration. The active influx is written as $G\, c_o(1 - k\, c_i)$. The differential equation describing the rate of change of internal concentration of the metabolite is

$$\frac{dc_i}{dt} = G\, c_o(1 - k\, c_i) - (P+E)c_i$$

The solution with the same boundary condition is similar to equation 1, and, as $t \to \infty$

$$c_i \to \frac{G\, c_o}{(P+E) + k\, G\, c_o} \qquad ..3$$

As in case 1, at higher external concentrations the internal concentration of the metabolite approaches a constant value $(1/k)$ which is therefore analogous to a set point. Again as in case 1, factors which tend to alter c_i or G will be compensated for by the system so that c_i remains nearly constant. On the other hand, when there is no separate set point, factors altering k (the sensitivity of the influx process to changes in c_i) would also alter c_i.

Relationships between flux and external and internal concentrations other than multiplicative ones might also occur, although this seems less likely. If the regulated influx were better represented as a function of the sum of the external concentration and the feedback signal, then it is simple to show that constancy of output concentration could be maintained either with or without a set point. A direct feedback, with a threshold above which c_i affects G, could take the place of a separate set point in giving constancy of output.

The significance of these properties becomes apparent when the system of Figure 1 (with or without a separate set point) is compared with the same system having no feedback. Here the differential equation describing the rate of change of the internal concentration of the metabolite is quite simple. Its solution with the same boundary condition is similar to equation 1, and, as $t \to \infty$,

$$c_i \to \frac{G\,c_o}{P+E} \qquad\qquad ..4$$

In this case the steady state internal concentration of the metabolite is proportional to the external concentration, and any change in G or (P+E) also causes a proportional change in c_i.

The important case to consider is that of a change in the activity of the enzyme catalysing the first step in the metabolism of the substance whose transport is being considered.

When the system of Fig. 1 has feedback, (with or without a separate set point) a change in (P+E) is compensated for, and c_i remains nearly constant. Consequently the rate of utilization of the metabolite (c_i E) changes in proportion to the change in enzyme activity. By contrast, when the system has no feedback a change in (P+E) leads to a proportional fall in c_i. Consequently, if utilization (E) is large compared with leakage (P), the fall in c_i will tend to offset the rise in E. The change in the rate of utilization may then change proportionally less than the change in enzyme activity. In other words, without feedback regulation of c_i there may be little change in the rate of metabolite utilization despite a large change in enzyme activity.

Intuitively it would appear that with non-linear functions for E the difference between the systems with and without feedback might not be so great. This would depend on whether the metabolite concentration was higher than or comparable with the K_m of the enzyme.

IMPLICATIONS FOR THE INVESTIGATION OF NUTRIENT UPTAKE BY PLANTS

There are many purely passive responses which can be represented in a similar manner to the feedback loop discussed above. For instance, if as the internal concentration rises the metabolite leaks out faster while the influx remains constant, then the net influx will fall. This response has the superficial appearance of negative feedback, and it must obviously be shown not to occur before regulatory negative feedback is invoked as a hypothesis to explain the fall in the rate of uptake of nutrients in nutrient-loaded cells or plants. (See also the discussion of "passive" or "intrinsic" feedback in Toates (1975)).

Among plant cell accumulatory systems, several have been shown to have a negative feedback from the output to a transport system (Cram 1976). Since homeostasis can be achieved without a separate set point it would seem that a search for a separate set point should only be undertaken with caution.

REGULATION OF THE UPTAKE OF MINERAL NUTRIENTS BY CELLS

At least three processes must be involved in the uptake of a metabolised nutrient into a cell. These are inflow across the plasmalemma, metabolism, and accumulation in the vacuole. Metabolism and vacuolar accumulation may be independent processes competing for a limited quantity of metabolite flowing in across the plasmalemma; or they both may be saturated by the supply of the nutrient, the excess flowing back out of the cell again. In the latter case the rate of supply across the plasmalemma would be matched to the demand of reduction and accumulation via changes in passive efflux. A less energy-dependent alternative is that influx might be matched to internal utilization via self-regulatory feedback signals.

The first necessity in investigating these hypotheses is to measure individual rates of transport and metabolism. There is no reason to expect that the quasi-steady influx which is what has invariably been measured will be to any extent "rate limited" by a single step, and hence interpretation of this flux at the level of individual membranes is difficult.

FEEDBACK REGULATION OF TRANSPORT

Many cells when starved of a metabolite show an increased

influx of that metabolite; and other cells that have been
allowed to accumulate a metabolite or its metabolic products
show a decreased influx of that metabolite (Cram 1976). Some
recent sets of experiments on higher plant tissues which pursue
this approach will be discussed in more detail here.

Sulphate.- Excised carrot tissue does not reduce $SO_4^=$ to
any extent. When washed carrot tissue is allowed to accumulate
$SO_4^=$ the quasi-steady influx of $SO_4^=$ falls. This could be due
to an increased efflux across the plasmalemma as well as to a
decreased influx; and therefore one cannot conclude that negative feedback regulation of $SO_4^=$ influx is operating. This
comment applies to the interpretation of all measurements of
quasi-steady influx.

When carrot tissue is allowed to accumulate methionine
or cysteine (the two first products of $SO_4^=$ reduction) $SO_4^=$
influx is not specifically inhibited. Hence neither plasmalemma nor tonoplast fluxes appear to be under any type of
metabolic end product inhibition.

Barley root tissue and tobacco pith cultures, on the other
hand, do show a reduced $SO_4^=$ influx after loading in methionine
or cysteine, as well as after loading in $SO_4^=$ (Ferrari and
Renosto 1972; Hart and Filner 1969; I.K. Smith 1975). In
tobacco pith cultures, on which the most extensive work has
been done, the internal $SO_4^=$ concentration and the concentrations of reduced S compounds have been altered by allowing the
tissue to grow in a limited quantity of $SO_4^=$. Sulphate influx
is then correlated with internal $SO_4^=$, rather than with the
concentrations of reduced S compounds. As emphasised above,
one cannot tell from measurements of quasi-steady influx
whether this is a change in a passive leak or in a regulated
influx.

Feeding cysteine to tobacco pith cultures leads to a
reduction in $SO_4^=$ influx. However, cysteine is partly converted
to $SO_4^=$, and the results suggest the possibility that it is the
change in $SO_4^=$ concentration, again, which leads to the change
in $SO_4^=$ influx. Some direct effect of internal cysteine on $SO_4^=$
influx is not excluded by the results, but it was concluded
that it is the change in internal $SO_4^=$ concentration which is
"the major determinant" of sulphate influx (Smith 1975).

Nitrate.- In carrot root tissue NO_3^- is taken up and may
be accumulated to 100 μmol. g f.wt.$^{-1}$. Individual fluxes
have not yet been measured, so it is not known if NO_3^- influx
is reduced by high internal NO_3^- concentrations. Loading with
K^+ malate does not reduce net NO_3^- influx in carrot, and loading with NH_4^+ also has no effect in net NO_3^- influx in storage
tissues that have been examined. The effects of amino acids
on net NO_3^- influx in storage tissues has not been examined

(Cram 1973, 1975).

In tobacco pith cultures NH_4^+ and amino acids have been shown to reduce the net influx of NO_3^- (Heimer and Filner 1971), suggesting that uptake may be subject to end product inhibition.

In addition to these putative feedback effects on NO_3^- influx, another type of regulation is evident in the apparent induction of NO_3^- influx by external NO_3^- (Heimer and Filner 1971; Jackson *et al.* 1973). As has previously been pointed out, this could be a *bona fide* induction, or a passive response to the product(s) of the induced NO_3^- reductase.

CONCLUSIONS

The approach of artificially or naturally altering the internal concentrations of substances or their reduction products has suggested the operation of negative feedback regulation of the uptake of some metabolised nutrients. It will be clear, however, that the evidence is very sketchy. This type of experiment suggests only the nature of the primary feedback signal (e.g. that it is a function of the concentration of SO_4^- in some intracellular compartment). What it does not do is to specify the physical nature or location of the pool or the process(es) concerned, or their quantitative relationships.

A nutrient may be both metabolised and accumulated (the latter sometimes being termed "luxury consumption"). Both processes may be regulated, and in either case a change in the quasi-steady rate of uptake would result. The plasmalemma influx may also be regulated. If the intracellular processes are to change independently of each other over a range of conditions, then the plasmalemma influx cannot be regulated via a signal from the output of either intracellular process, for this would tend to inhibit both accumulation and metabolism at the same time.

REGULATION OF THE UPTAKE OF NUTRIENTS BY THE WHOLE PLANT

The main elements in higher plants, after C,H and O, are N and K, followed by divalent cations, P, and S. These must be taken up across the root system, generally as NO_3^-, K^+, etc. Some factors influencing the uptake by the whole plant of NO_3^- and K^+ will be considered here.

THE UPTAKE OF K BY BARLEY SEEDLINGS

The concentration of K^+ in the shoots of barley seedlings growing at different rates is fairly constant, and the rate of uptake by the root is, as expected, proportional to the rate of growth of the shoot (Pitman 1972).

One possibility to account for this is that growth is determined by the supply of K^+. Environmental conditions permitting greater growth might signal an increased rate of K^+ uptake by the root, and the greater K^+ supply might determine the greater rate of expansion of the shoot. The inbuilt response to K^+ supply would be the factor ensuring constancy of internal K^+ concentration. On this hypothesis, to the extent that K^+ uptake is regulated, it is in an "open loop" manner, a "steering" by external factors rather than an internal feed-back.

The alternative possibility is that a tendency for shoot K^+ concentration to fall signals greater K^+ uptake by the root. This would constitute a negative feedback regulation of K^+ influx, functioning simply to match K^+ supply to growth.

The first hypothesis suggests that the K^+ concentration in the shoot would be somewhat higher at greater rates of growth, while the second hypothesis suggests that the opposite would hold. In barley the shoot K^+ concentration in fact rises by a small but significant amount at higher growth rates (Pitman 1972), which supports the first hypothesis. The time courses of the responses of growth and of K^+ uptake to growth stimuli would also serve to distinguish between the two hypotheses.

While they are certainly over-simple, and ignore other than proportional control, these hypotheses do at least represent the two extremes of the possibilities that might account for the constancy of internal K^+ concentration.

THE UPTAKE AND DISTRIBUTION OF NO_3^- BY WHOLE PLANTS

A picture of the flow of N and of related flows of K^+ and organic acid anions has been proposed by Dijkshoorn (1958, 1971) and amplified by Lips *et al.* (1971). During the reduction of NO_3^- its negative charge is transferred to organic acid anions (refs. in Ben Zioni *et al.* 1970) which in the short term can accumulate in equivalent amounts. In the long term, however, the total N taken up by the plant is often greater than the organic acid produced and than the equivalent cation concentration. Dijkshoorn (1958) originally suggested that the NO_3^- uptake in excess of accompanying cations was ultimately in exchange for the negative charge on organic acids produced during NO_3^- reduction. A corollary of this hypothesis was that since NO_3^- reduction was taken to occur predominantly in the shoot, the organic acid anions would have to flow back to the root before exchanging with NO_3^-. The flow would be in the phloem with K^+ as the counter ion, and the K^+ would then be available to flow back to the shoot with fresh NO_3^-.

QUANTITATIVE ANALYSIS OF THE COMPONENT PROCESSES OF THIS PICTURE

1. *The relative amounts of NO_3^- taken up by exchange or with cations.*- From the ratio of organic acid anions to total N in a plant it appears that in Gramineae 50 - 60% of the NO_3^- is taken up by exchange. In other taxa uptake by exchange varies from 0 (*Nicotiana, Fagopyrum*) to 75% (*Brassica*)of the total (e.g. Pierre and Banwart 1973). The above picture obviously does not apply to those plants which appear to take up all their NO_3^- with cations.

2. *Location of NO_3^- reduction.*- In some plants NO_3^- reduction appears to take place principally in the shoot, but in others, an appreciable or major fraction may take place in the root. Means of estimating this fraction by examination of xylem exudate are uncertain (cf. Pate, this volume), since excision of the shoots may alter the fraction, but the estimate correlates with the distribution of extractable nitrate reductase (Wallace and Pate 1967). On this basis barley, a case under discussion here, would appear to reduce about 50% of its NO_3^- in the root (Wallace and Pate 1967).

It immediately follows from 1 and 2 that in barley the NO_3^- taken up without cations could exchange for the negative charges left in the root after the reduction of NO_3^- there. Consequently no recycling of K^+ and organic acid is necessary to give charge balance in the whole barley plant.

3. *Downwards flow in the phloem.*- K^+ and organic acid anions are major components of phloem exudates (Pate, this volume), which appears to be consistent with the downwards flow of considerable quantities of these substances. However, direct measurement does not support this contention.

In Gramineae, if all the NO_3^- is reduced in the shoot, then the rate of organic acid production in the shoot must equal the rate of flow of NO_3^- upwards to the shoot. Since only half as many organic acid equivalents as N are accumulated in the shoots, half the organic acid equivalents produced during NO_3^- reduction must flow downwards again. K^+ must balance the downwards flow in the phloem (as pointed out by Ben Zioni *et al.* 1971), while it will accompany some or all of the NO_3^- (which is the major anion) flowing up in the xylem. Therefore the flux of K^+ to the root should be about 50% of the flux to the shoot. However, in barley, at least, this is not so. The K^+ flux down to the root is nearly zero (Pitman; see Pitman and Cram 1973). This does not invalidate the concept of NO_3^- - organic acid anion exchange, but it is a serious experimental objection to the K^+ recycling picture.

This evidence supports the conclusion above that in barley roots any organic acid anions involved in exchange with

external NO_3^- do not come from the shoot. It also points again
to the critical necessity for determining where NO_3^- is reduced in the plant.

4. *Factors influencing the uptake of NO_3^- by exchange for
organic acid anion equivalents in the root.* - Ben Zioni *et al.*
(1971) have suggested that the malate concentration in the
root may limit the uptake of NO_3^- by exchange, and hence that
the influx of NO_3^- unaccompanied by cations should fall after
NO_3^- is removed from the medium bathing the plant.

In qualitative agreement with this prediction Ben Zioni
et al. (1971) found that in *Nicotiana rustica* grown in full
culture solution net NO_3^- uptake from 1 mM KNO_3 is greater than
net K^+ uptake from the same solution; and that after transfer
to $CaSO_4$ the net NO_3^- influx falls and after 5 days becomes
equal to the net K influx. These observations are difficult
to interpret, as NO_3^- uptake appears to be inducible by
external NO_3^- in *Nicotiana* (Jackson *et al.* 1972), and
Nicotiana is a genus in which the content of organic acid
equivalents is equal to the content of organic N (Pierre and
Banwart 1972), so that no organic acid - NO_3^- exchange would
appear to occur naturally. It has also been found that there
is no effect on NO_3^- influx of increasing the internal K
malate concentration in carrot and barley root tissue (Cram
1973, Smith 1973).

However, intracellular malate does have a subtle effect
on NO_3^- influx, as described by F.A. Smith (1973). In excised
barley roots loading with K malate did not alter the net NO_3^-
influx, but in the K malate loaded roots NO_3^- influx was no
longer dependent on the presence of K^+ in the external medium,
as though the fraction of NO_3^- that was taken up by exchange
had increased to 100%.

THE OCCURRENCE AND SIGNIFICANCE OF NITRATE - ORGANIC ACID
EXCHANGE

The quantity of NO_3^- taken up by exchange is indicated by
the total organic N which is not balanced by organic acid
anions, and this is well documented. The necessity for the
flow of organic acid anions from shoot to root could only arise
when the fraction of NO_3^- uptake by exchange exceeds the fraction of NO_3^- which is reduced in the root, and this is not so
well documented. In *Hordeum* as much NO_3^- is reduced in the
roots as is taken up by exchange, and therefore no recycling
from the shoot need be postulated. In *Nicotiana* organic N is
fully balanced by organic acid anions, and no organic acid -
NO_3^- exchange need be postulated. Thus recycling of organic
acids and K^+ is not a universal phenomenon of plant growth.

Even if organic acids do not flow extensively from shoot
to root, quantitatively equivalent flows of carbon compounds

from the shoot must occur when NO_3^- is reduced in the roots, since carbon skeletons are needed for the amino acids that would be produced.

The significance of an antiport as against a symport mechanism of NO_3^- uptake has been suggested previously to be related to osmotic pressure generation in the plant (Cram 1976). If all N were taken up as KNO_3 and one K^+ and organic acid anion produced per N reduced to organic form, the osmotic pressure of cells of Gramineae might be up to twice the value observed. Hence the switch to NO_3^- uptake by exchange appears to keep internal osmotic pressure from rising, and may be part of the system(s) regulating internal osmotic pressure and hence turgor.

REFERENCES

Ben Zioni, A., Vaadia, Y. and Lips, S.H. (1970). Correlations between nitrate reduction, protein synthesis and malate accumulation. *Physiol. Plant. 23*, 1039-1047.

Ben Zioni, A., Vaadia, Y. and Lips, S.H. (1971). Nitrate uptake by roots as regulated by nitrate reduction products of the shoot. *Physiol. Plant. 24*, 288-290.

Cram, W.J. (1973). Internal factors regulating nitrate and chloride influx in plant cells. *J. Exp. Bot. 24*, 328-341.

Cram, W.J. (1975). Storage tissues. In "Ion Transport in Plant Cells and Tissues" ed. by D. Baker and J.L. Hall, Elsevier, Amsterdam.

Cram, W.J. (1976). Negative feedback regulation of transport in cells. In "Encyclopedia of Plant Physiology" New Series, Vol. II A. ed by M.G. Pitman and U. Lüttge. Springer Verlag, Berlin, Heidelberg, New York (in press).

Dijkshoorn, W. (1958). Nitrogen, chlorine and potassium in perennial ryegrass and their relation to the mineral balance. *Neth. J. Agric. Sci. 6*, 131-138.

Dijkshoorn, W. (1971). Partition of ionic constituents between organs. In "Recent Advances in Plant Nutrition" ed. by R.M. Samish, Gordon and Breach, New York pp. 447-476.

Ferrari, G. and Renosto, F. (1972). Regulation of sulfate uptake by excised barley roots in the presence of selenate. *Plant Physiol. 49*, 114-116.

Halse, N.J., Greenwood, E.A.N., Lapins, P. and Boundy, C.A.P. (1969). An analysis of the effects of nitrogen deficiency on the growth and yield of a Western Australian wheat crop. *Aust. J. Agric. Res. 20*, 987-998.

Hardy, J.D. (1966). The "set-point" concept in physiological temperature regulation. In: "Physiological Controls and Regulations" ed. by W.S. Yamomoto and J.R. Brobeck, Saunders, Philadelphia pp. 98-116.

Hart, J.W. and Filner, P. (1969). Regulation of sulfate uptake by amino acids in cultured tobacco cells. *Plant Physiol. 44*, 1253-1259.

Heimer, Y.M. and Filner, P. (1971). Regulation of nitrate assimilation pathway in cultured tobacco cells. III. The nitrate uptake system. *Biochim. Biophys. Acta. 230*, 362-372.

Jackson, W.A., Flesher, D. and Hageman, R.H. (1973). Nitrate uptake by darkgrown corn seedlings. Some characteristics of apparent induction. *Plant Physiol. 51*, 120-127.

Jackson, W.A., Volk, R.J. and Tucker, T.C. (1972). Apparent induction of nitrate uptake in nitrate-depleted plants. *Agron. J. 64*, 518-521.

Lips, S.H., Ben Zioni, A. and Vaadia, Y. (1971). K^+ recirculation in plants and its importance for adequate nitrate nutrition. In "Recent Advances in Plant Nutrition" ed. by R.M. Samish, Gordon and Breach, New York pp. 207-214.

Pierre, W.H. and Banwart, W.L. (1973). Excess-base and excess-base/nitrogen ratio of various crop species and parts of plants. *Agron. J. 65*, 91-96.

Pitman, M.G. (1972). Uptake and transport of ions in barley seedlings. III. Correlation between transport to the shoot and relative growth rate. *Aust. J. Biol. Sci. 25*, 905-919.

Pitman, M.G. and Cram, W.J. (1973). Regulation of inorganic ion transport in plants. In "Ion Transport in Plants" ed. by W.P. Anderson, Academic Press, London pp. 465-481.

Smith, F.A. (1973). The internal control of nitrate uptake into excised barley roots with differing salt contents. *New Phytol. 72*, 769-782.

Smith, I.K. (1975). Sulfate transport in cultured tobacco cells. *Plant Physiol. 55*, 303-307.

Toates, F.M. (1975). Control Theory in Biology and Experimental Psychology. Hutchinson, London.

Wallace, W. and Pate, J.S. (1967). Nitrate assimilation in higher plants with special reference to the Cocklebur (*Xanthium pennsylvanicum* Wallr.) *Ann. Bot. 31*, 213-228.

Williams, R.F. (1955). Redistribution of mineral elements during development. *Ann. Rev. Plant Physiol. 6*, 25-42.

CHAPTER 11

The Electrophysiology of Higher Plant Roots

W.P. Anderson

Research School of Biological Sciences, Australian National University, Canberra. A.C.T. 2601. Australia

Throughout the last decade there has been an increasing application of electrophysiological techniques to the study of ion transport in higher plant roots. In this paper I shall attempt to summarise what seem to me to be the more salient features of these developments and I may describe three, rather separate areas of interest.

EXCISED ROOT XYLEM EXUDATE POTENTIALS

Many higher plant root systems, excised cleanly from their aerial parts and bathed in sufficiently dilute nutrient solution, will exude by root pressure. The hydraulics of this phenomenon has been extensively discussed (Arisz et al. 1951; House and Findlay 1966; Anderson et al. 1970; Fiscus and Kramer 1975) and is essentially simple; salts (predominantly) are accumulated from the external solution into the xylem stream across some semi-permeable membrane and the resultant osmotic pressure gradient then drives water flow into the xylem. The consequent exudate may be collected in some suitably fitted tube, and subjected to chemical, or other, analysis.

Here I shall deal only with the electrical observations reported for this excised root system. The exudate electropotential is commonly measured by inserting a saturated KCl salt bridge, coupled to a Ag/AgCl reversible electrode, into the xylem fluid, with a similar probe into the solution bathing the root system, and a high impedance voltmeter connected in series. Such measurements were among the earliest reported electrophysiological observations on higher plant root systems (e.g. Bowling and Spanswick 1964) and there is now a considerable body of evidence on several species.

These data, in conjunction with ion concentrations for

exudate fluid and bathing solution, have been used to decide
whether an ion is actively transported to the xylem, by
evaluating the electrochemical potential $\bar{\mu}$, given in the usual
notation by

$$\bar{\mu} = \bar{\mu}^o + RT\ln\gamma C + zF\phi \qquad \ldots\ldots(1)$$

where $\bar{\mu}^o$ is the standard electrochemical potential, R is the
gas constant, T is the temperature (^{o}K), γ is the activity
coefficient, C the concentration, z the ion valency, F the
Faraday constant, and ϕ the phase electrical potential. With
appropriate units for these parameters, $\bar{\mu}$ will be in J mole^{-1}
and the difference in $\bar{\mu}$ for an ion in the xylem fluid and
bathing solution, will be a measure of the work done in reversibly transferring the ion from one phase to the other.

It seems worthwhile pointing out that a non-zero electrochemical potential difference in the situation just described,
does not by itself show the existence of active, passive or
any transport. The appropriate test with independent ion
movement is the Ussing-Teorell flux ratio equation for each
ion,

$$RT\ln \frac{J_{in}}{J_{out}} \underset{>}{\overset{<}{}} \bar{\mu}_{out} - \bar{\mu}_{in} \qquad \ldots\ldots(2)$$

where J_{in} is the ion influx, J_{out} is the ion efflux. If the
LHS is greater, then we have active transport of the ion. The
implicit assumption in all the reported tests for active
transport from exudate potential measurements is that J_{in}
$> J_{out}$, so that when the RHS of equation (2) is negative,
active transport is indicated. This assumption is un-
doubtedly justified for the major ions of the xylem stream,
but it would have been better if the authors had made the
situation explicit at the time.

The origin of the exudate potential has also been much
discussed (e.g. Davis and Higinbotham 1969, Shone 1969). The
equation commonly used (e.g. Davis and Higinbotham 1969) to
describe the exudate electropotential is the Goldman-Hodgkin-
Katz equation, originally developed for animal cell electrophysiology and used extensively in single plant cell
electrophysiology.

$$E_G = \frac{RT}{F} \ln \left(\frac{P_K[K_e] + P_{Na}[Na_e] + P_{Cl}[Cl_x]}{P_K[K_x] + P_{Na}[K_x] + P_{Cl}[Cl_e]} \right) \qquad \ldots\ldots(3)$$

In this equation the subscripts "x" and "e" refer to the
xylem fluid and external solution respectively. Some of the
parameters are as previously described, the P's are permeability coefficients for the ions indicated by the subscripts,
and the square bracket terms are concentrations.

There are two shortcomings in the use of equation (3).

Firstly it applies only to the potential difference across a single membrane, which the exudate electropotential most certainly is not;   secondly it is incomplete in that H^+ and Ca^{++} are likely to be important potential determining ions, and are not included in the equation as it has been most often applied.   It is easy to add the H^+ term, but inclusion of Ca^{++}, being divalent, requires considerable modification of equation (3).

In any event, it has been suggested that the exudate electropotential is not simply a diffusion potential, given by the Goldman equation, but contains an additional component due to the electrogenic transport of one or more ions, so that we should write,

$$E_x = E_G + E_P \quad \ldots \ldots (4)$$

where E_P is the electrogenic pump potential.   This topic will be taken up more fully later, for the case of single cell potentials where the evidence is more direct.   With exudate potentials the evidence for an electrogenic component (Davis and Higinbotham 1969;   Shone 1969) is essentially that there is a depolarisation following the application of the un-coupler di-nitrophenol or the inhibitor cyanide (Fig. 1).

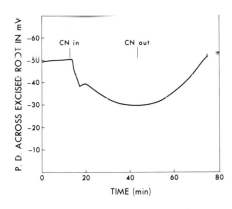

Fig.1. - The effect of cyanide on the xylem exudate potential of excised maize primary roots (redrawn from Davis and Higinbotham 1969).

ROOT TISSUE RESISTANCES

The second type of electrical measurement which has been reported for root tissue, as distinct from single cells, involves determination of electrical resistances.   The electrodes are usually Ag/AgCl wires.   The method of measure-ment is extremely simple, being basically the application of Ohm's law.   Usually the current polarity is switched at regular intervals, every few seconds, to avoid polarising the electrodes, and the current magnitude is kept as low as is consistent with accurate measurement, generally not

exceeding a few hundred nanoamperes. A more complete account
of the method used can be found in Anderson and Higinbotham
(1976). Figure 2, taken from that paper, shows how the longitudinal resistance of maize root segments varies with distance from the apex. Note that there is a definite trend, with
the older segments showing a lower resistance, although perhaps
surprisingly, the 0-2 cm segment has a lower resistance than
the 1-3 cm segment.

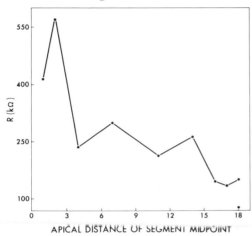

*Fig.2. - Variation in the
longitudinal resistance
of 2 cm maize root segments
at differing distances
from the root tip (from
Anderson and Higinbotham
1976).*

The two chief anatomically distinct regions of a cereal
root, the outer cortex and the central stele, can easily be
separated in maize and because of this, one can quite simply
determine which region carries the bulk of the longitudinal
electrical current. Measurements on isolated steles give
resistance values of around 750 $k\Omega$ cm^{-1}, while for entire
root segments the resistance is around 150 $k\Omega$ cm^{-1}. Thus 80%
of the applied current, (which is of course the longitudinal
net ion flux induced in the tissue) flows through the cortex,
while only 20% flows through the stele. However, the cross-
sectional area ratio of cortex to stele in maize is 8:1, while
the current ratio is 4:1, so that the resistivity of the
stele is only 0.5 that of the cortex, presumably due to the
vascular elements providing a low electrical resistance
pathway.

There are two principal pathways for electric current
flow in the cortex, the first through the apoplasm and the
second through the symplasm. Ginsburg and Laties (1973)
showed that the apoplasm pathway in a normal young maize
primary root, has a resistance of about 10 $M\Omega$ cm^{-1}, which
must be contrasted with the measured value of about 150 $k\Omega$
cm^{-1}. By application of the usual rule for resistances in

parallel, the symplasmic resistance is therefore 152 kΩ cm^{-1}; in other words the observed electrical resistance is essentially the symplasmic resistance.

The radial resistance of maize root cortex can also be measured by carefully inserting an electrode along the central cavity of a cortical sleeve and measuring between it and the external solution. As just discussed, the measured resistance will again be the symplasmic resistance; the mean recorded value of radial resistivity of the cortex (the resistance over unit cross-sectional area per unit length) is 324 kΩ cm, varying between 504 kΩ cm at the 2-4 cm segment, to 252 kΩ cm at the 10-12 cm segment. These values are 10 to 20 times larger than membrane resistivity values measured by intracellular probes into single maize root cells (see later), and imply that the major resistance encountered in the symplasm pathway is in ions entering (and perhaps leaving) the system by membrane transit.

Finally, these root resistance measurements have implications for the interpretation of exudate electropotentials. As has been pointed out by Ginsburg (1972), and as the data presented here and in Anderson and Higinbotham (1976) also show, an excised maize root behaves electrically like a very leaky cable. Thus, the exudate electropotential recorded by the usual method is generated only over the top centimetre or so of root, and in the case of a 10 cm long root segment, say, gives no information of the electropotential in the young regions where it is known from tracer experiments that the chief ion uptake occurs. The leaky cable model also explains the observation by Davis (1968) that an electropotential is maintained between the xylem exudate and the bathing solution, even when successive segments are removed from the root tip.

SINGLE ROOT CELL ELECTROPHYSIOLOGY

Finely drawn glass micropipettes, filled with 3M KCl, can be inserted by micromanipulator under a microscope into single cells of plant roots. Coupled through Ag/AgCl electrodes to a high inpedance voltmeter, these can be used to record cell membrane potentials. Because higher plant cells are highly vacuolated, with the cytoplasm being only 1 µm or so thick, the first question is whether the microelectrode is lodged in cytoplasm or vacuole. It seems likely that in cereal root cells, one can reproducibly achieve either condition. What follows here should be taken on the assumption that the electrode is in the cytoplasm.

First, there is a ramification of the earlier discussions of symplasmic resistance, for single cell work. As suggested in the previous section, and as the cell-to-cell electrical coupling observations by Spanswick (1972) demonstrate, the cells of plant roots are inter-connected by the low resistance junctions of the symplasm. Therefore, a probe in the cytoplasm of one cell will record the mean of the electrical properties of a large number of cells. This will be less true if the probe is in a cell vacuole, because the tonoplast resistance will help isolate the probe from the other cells of the tissue. However, the reported values of single cell membrane resistivity (e.g. Anderson et al. 1974) are likely to be considerable underestimates because of this effect; the injected current to record resistance will be dissipated through the membranes of several cells, so that the membrane area used to calculate the resistivity will not be simply that of the cell into which the probe is inserted, but should be some weighted mean of all the near-neighbour cells.

The most commonly observed parameters with single cell probes are the membrane potential and the membrane resistivity. The problems associated with estimation of the latter have just been described, and it seems possible that the published values, of several $k\Omega$ cm, might be as much as 10 to 20 times too low. The later discussion on changes in membrane resistance following inhibitor application is not really affected by this consideration, because all the cells in the root cortex are likely to be equally altered by the inhibitors.

The membrane potential in all the plant root cells so far studied, contains two separate, additive components, as has been earlier described for the exudate potential. First, there is a diffusion potential which arises from the differing mobilities of the ions in passing across the membrane, and it will be given by the Goldman equation (3). Second, there is an electrogenic pump potential, arising from the direct coupling of cellular exergonic metabolism to the transport of net electric charge across the membrane. Thus the important, distinctive property of an electrogenic ion pump is that its action contributes directly to the cell membrane potential. There are two common mechanisms by which this may occur: (i) a uni-directional pump, carrying an ion one way across the membrane, and "returning empty" (e.g. the H^+ ATPase of mitochondria, chloroplasts, bacteria and plant cells) (ii) an exchange pump which is not 1:1 coupled (e.g. the Na^+ : K^+ ATPase of the red cell, where the stoichiometry is thought to be 3:2.

As early as 1941, this possibility and its electrical consequences were recognised by Dean (1941), but the best, early experimental demonstration was not reported until

Slayman (1965) showed the presence of an electrogenic pump in *Neurospora crassa*. In fact, Etherton and Higinbotham (1960) had shown that DNP caused depolarisation in *Avena* coleoptile cells, and further that the normal resting potential was higher than the Goldman equation prediction. However, the joint possibilities that the uncoupler might be causing membrane permeability alterations and that ions other than those commonly written into the Goldman equation were important for potential determination, were together taken as sufficient grounds for continued resistance to the idea that an electrogenic ion transport mechanism was making a significant contribution to the measured membrane potential in plant cells. Therefore it was not until Kitasato (1968) published his work on *Nitella clavata*, that it was generally accepted that there was an electrogenic pump potential in plant cells.

Higinbotham et al. (1970) produced cyanide inhibition and DNP uncoupler data to demonstrate an electrogenic component in *Pisum* epicotyl cells, but the resultant depolarisations, although faster than might be expected for a diffusion potential run-down, were not dramatically fast (10 mins or so), and a purist might argue that permeability changes were occurring. The most recent demonstration in higher plant cells seems quite unequivocal (see Figure 3). Anderson et al (1974), using a technique developed to record both membrane potential and

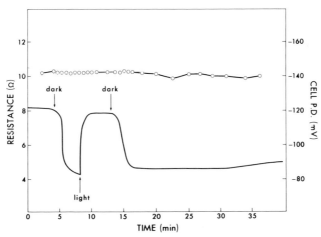

Fig.3. - The effect of carbon monoxide on the membrane potential, and membrane resistance of pea root epidermal cells. Inhibition commences when the light is switched off. The open circles are resistances.

membrane resistance, were able to show in etiolated, non-photosynthetic pea epicotyl, and in wheat root epidermal and cortical cells, that there is rapid depolarisation-repolarisation of membrane potential with no change in membrane conductance (permeability) following respiratory inhibition-release by carbon monoxide.

Rapid and reversible membrane depolarisation upon respiratory inhibition with no concomitant membrane conductance change is the best test of an electrogenic pump potential.

Although I do not plan to go into the theory of electrogenic pumps here, and anyone interested should consult Rapoport (1970) or Higinbotham and Anderson (1974), it may be useful to briefly describe the system. The electrogenic pump, driven by the cell metabolism, carries electric current across the membrane, and therefore charges or discharges the cell membrane capacitance. (An outward cation, or an inward anion electrogenic pump will have the effect of making the cell interior more negative, the observed direction of the effect in all cells so far studied). The membrane potential will therefore be displaced, and will then drive a passive current flow across the membrane, to balance the current in the electrogenic pump. Within two or three resistance-capacitance time constants for the membrane, there must be zero net electric current flow. The electrogenic pump potential will therefore be given as

$$E_p = I_p R_M$$

where I_p is the current flow through the pump and R_M is the passive membrane resistance.

There is to date no definite information of which ion is electrogenically transported in higher plants. In *Nitella clavata* the H^+ is extruded (Kitasato 1968), in *Nitella translucens* it is also H^+ (Spanswick 1972) and in *Chara corallina* Walker and Smith (1975) have shown a remarkable correlation between measured membrane potential and the expected potential if $2H^+$ are electrogenically extruded per ATP molecule hydrolysed. It is also known that H^+ is electrogenically transported in bacteria, mitochondria and chloroplasts. It therefore seems highly likely that H^+ ions are involved in the electrogenic effect in the higher plant cells.

There is one further type of single cell probe experiments which is unique to root cells. Attempts have been made to record membrane potentials, and indeed chemical concentrations, across the radius of young roots, by pushing probes continuously through the tissue, from the epidermis all the way through to the xylem parenchyma (Dunlop and Bowling 1971; Dunlop 1974). Recently Anderson and

Higinbotham (1975) have urged caution in the application of this technique because they see several possible objections to the approach. Low resistance electrodes, such as seem to have been used, frequently become plugged with debris after only 2 or 3 cells have been probed and no longer give stable membrane potentials; electrode tip distortion, likely as the probe is driven into the tissue, can *per se* produce voltage signals; ruptured cells behind the probe tip may increase the local free space ion concentrations and thus distort the measured values especially if these values are only recorded for a short time. However, the general situation found by these radial probes across a root, namely that the symplasm is a low resistance pathway (all the cells have the same membrane potential), is exactly what is expected from other sources of evidence.

The ion selective electrode work by these same authors may also be criticised. Such electrodes are sensitive to electrochemical potential, and therefore are affected by the membrane potential when the probe is inserted into a cell. The commonest approach has been to assume that a cell, probed by an ion selective electrode, has a measurable membrane potential equal to that of the mean potential value recorded on KCl micro electrodes. Because there is an exponential dependance on potential, there may be rather large errors (tens of millimoles/l) associated with these estimates.

REFERENCES

Anderson, W.P., Aikman, D.P. and Meiri, A. (1970). Excised root exudation - a standing gradient osmotic flow. *Proc. Roy. Soc. Lond. B. 174*, 445-458.

Anderson, W.P., Hendrix, D.L. and Higinbotham, N. (1974). The effect of cyanide and carbon monoxide on the electrical potential and resistance of cell membranes. *Plant Physiol. 54*, 712-716.

Anderson, W.P. and Higinbotham, N. (1975). A cautionary note on plant root electrophysiology. *J. Exp. Bot. 26*, 533-535.

Anderson, W.P. and Higinbotham, N. (1976). Electrical resistances of corn root segments. *Plant Physiol.* (in press).

Arisz, W.H., Helder, R.J. and Van Nie, R. (1951). Analysis of the exudation process in tomato plants. *J. Exp. Bot. 2*, 257-297.

Bowling, D.J.F. and Spanswick, R.M. (1964). Active transport of ions across the root of *Ricinus communis*. *J. Exp. Bot. 15*, 422-427.

Davis, R.F. (1968). Ion transport across excised corn roots.

Doctoral Thesis, Washington State University, Pullman, Wash.

Davis, R.F. and Higinbotham, N. (1969). Effects of external cations and respiratory inhibitors on electrical potential of the xylem exudate of excised corn roots. *Plant Physiol.*, *44*, 1383–1392.

Dean, R.B. (1941). Theories of electrolyte equilibrium in muscle. *Biol. Symp. 3*, 331–348.

Dunlop, J. (1974). The transport of potassium to the xylem exudate of ryegrass. II. Exudation. *J. Exp. Bot. 25*, 1–10.

Dunlop, J. and Bowling, D.J.F. (1971). The movement of ions to the xylem exudate of maize roots. I. Profiles of membrane potential and vacuolar potassium activity across the root. *J. Exp. Bot. 22*, 434–444.

Fiscus, E.L. and Kramer, P.J. (1975). General model for osmotic and pressure-induced flow in plant roots. *Proc. Nat. Acad. Sci. USA. 72*, 3114–3118.

Etherton, B. and Higinbotham, N. (1960). Transmembrane potential measurements of cells of higher plants as related to salt uptake. *Science 131*, 409–410.

Ginsburg, H. (1972). Analysis of plant root electropotentials. *J. Theor. Biol. 37*, 389–412.

Ginsburg, H. and Laties, G.G. (1973). Longitudinal electrical resistance of maize roots. *J. Exp. Bot. 24*, 1035–1040.

Higinbotham, N. and Anderson, W.P. (1974). Electrogenic pumps in higher plant cells. *Can. J. Bot. 52*, 1011–1021.

Higinbotham, N., Graves, J.S. and Davies, R.F. (1970). Evidence for an electrogenic ion transport pump in cells of higher plants. *J. Membrane Biol. 3*, 210–222.

House, C.R. and Findlay, N. (1966). Water transport in isolated maize roots. *J. Exp. Bot. 17*, 344–354.

Kitasato, H. (1968). The influence of H^+ on the membrane potential and ion fluxes of *Nitella*. *J. Gen. Physiol. 52*, 60–87.

Rapoport, S.I. (1970). The sodium-potassium exchange pump: relation of metabolism to electrical properties of the cell. *Biophys. J. 10*, 246–259.

Shone, M.G.T. (1969). Origins of the electrical potential difference between the xylem sap of maize roots and the external solution. *J. Exp. Bot 20*, 698–716.

Slayman, C.L. (1965). Electrical properties of *Neurospora crassa*: respiration and the intracellular potential. *J. Gen. Physiol. 49*, 93–116.

Spanswick, R.M. (1972). Evidence for an electrogenic pump in *Nitella translucens*. I. The effects of pH, K^+, Na^+, light, and temperature on the membrane potential and resistance. *Biochim. Biophys. Acta. 288*, 73–89.

Walker, N.A. and Smith, F.A. (1975).   Intracellular pH in
     Chara corallina measured by DMO distribution.   *Pl. Sci.
     Letters*. *4* 125-132.

CHAPTER 12

The Regulation of Potassium Influx into Barley Roots: An Allosteric Model

A.D.M. Glass

Department of Botany and Zoology, Massey University, Palmerston North, New Zealand

The rates of absorption of several inorganic ions by the roots of higher plants have been shown to be negatively correlated with their respective internal concentrations (Pitman and Cram 1973). As a result of such observations it has been suggested that the regulation of ion uptake is achieved by some form of negative feedback mechanism (Pitman and Cram 1973, Glass 1975). In microorganisms transport systems are subject to induction or repression, as well as negative feedback inhibition, by specific molecules in a manner analogous to that of enzyme systems. Through the operation of such processes nutrient uptake may be regulated to satisfy the metabolic requirements of the cell. To date there appears to be little evidence to demonstrate the operation of similar mechanisms for control of ion uptake in higher plants.

In barley roots, plasmalemma influx of potassium is negatively correlated with the internal concentration ($[K^+]_i$) of this ion. In order to arrive at a model which might account for the regulation of K^+ influx in this plant it was decided to examine what might appear, a priori, to be the most likely mechanisms for control, namely:-

(1) that K^+ influx is regulated through repression of the transport system by internal K^+, and/or

(2) that K^+ influx is regulated via direct inhibition of the K^+ transport system by internal K^+.

Now if influx be regulated by repression alone, it might be expected that increasing $[K^+]_i$ might reduce the V_{max} for K^+ influx without modifying the affinity ($1/K_m$) of the transport system for K^+. Indeed Young, Jefferies and Sims (1970) showed that in *Lemna minor* V_{max} was negatively correlated with internal K^+ concentrations, although K_m values were unaltered by increasing $[K^+]_i$. However, in barley it has been shown that above about 35 μmoles K^+ g^{-1} the K_m for influx is positively correlated with $[K^+]_i$ in plants which have been acclimated

for several days in KCl solutions (Glass 1976). Similarly when
plants, previously grown in $CaSO_4$ solution, are loaded with
K^+ by pretreatment in 50mM KCl for various intervals of time,
K^+ influx from solutions in the System I concentration range
decline rapidly, and significant changes in the K_m for influx
occur within 3 hr of pretreatment. Fig. 1 shows the relation
between K_m for influx and $[K^+]_i$ obtained in the latter experiments. Those higher plant enzymes which have been studied in
detail appear to turn over relatively slowly. Furthermore the
use of various inhibitors has revealed that carriers involved
in K^+ accumulation in barley roots are relatively long-lived

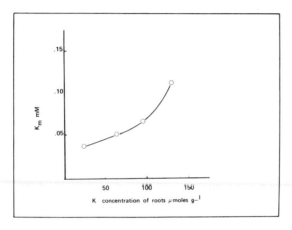

*Fig.1.- The relation-
ship between Km for K^+
influx and internal K^+
concentrations for
barley roots in which
internal K^+ concen-
trations were increased
rapidly by pretreat-
ment in 50 mM KCl.*

(Schaefer, Wildes and Pitman 1975). Thus considering the
correlation between K_m and $[K^+]_i$ and the time course of these
changes it would seem unlikely that regulation proceeds
through repression of the transport system.

However, to assert that a given regulatory process operates at the genomic level it would seem to be essential to
demonstrate that the operation of the regulatory system is
contingent upon DNA dependent syntheses. Conversely if the
regulatory process can be shown to operate in the absence of
DNA, RNA or protein synthesis then the hypothesis of regulation by repression can almost certainly be rejected. To
evaluate this hypothesis barley roots were allowed to
accumulate K^+ from 50 mM KCl solutions in the presence or absence of various inhibitors of protein synthesis. Following
this treatment K^+ influx was measured from 0.05 mM KCl solutions. The data of Table 1 indicate that none of the
inhibitors prevented the reduction of K^+ influx which is
associated with increased K^+ status. These observations
suggest that the regulation of K^+ influx is not mediated via
DNA dependent syntheses. Thus it seems unlikely that the

TABLE 1. Influence of various inhibitors of protein synthesis upon the reduction of K^+ influx associated with increased K^+ status. Barley roots were pretreated for 6 hr. in $CaSO_4$ or 50 mM KCl plus or minus inhibitor. After this time K^+ influx was measured from 0.05 mM KCl. Inhibition of influx is a measure of the reduction of influx associated with the accumulation of K^+ during pretreatment.

Pretreatment	Influx (μmoles $g^{-1}hr^{-1}$)	% inhibition
1. $CaSO_4$	2.7 ± 0.19	–
$CaSO_4$ + KCl	1.3 ± 0.02	52%
$CaSO_4$ + fluorodeoxyuridine (5 x 10^{-4}M)	3.8 ± 0.13	–
" " + KCl	1.2 ± 0.03	68%
$CaSO_4$ + actinomycin D (20 μg/ml)	2.9 ± 0.06	–
" " + KCl	1.1 ± 0.06	62%
2. $CaSO_4$	1.9 ± 0.03	–
$CaSO_4$ + KCl	1.0 ± 0.04	54%
$CaSO_4$ + cycloheximide (10 μg/ml)	1.4 ± 0.02	–
" " + KCl	0.8 ± 0.02	43%
$CaSO_4$ + fluorophenyl-alanine (2 mM)	1.9 + 0.05	–
" " + KCl	0.7 ± 0.03	63%

repression of transport capacity, in the short term at least, is of major significance in regulating K^+ influx in barley roots. Considering the longevities of higher plant cells as compared to microbial cells it is not surprising to find such differences in mechanisms of regulation.

Detailed examination of the relationship between K^+ influx and $[K^+]_i$ in barley roots whose K^+ status was changed rapidly by immersion in 50 mM KCl has revealed the existence of a negative sigmoidal relationship (Glass 1976). Fig. 2 demonstrates this relationship between influx from 0.16, 0.08, 0.04 and 0.02 mM KCl and $[K^+]_i$. Sigmoidal kinetics are highly characteristic of allosteric enzymes such as haemoglobin, which was probably the first such system to be examined in detail. A.V. Hill developed an empirical equation to describe such kinetics, which in its linear transformation:-

$$\log (v/V_{max}-v)) = n \log [S] - \log K'$$

has been employed by enzymologists to evaluate the number of interaction sites (n) for allosteric modifiers which may

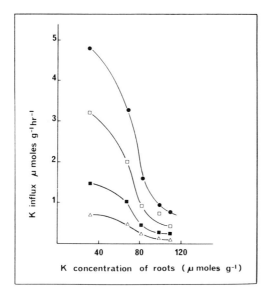

Fig.2.- K^+ influx from solutions containing 0.16 (●), 0.08 (□), 0.04 (■), and 0.02 mM KC1 solutions (△) plotted against internal K^+ concentration.

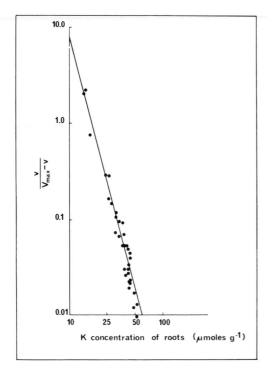

Fig.3.- Hill plot (v/V_{max} $-v$) against internal K^+ concentration plotted on a log scale) of influx data obtained from many separate experiments. n (slope of Hill plot) = -3.7 ± 0.2. Correlation coefficient for the regression line = -0.96.

activate or inhibit enzyme activity. Applying this methodology to K^+ influx data obtained from barley roots, Hill plots
(Fig. 3) gave linear transformations with a high index of
correlation (-0.96) and n values of 3.7 ± 0.2. It is therefore attractive to propose that the K^+ carrier may consist
of several (perhaps 4) 'internal' sub-units, each of which
possesses an allosteric binding site for internal K^+. Saturation of these sites may cause conformational changes in the
carrier which reduce the affinity of the external binding site
for external K^+. By such a mechanism the observed decreases
of affinity ($1/K_m$) associated with increased $[K^+]_i$ may be
accounted for. The ideas of this model are summarized in
Fig.4.

LOW SALT ROOTS

outside   membrane   inside

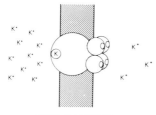

HIGH SALT ROOTS

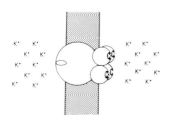

*Fig.4.- Allosteric model for the
regulation of K^+ influx. The K^+
carrier possesses 4 sub-units, each
bearing an allosteric binding site
for internal K^+. In 'low-salt' roots
the sites are vacant and measured
K^+ influx rates are high. As K^+
accumulates (high-salt roots) the
allosteric sites become saturated
bringing about conformational changes
in the carrier reducing its affinity
for external K^+.*

It is perhaps hazardous to extrapolate from the kinetics
of simple enzyme reactions to the kinetics of whole root
influx. Furthermore n values obtained from Hill plots of the
oxygen-haemoglobin dissociation curve do not necessarily give
values equal to 4, the number of sub-units in the haemoglobin
molecule. Nevertheless the vast majority of regulatory
enzymes have been found to be tetrameric (Schachman 1963).
Furthermore, as a mechanism for regulation, a subunit model
displaying sigmoidal kinetics does provide an extremely
efficient means of regulating influx since, by the very nature
of the sigmoidal response, dramatic changes of influx are
achieved by a relatively small change of internal K^+

concentration. Biochemical systems are essentially conservative at the molecular level and hence there is no reason, a priori, why a sub-unit mechanism which so effectively provides the requirements for regulation of enzyme action should not also serve the requirements for the regulation of carrier mediated transport.

The K^+ influx data obtained in the experiments described above were all obtained at external concentrations between 0.01 and 0.32 mM KCl. Under natural conditions (i.e., in the soil) plant roots are only rarely exposed to potassium concentrations that are in excess of these values. Thus it is interesting to observe that in the high concentration range (1 - 50 mM) which Epstein has called System II, there appears to be little correlation between K^+ influx and $[K^+]_i$. Influx data for $CaSO_4$ and 5 mM KCl-grown roots differ much less in the System II than in the System I range. When the differences between influx values from System II concentrations for $CaSO_4$ and KCl-grown roots were expressed as a % inhibition of influx, it became evident that with increasing external K^+ concentration the % inhibition of influx declined from 80% at 1.56 mM to 28% at 50 mM. The question of whether System II influx kinetics can be explained by a passive diffusive flux driven by the electrochemical potential gradient across the plasmalemma rather than by a carrier mediated flux has frequently been raised in recent years (Clarkson 1974). The apparent absence of regulation at high external concentrations of KCl might be interpreted as support for the former interpretation of System II influx kinetics.

REFERENCES

Clarkson, D. (1974). In "Ion Transport and Cell Structure in Plants", ed. by D. Clarkson, McGraw-Hill Book Co. Ltd. (U.K.). pp.292-296.

Glass, A.D.M. (1975). The regulation of potassium absorption in barley roots. *Plant Physiol.* 56, 377-380.

Glass, A.D.M. (1976). The regulation of potassium absorption in barley roots. II. relationship between K_m for uptake and potassium status. *Plant Physiol.* (In press).

Pitman, M.G. and Cram, W.J. (1973). Regulation of inorganic ion transport in plants. In: "Ion Transport in Plants", ed. by W.P. Anderson, Academic Press, London, pp. 465-481.

Schachman, H.K. (1963). Considerations on the tertiary structure of proteins. *Cold Spring Harbour Symp. Quant. Biol.* 28, 409-430.

Schaefer, N., Wildes, R.A. and Pitman, M.G. (1975). Inhibition by p-fluorophenylalanine of protein synthesis and ion transport across the roots in barley seedlings. *Aust. J. Plant Physiol.* 2, 61–73.

Young, M., Jefferies, R.L. and Sims, A.P. (1970). The regulation of potassium uptake in *Lemna minor*. *Abh. Deut. Akad, Wiss. Berlin, Kl. Med. Bd.* b 67–82.

CHAPTER 13

Genotypic Differences in Potassium Translocation in Ryegrass

J. Dunlop and B. Tomkins
Grasslands Division, DSIR, Palmerston North, New Zealand

INTRODUCTION

Most plant physiologists who have investigated ion uptake and translocation have encountered genetic variation. This necessitates considerable replication of plants or alternatively use of genetically uniform plants in nutrition studies. However, because plant breeding programmes may be able to capitalise on genetic variation to produce new varieties of agricultural plants that are less dependent on increasingly costly fertiliser inputs (Cooper 1973) the phenomenon deserves study in its own right. There are numerous reports of differences between cultivars in their yield response to fertilisers (see Vose 1963) but the literature on the physiological basis of this is much less extensive. Differences in the rate of potassium translocation by excised seminal roots of ryegrass cultivars have been reported by Dunlop (1975). However, because seminal roots are only a temporary component of the root system the practical relevance of these results may be limited. This paper describes the variation in potassium translocation in nodal roots and intact plants of four varieties of ryegrass.

METHODS

The plants used were the *Lolium multiflorum* Lam. cultivars 'Grasslands Paroa' ('G. Paroa') and 'S22', and the *Lolium perenne* L. cultivars 'Medea' and 'Grasslands Ruanui' ('G. Ruanui'). Two experiments were conducted. Experiment one involved all four cultivars and the data were examined for differences among cultivars. Experiment two involved four genotypes of 'G. Paroa' only and the data were examined for genotypic differences within this cultivar.

Germinated seedlings were grown in a complete nutrient solution which contained potassium at a concentration of 0.4 mM until they had produced 10 to 15 tillers. For experi-

ment two, plants were grown to this stage and then separated
into tillers and the tillers grown separately to provide replication of each genotype. The plants were grown in a glasshouse, where reasonable control over the environment was maintained. The plants grew well and displayed no visual symptoms
of mineral deficiency. The plants were used when the tiller
number was between 10 and 15.

In each experiment translocation of potassium was determined in both excised roots and intact plants. For excised
roots this was determined as the amount of potassium moving
to the xylem exudate. The distal 6.0 cm of unbranched primary
nodal roots were excised and the cut ends sealed into micro-
capillaries. The roots were then bathed in nutrient solution
maintained at 25°C and exudation was allowed to proceed for
6 hours when the volume of exudate and its potassium concentrations were determined.

Translocation by roots of intact plants was estimated by
transferring the plants to nutrient solution in which the
potassium was replaced by rubidium. It was assumed that
rubidium acts as a tracer for potassium. During these measurements the temperature was maintained at 25°C. After 24 hours
the roots were rinsed in ice-cold 0.4 mM K_2SO_4 plus 0.4 mM
$CaCl_2$ for 15 minutes and briefly rinsed in deionised water.
The roots and shoots were separated, weighed and analysed for
rubidium. In experiment one transpiration was measured by
weighing the beakers containing the experimental solution
before and after uptake. Controls showed that evaporation
from the solution was negligible.

RESULTS

The rates of translocation by excised roots and in intact
plants of the four cultivars are reported in Table 1. There
were significant differences among cultivars in values of
translocation for both excised roots and intact plants. For
both measurements 'G. Ruanui' gave low values, while 'G. Paroa'
gave values that were in the middle of the range. However,
'Medea', which gave the lowest value for the excised roots,
had the highest value for the intact plants, while 'S22' re-
acted the inverse way. For intact plants the amount of
rubidium translocated as a percentage of the total amount
which entered the root during the measurement period is included in Table 1. There were significant differences among
cultivars in this respect also. In intact plants Medea,
which had the highest rate of rubidium translocation, also
transported the greatest proportion of the rubidium absorbed
to the aerial parts. There were large differences in rates
of transport between excised roots and intact plants. Trans-

location in excised roots was only 7 to 20% of the rates found in intact plants.

TABLE 1. Potassium-rubidium translocation by four ryegrass cultivars. Rates are expressed on the basis of grams dry weight (gdw) of roots. Values followed by a different letter differ significantly at p < .05.

Cultivar	Excised roots	Intact plants	
	K^+ translocated μg ion/gdw/h	Rb^+ translocated μg ion/gdw/h	% trans-located
'G. Paroa'	4.16±0.53 ab	35.5±4.0 ab	49.7±1.0 a
'S22'	5.43±0.50 a	26.8±2.4 b	46.1±1.6 a
'Medea'	3.19±0.50 b	46.8±4.9 a	54.7±1.0 b
'G. Ruanui'	3.35±0.60 b	22.9±1.3 b	49.3±0.9 a
Level of significance	*	***	***
Number of measurements/ cultivar	20	12	12

* indicates significance at p < .05, *** p < .001.

The results from experiment two which involved four genotypes of 'G. Paroa' are presented in Table 2. These resemble the data of Table 1 in a number of ways e.g. there are significant differences among genotypes in both excised roots and intact plants but the two sets of data do not correlate with each other. The percentage of rubidium absorbed which was subsequently translocated could be ranked in the same order as rates of translocation in the intact plants for the genotypes studied. As in the previous experiment, rates of translocation were higher for intact plants than for excised roots. However in this experiment differences were smaller, and rates for excised roots ranged from 27% to 80% of the values for intact plants.

In previous studies potassium translocation by excised ryegrass roots was found to be strongly correlated with water flow into the exudate (Dunlop 1975). The present data were examined for correlations between these two parameters, and a summary is presented in Table 3. The data indicate that in both experiments one and two there were strong correlations between water movement and potassium translocation in the excised roots. However in intact plants in experiment one there was no corresponding correlation. Measurements of transpiration were not made in experiment two.

147

TABLE 2. Potassium-rubidium translocation by four genotypes of the ryegrass cultivar 'G. Paroa'. Rates are expressed on the basis of grams dry weight (gdw) of roots. Values followed by a different letter differ significantly at p < .05.

Genotype	Excised roots	Intact plants	
	K^+ translocated µg ion/gdw/h	Rb^+ translocated µg ion/gdw/h	% trans- located
1	8.15±0.88 a	19.8±2.6 a	77.5±1.2 a
2	10.34±0.82 a	12.7±1.4 ab	73.3±2.6 ab
3	2.79±0.29 b	7.9±2.3 b	65.8±2.5 b
4	4.4 ±0.43 b	16.2±3.0 ab	75.0±1.2 a
Level of significance	***	*	**
Number of replicates/ genotype	10	4	4

* indicates significance at p < .05; ** p < .01, *** p < .001.

TABLE 3. Correlation coefficients[+] for the linear regression of potassium translocation with the movement of water to the exudate or, in the case of intact plants, with transpiration.

Experiment one			Experiment two	
Cultivar	Excised roots	Intact plants	Genotype	Excised roots
G. Paroa	.84***	-.53 ns	1	.94***
S22	.60**	-.04 ns	2	-.19 ns
Medea	.68***	.23 ns	3	.79**
G. Ruanui	.90***	.38 ns	4	.85**
Overall	.78***	.02 ns	Overall	.69***

[+] ns indicates that the regression was non-significant;
* indicates significance at p < .05; ** p < .01; *** p < .001

Fig. 1. - The relationship between rubidium translocation and rubidium absorption. Open symbols indicate individual data points; closed symbols indicate means. Values for r, the correlation coefficient, are given on the figures; A, experiment one; B, experiment two.

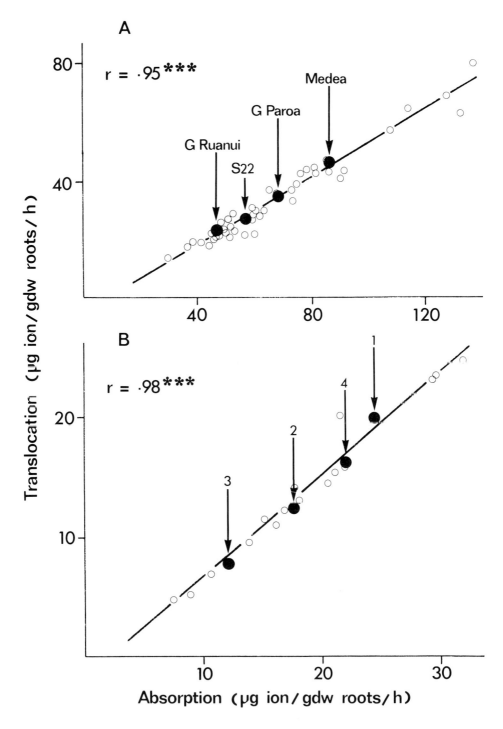

The data for intact plants in both experiments showed strong linear correlations between the rate of rubidium absorption by roots and the rate of rubidium translocation to shoots (Fig. 1). The regression lines for these correlations do not pass through the origin.

DISCUSSION

This study does not give a complete indication of the extent of differences among cultivars and within a cultivar in potassium - rubidium transport characteristics. There are many ryegrass cultivars and ecotypes and the sample studied in experiment one represents a mere fragment of the entire population. Likewise the four genotypes studied in experiment two cannot be taken to indicate the range of differences in the cultivar 'G. Paroa'. In spite of these deficiencies the data indicate that there are significant differences among cultivars and within a cultivar (200% in experiment one; 250% in experiment two). Variation in potassium translocation within a cultivar measured in experiment two is more impressive when it is considered that these differences were obtained without undertaking an extensive search.

The rates of rubidium translocation for intact plants of 'G. Paroa' in experiment one were more than twice as great as the corresponding rates reported in experiment two. Although this discrepancy was not investigated in detail, it probably relates to the environment in which the plants were grown. Plants used in experiment one were grown during winter while the plants for experiment two were grown during late spring and early summer. There were large differences in temperature and light intensity during the growth period between the two experiments. Both of these factors were found by Pitman et al. (1974) to influence properties of rubidium translocation in barley. In an experiment not reported here, potassium translocation by excised roots of the four genotypes used in experiment two was measured for plants grown during the winter. The ranking of genotypes on the basis of rate of translocation was different from that obtained in experiment two. It seems likely that the differences found in this study may be subject to an interaction with climate.

The use of exuding nodal roots could be useful if they accurately reflected the situation in intact plants. However, the quantitative effects of excision on potassium translocation are large and vary from cultivar to cultivar, and from genotype to genotype, making extrapolation from excised roots to intact plants impossible. In intact plants rubidium translocation was not only more rapid but it was not correlated with water flow as in the excised roots. This suggests that the linear relationship between potassium translocation and water flow reported by a number of workers for both excised roots and intact plants (see Dunlop 1974a) is valid only at the slower rates of translocation. In connection with this it is interesting to note that in genotype 2 in experiment two, the only batch of excised roots where there was no correlation between water flow and potassium translocation, the excised roots translocated potassium at a rate approaching that of rubidium measured in intact roots. This observation has implications for the controversy concerning the mechanism of potassium translocation (Dunlop 1974b). Measurements made on excised roots translocating potassium at a slow rate may represent a totally different situation from roots with a more active translocation mechanism as used by Läuchli et al. (1971).

The strong correlations between rates of rubidium translocation and rubidium uptake in intact plants indicate that there is a very close relationship between these two factors. In each experiment data for all cultivars or genotypes conformed to the same relationship which indicates that the differences in rates of translocation arise almost solely from differences in rates of uptake. However, because the cultivar or genotype means were spaced at significantly different points along the regression lines, which did not pass through the origin, there were significant differences in the percentages of rubidium entering the root which were subsequently translocated. While these differences were only of the order of 10% they could be quite important in practice. In pasture plants growing in competition with each other differences of this order could be significant.

While this study does not indicate a practical method for screening plants for potassium translocation it has shown that there exists within the ryegrass population genetic variability which a breeding programme could utilise.

REFERENCES

Cooper, J.P. (1973). The use of physiological criteria in grass breeding. Rep. Welsh Pl. Breed. Stn, for 1973. pp.95-102.

Dunlop, J. (1974a). A model of ion translocation by roots which allows for regulation by physical influences. In "Mechanisms of Regulation of Plant Growth" Bull. 12 Roy. Soc. New Zealand. Ed. by R.L. Bieleski, A.R. Ferguson, and M.M. Cresswell, Wellington. pp.133-138.

Dunlop, J. (1974b). The transport of potassium to the xylem exudate of ryegrass. II. Exudation. *J. Exp. Bot. 25*, 1-10.

Dunlop, J. (1975). Differences in xylem exudation by seminal roots of *Lolium* varieties. *New Phytol. 74*, 19-23.

Läuchli, A., Spurr, A.R. and Epstein, E. (1971). Lateral transport of ions into the xylem of corn roots. II. Evaluation of a stelar pump. *Plant Physiol. 48*, 118-124.

Pitman, M.G., Luttge, U., Läuchli, A. and Ball, E. (1974). Action of abscisic acid on ion transport as affected by root temperature and nutrient status. *J. Exp. Bot. 25*, 147-155.

Vose, P.B. (1963). Varietal differences in plant nutrition. *Herb. Abs. 33*, 1-13.

CHAPTER 14

Auxin Transport in Oats: A Model for the Electric Changes

I.A. Newman and Julie K. Sullivan

Physics Department, University of Tasmania, Hobart, Tasmania

INTRODUCTION

For many years it has been known that auxin transport has associated with it electric changes measurable very simply with wet contacts on the plant surface. Early work was done by Clarke (1935). Observations in greater detail, using more reliable techniques, have been made by others since, for example Newman (1963) and Woodcock and Wilkins (1970). They reported electric potential changes associated with longitudinal auxin transport and lateral geotropic redistribution of auxin respectively.

The contact systems developed by Newman (1963, Fig. 1) ensured constant concentration of electrolyte (10 mM KCl) bathing the cells in the small region of contact with an oat coleoptile. The coleoptile had been gently scraped to penetrate the cuticle at the point of contact so, during the 15 minutes required for the potential to become steady, the electrolyte would have permeated the free-space around the epidermal and nearby parenchymal cells. The electric changes observed were in the form of a potential wave which moved down the coleoptile in presumed association with a change in auxin concentration. A wave was initiated either by apical illumination of an intact etiolated coleoptile or by Indole Acetic Acid (IAA) application to a decapitated coleoptile (Newman 1963, Figs. 10 and 9).

The association of the electric wave with auxin transport was confirmed by transport studies using IAA labelled with carbon-14 (Newman 1970) which showed in particular a well-defined front to the IAA stream. The electric wave showed an initial rise in potential of about 5 mV at each contact at the time that this front passed the region of the contact.

The electric waves, and the other observations, have been useful in investigating auxin transport speeds and time sequences of events in tropic auxin redistributions. However, it is of more fundamental interest to understand the cellular changes of which the reported electric changes are the manifestations on the plant surface. To put it another way, there is need to specify the electric path(s) through the plant between the contacts. It is known that auxin has effects on the plasmalemma and cell wall (e.g. Etherton 1970). Can these be related to the observations with plant surface contacts? This paper suggests a model which may help to throw light on the matter.

EXPERIMENTAL

Evidence for our model comes from studies involving, for convenience, the faster electric responses of the etiolated *Avena sativa* coleoptile to transformation of phytochrome. There is evidence from a variety of viewpoints that phytochrome, like auxin, interacts with cell membranes, the plasmalemma in particular (e.g. Quail et al. 1973).

Surface Measurements.- Electric responses to phytochrome transformation (Newman and Briggs 1972) have been observed using the same electric contact systems as were used for the auxin wave (Newman 1963). The same questions arise about the location of the changes and the electric path in the coleoptile, and the answers could be very similar.

Recent observations, with three doses of red light, are shown in Fig. 1. The main features of the responses are a rapid rise (beginning within 10 s of the brief light treatment) to a peak at 60 s. For the two saturating doses the peak height is about 4 mV. Subsequent changes depend on light dose and duration partly because of cycling of the phytochrome. Restricting the illumination to narrow bands shows that the responses, as in the auxin case, are produced close to, though not necessarily at, the liquid drop contact.

Microelectrode Measurements.- Microelectrode measurements of vacuolar potential changes were undertaken in order to relate them to the changes on the coleoptile surface.

Microelectrodes having tip diameter < 1 μm were prepared and filled with 3 M KCl using standard techniques. Resistances ranged from 30 to 45 MΩ and negative tip potentials in the bathing solution were smaller than 15 mV. The bathing medium was 1.0 mM KCl, 1.0 mM Ca $(NO_3)_2$, 0.25 mM $MgSO_4$, 0.904 mM NaH_2PO_4, 0.048 mM Na_2HPO_4 (after Higinbotham et al. 1964). The intact coleoptiles were placed horizontally in the bathing medium and the microelectrode was adjusted to have its tip just outside the coleoptile 5 min before impalement. These manipulations were done under dim green safelight with brief exposure to bright green microscope light.

It is not possible to see the position of the tip inside

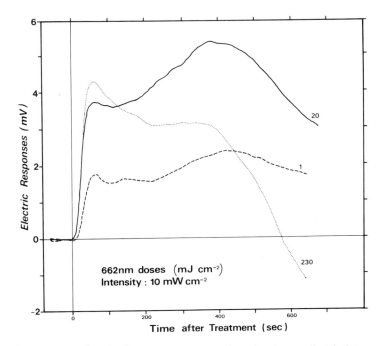

Fig. 1. - *Electric responses of oats to red light.*
These average responses from intact, etiolated seedlings
were measured by a wet cotton contact on the coleoptile
surface. The reference contact was in the medium bathing
the roots. Time is after commencement of red light last-
ing 0.1, 2.0 or 23 s. The peaks occur at 60 s in each
case.

the intact coleoptile. The tip was inserted into the parenchyma by manually advancing the micromanipulator a measured
distance. The presence of the tip in the vacuole of a cell
was inferred from the sudden fall in potential on penetration.
It is thought unlikely that the tip was in the cytoplasm in
these highly vacuolated cells. Measurements were only recorded
from cells which had potentials drifting less than 1 mV per
min within 15 min of insertion.

The vacuolar potentials of parenchyma cells, measured
with respect to the bathing medium, were -93 ± 5 mV ($\pm$ s.e.).
These may be compared with about -105 mV observed by Etherton
(1970) in a similar medium.

Representative responses of the vacuolar potential of
these cells to saturating red (and to far-red) light are shown
in Fig. 2. (Additional work has shown that the responses are
red, far-red reversible as is required for phytochrome.) Note,
as in Fig. 1, the rapid rise of the graph to a peak at $1\frac{1}{2}$ min,

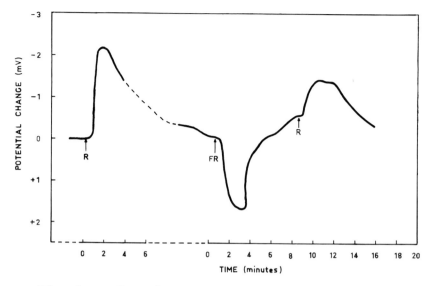

Fig. 2. - Electric response of an oat cell to red (and far-red) light. These representative vacuolar potential changes of one parenchyma cell of an intact etiolated coleoptile were measured with an inserted microelectrode. The peak in the response occurs at about 80 s after the start of the initial red light which lasted 2 s.
R: treatment by 50 mJ cm^{-2} red light.
FR: treatment by 50 mJ cm^{-2} far-red light.
The potential axis has been drawn inverted to allow more ready comparison of the response shape with that of Fig. 1 (from Sullivan 1974).

though the magnitude is 2 rather than 4 mV. The mean magnitude of the peak for four cells is 2.1 ± 0.3 mV. The mean time to the peak is 1.3 ± 0.1 min. The striking thing about the response to red in Fig. 2 is that, although having a similar initial time course, it is of opposite polarity to the responses shown in Fig. 1. The membrane (tonoplast + plasmalemma) change is a hyperpolarisation.

DISCUSSION

A Model. - The potential changes, following red light, shown in Fig. 2 are presumably occurring predominantly at the plasmalemma of the parenchyma cell, since tonoplast changes are generally very small. Thus $V_{in} - V_{out}$ is displayed, where V_{in} represents the potential of the inside of the plasmalemma (the cytoplasm) and V_{out} represents the potential of the bathing medium around the cell. It follows that a reasonable interpretation of the opposite polarity changes of Fig. 1 is that they display $V_{out} - V_{in}$. Since the measuring contact for those plant surface observations is indeed bathing the nearby cells, it will represent V_{out}. Hence the reference contact, which is in the solution around the seedling's roots, must be effectively connected more or less directly to the inside of the coleoptile parenchyma cells in order to represent V_{in}.

A simple model depicting a possibility for this electric path through the plant is illustrated in Fig. 3. The vascular

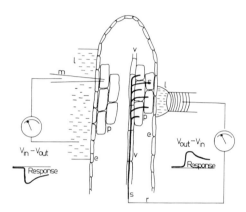

Fig. 3. - A diagram to illustrate possible electric paths in the oat coleoptile. On the left, a microelectrode measures the potential of the vacuole, V_{in}, of a cell with respect to the potential, V_{out}, of the bathing medium. The change in $V_{in} - V_{out}$, in response to red light, is a negative one, a hyperpolarisation as displayed in Fig. 2.
On the right, the small liquid drop bathes the cells, measuring V_{out}. V_{in} is measured by the reference connection via the roots, vascular tissue and symplasm which is shown as a thick line. The response to red light appears as a positive one since the change in $V_{out} - V_{in}$ is displayed as in Fig. 1.
e: epidermal cells
p: parenchyma cells
l: liquid drop or bathing medium which permeates nearby cell walls
m: microelectrode
r: path via roots or base of coleoptile
v: vascular tissue
s: possible symplastic pathway to the insides of the cells

tissue, particularly as the xylem in these coleoptiles is

living (Higinbotham, personal communication), is the suggested pathway from the roots up to the level of the contact. The model then presumes that the symplastic connections from vascular tissue to the cytoplasm of the parenchyma cells have lower resistance than the apoplastic path through cell walls and cell membranes. The elegant experiments of Goldsmith et al. (1972) seem to argue against this but their measurements were on cells which had reduced turgor pressure and this may have affected conduction properties of the plasmodesmata. Williams and Pickard (1974) give evidence that plasmodesmata, although in a very different species, can allow action potential transfer between cells. Figure 3 shows how a particular change of membrane potential can appear as having one polarity when observed by a microelectrode but as having the opposite polarity for a contact on the plant surface with a reference connection through the roots.

Deductions from the Model.- The initial 5 mV rise in potential observed (Newman 1963, Fig. 9) on the coleoptile surface is interpreted as being due to hyperpolarisation of the plasmalemma of underlying parenchyma cells as the IAA stream reaches them. This is consistent with the hyperpolarisation of up to 35 mV reported by Etherton (1970) for cells of oat coleoptile segments bathed in physiological concentrations of IAA.

The model suggests the following as the electric path between contacts placed on the surface on each side of a coleoptile as used by Woodcock and Wilkins (1970): from one contact in through the plasmalemma of adjacent cells, across (or around) the coleoptile via the symplasm and out through the plasmalemma of the cells near the other contact. If one side of the coleoptile has an increased auxin concentration the cells on that side will become hyperpolarised and the contact outside them will become positive to the symplastic connection and the contact on the other side. Thus if IAA is applied asymmetrically to a decapitated coleoptile, the side under the IAA donor should become positive; if a coleoptile is turned on its side, the underneath, to which the auxin is redistributed, should become positive. Just these changes have indeed been observed by Woodcock and Wilkins (1970).

CONCLUSIONS

If the simple model presented here is valid, it means that some kinds of investigations of oat coleoptile parenchyma plasmalemma properties may be made with external contacts - a technique very much simpler than microelectrode impalement. Metabolic inhibitors or other substances affecting plasmalemma properties may be applied in the contact solution. The electric changes measured would then be precisely on those cells affected. Such studies could also provide necessary further testing of the model.

REFERENCES

Clark, W.G. (1935). Note on the effect of light on the bioelectric potentials in the *Avena* coleoptile. *Proc. Nat. Acad. Sci. Wash. 21*, 681-684.

Etherton, B. (1970). Effect of indole-3-acetic acid on membrane potentials of oat coleoptile cells. *Plant Physiol. 45*, 527-528.

Goldsmith, M.H.M., Fernandez, H.R. and Goldsmith, T.H. (1972). Electrical properties of parenchymal cell membranes in the oat coleoptile. *Planta (Berl.) 102*, 302-323.

Higinbotham, N., Hope, A.B. and Findlay, G.P. (1964). Electrical resistance of cell membranes of *Avena* coleoptiles. *Science 143*, 1448-1449.

Newman, I.A. (1963). Electrical potentials and auxin translocation in *Avena*. *Aust. J. Biol. Sci. 16*, 629-646.

Newman, I.A. (1970). Auxin transport in *Avena*. I. Indole-acetic acid-^{14}C distributions and speeds. *Plant Physiol. 46*, 263-272.

Newman, I.A. and Briggs, W.R. (1972). Phytochrome-mediated electric potential changes in oat seedlings. *Plant Physiol. 50*, 687-693.

Quail, P.H., Marmé, D. and Schäfer, E. (1973). Particle-bound phytochrome from maize and pumpkin. *Nature (New Biology) 245*, 189-191.

Sullivan, J. (1974). The effect of red and far red light on *Chara corallina* and *Avena sativa*. Honours Thesis, University of Tasmania.

Williams, S.E. and Pickard, B.G. (1974). Connections and barriers between cells of *Drosera* tentacles in relation to their electrophysiology. *Planta (Berl.) 116*, 1-16.

Woodcock, A.E.R. and Wilkins, M.B. (1970). The geoelectric effect in plant shoots. III. Dependence upon auxin concentration gradients and aerobic metabolism. *J. Exp. Bot. 21*, 985-996.

CHAPTER 15

Lateral Transport of ^{14}C-labelled IAA in Maize Roots

P. Jarvis and R.J. Field

Department of Plant Science, Lincoln College, New Zealand

INTRODUCTION

Solutes enter the plant by passing radially across the root from the external solution to the xylem vessels or tracheids. Two major pathways exist. There is a free space pathway through the cortex up to the endodermal cells which constitute a barrier to the free space movement of solutes into the stele. Alternatively, solutes can be taken up into the cytoplasm of the cells and pass in the symplasmic pathway to the stele. In the latter case the solutes traverse the endodermal barrier. The precise mechanism for the movement of the solutes from the symplasm to the xylem vessels has not yet been defined. The possible means by which this might occur are discussed by Epstein (1972).

Auxin may be rapidly metabolised by roots (see references quoted by Scott (1972)). Greenwood et al. (1973) investigated the uptake and metabolism of auxin in isolated maize roots but did not attempt to analyse root exudate or to elaborate on the pathway of lateral movement of auxin.

MATERIALS AND METHODS

Primary roots of maize (*Zea mays* cv. Northrup King) were grown as described by Jarvis and House (1967). The roots were excised at 10 cm in length and inserted into 100 µl capillaries and placed in a saline solution (1 mM KCl and 0.1 mM $CaCl_2$) containing 3-Indolyl 1-^{14}C - acetic acid (IAA). Two concentrations of IAA, 2.4×10^{-5}M (Specific activity 1 µCi/102 µM) and 2.4×10^{-6}M (specific activity 1 µCi/137 µM) were used. The radioactivity present in the load solutions, root exudate and in extracts of root tissue was determined over periods of up to 24 hours. Whole roots or roots separated into stele and cortex were extracted in 2 ml of 70% ethanol. Aliquots of samples were counted on a Packard 3390 scintillation spectrometer or analysed by descending paper chromatography using Isopropanol: Ammonia : Water (10:1:1) or n-Butanol :

Acetic acid : Water (5:1:2.2) as the solvents. The developed chromatograms were analysed using a Packard 7201 radiochromatogram scanner.

Initial experiments showed that the loss of activity from the load solutions was considerable, with 90% disappearing in a 24 h incubation period. This loss appeared to be due to bacterial activity, so chloramphenicol (25 mg/l) is now routinely incorporated into all load solutions. At the end of a 24 h experiment, under these conditions, at least 95% of the radioactivity present initially could be accounted for and the IAA in the load solution was not converted to metabolites of IAA. The chloramphenicol at this concentration did not appear to have a marked effect on the root, as roots exuded in the absence of IAA, but in the presence or absence of chloromphenicol had similar exudation rates.

RESULTS AND DISCUSSION

Preliminary experiments showed that there was a linear uptake of activity into the roots over a 24 h period. Incubation of excised roots in a solution of IAA (2.4×10^{-5}M) for 12 h showed that the roots rapidly metabolised the IAA. Estimates of the activity in IAA and IAA metabolite fractions (Table 1) indicate that 78% of the activity in the whole root extract was in the form of the metabolites. Although the con-

TABLE 1. Concentration of activity in root extract and exudate following a 12 h incubation in 2.4×10^{-5}M IAA. Figures in parenthesis represent the percentage distribution of activity in each fraction. Chromatographic separation in isopropanol : ammonia : water.

	Concentration (cpm/mg)	
Rf	0.35	0.03
	IAA	Metabolites
Root extract	1080 (21.6%)	3768 (78.4%)
Exudate	139.4 (90.6%)	14.4 (9.4%)

centration (cpm/mg) of IAA in the exudate was low it represented at least 90% of the total activity in that sample. The metabolites appear relatively immobile with a very low concentration detected in the exudate.

In order to examine the movement of IAA and its metabolites more closely a further series of experiments was performed in which the stele and cortex were separated at the end of the experiment before extraction. Table 2 shows the concentration of activity found in extracts of the stele and cortex and in the exudate at the end of a 24 h incubation period. By using the n-butanol : acetic acid : water solvent system it

TABLE 2. Concentration of activity in extracts of isolated steles and cortices and root exudate following a 24 h incubation in 2.4 x 10^{-5}M IAA. Figures in parenthesis represent the percentage distribution of activity in each fraction. Chromatographic separation in n-butanol : acetic acid : water.

Rf	Concentration (cpm/mg)		
	0.85	0.52	0.73
	IAA	Metabolites	
Cortex	312 (7.1%)	2084 (47.5%)	1996 (45.5%)
Stele	3632 (26.7%)	5384 (39.6%)	4596 (33.8%)
Exudate	42.0 (85.2%)	2.0 (5.9%)	4.4 (8.9%)

is clearly shown that IAA runs at a Rf of 0.83 to 0.88. There are two major metabolites, one running at a Rf of 0.45 to 0.56 and the other at a Rf of 0.66 to 0.76. It appears reasonably certain that the latter metabolite is indoleacetyl aspartate as it co-chromatographs with marker spots of this compound. The other compound has not yet been positively identified but it is possibly another conjugate compound, indoleacetyl glucose. This compound has been isolated from other plant tissues and runs with a Rf of 0.56 in this solvent (Zenk 1961). Both metabolites behave in a similar manner and their concentrations within the tissue or exudate fractions appear to run in parallel.

As in the previous experiment there was a high percentage of activity in the exudate in the form of IAA. However the concentration of IAA in the exudate was less than that in the external medium when that contained 2.4 x 10^{-5}M of IAA. There is a low concentration of metabolites in the exudate where together they only account for 14.8% of the total activity.

It appears that both the cortex and stele accumulate IAA, as its concentration there was greater than that in the load solution. The separate analysis of the cortex and stele indicates that there are major differences in the distribution of radioactivity among different fractions within these tissues. There is a much higher concentration and proportion of the total activity found in the form of IAA in the stele. The figure of 26.7% in these experiments would appear to be a conservative figure as in other experiments, the proportion of label in IAA in the stele has reached 50%.

In the cortex, there was a low concentration of IAA representing only 7.1% of the total label in that extract. Obviously both of the tissue extracts have much higher concentrations of metabolites than are found in the exudate.

The maize root can therefore be thought of in terms of a three compartment model with respect to the concentration of

IAA and IAA-metabolites and in terms of the distributional pattern of these compounds (Fig. 1). IAA appears to be able

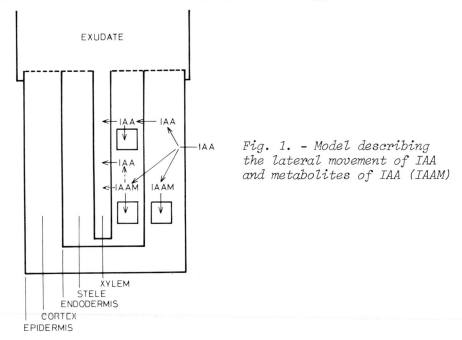

Fig. 1. - Model describing the lateral movement of IAA and metabolites of IAA (IAAM)

to move from the external solution, across the epidermis, cortex, endodermis, stelar parenchyma and into the xylem exudate. The IAA is concentrated in the stele but at higher external concentrations of IAA (2.4 x 10^{-5}M) there is a lack of evidence that IAA in the exudate is more concentrated than in the external medium. IAA is however more concentrated in the xylem exudate when the external concentration is lowered to 2.4 x 10^{-6}M.

There is a marked difference in the IAA concentration between the cortex and the stele with at least a ten-fold increase in concentration in the stele. This suggests that the endodermis must be considered as a barrier to the free diffusion of IAA. Ferguson and Clarkson (1975) have indicated that in maize each endodermal cell, further than 1 cm from the root tip, has a fully developed Casparian strip which will effectively prevent a free space movement of solutes and water into the stele. The difference in concentration of IAA between the cortex and the stele suggests that IAA may be accumulated in a "pool" in the stele. The possibility that this "pool" is the vacuoles of the stelar parenchyma cannot be ruled out. Alternatively the results may simply relate to the differing abilities of the cortical and stelar cells to metabolise the IAA.

The position of the metabolites of IAA in this model is unclear. The high concentration of metabolites in the cortex may suggest that these compounds are immobilised in a vacuolar compartment. If the metabolites are conjugates of IAA, this does not represent a new suggestion as these compounds have been reported to be immobile in plants (Eschrich 1968; Morris et al. 1969). Additional results obtained from a pulse-chase experiment support this hypothesis. Roots were loaded with radioactive IAA (2.4 x 10^{-6}M) for an initial period of 4 h followed by incubation for a further 24 h in equimolar un-labelled IAA. The results indicate that the metabolites form-ed after 4 h appeared to be immobile but that the free IAA in the stele and cortex is redistributed into metabolites or exuded in the exudate stream. The possibility that the IAA metabolites can be converted back to free IAA can not be dis-counted (Field and Peel 1971) but further experiments are re-quired for this to be determined in maize roots.

The fact that IAA and the metabolites are found in only low concentrations in the exudate, compared to the stelar cells is indicative of a barrier between the living cells in the stele and the dead cells in the xylem through which the exudate passes. One suggestion for the control of solute flow into the exudate is that solutes are first accumulated in the cytoplasm and vacuoles of developing xylem vessels and sub-sequently released into exudate stream as the xylem vessel matures (Hylmö 1953).

In view of the very low concentrations of IAA metabolites in the exudate it would appear that this proposal is not valid. Were this the case a considerably higher concentration of metabolites would be expected. The results seem consistent with the proposals suggested for ions, that IAA is either leaked passively into the exudate (Laties and Budd 1964) or is actively pumped into the exudate stream by the stelar parenchyma cells (Steward and Sutcliffe 1959).

REFERENCES

Epstein, E.E. (1972). Mineral nutrition of plants. Princi-ples and perspectives. Wiley & Sons New York pp 151-189.

Eschrich, W. (1968). Translokation radioactiv markiere indolyl-3-essigsaüre in Siebröhren von *Vicia faba*. *Planta (Berl.)* *78*, 144-157.

Ferguson, I.B. and Clarkson, D.T. (1975). Ion transport and endodermal suberization in the roots of *Zea mays*. *New Phytol*. *75*, 69-79.

Field, R.J. and Peel, A.J. (1971). The movement of growth

regulators and herbicides into the sieve elements of willow. *New Phytol. 70*, 997–1003.

Greenwood, M.S., Hillman, J.R., Shaw, S. and Wilkins, M.B. (1973). Localization and identification of auxin in roots of *Zea mays. Planta (Berl.) 109*, 369–374.

Hylmö, B. (1953). Transpiration and ion absorption. *Physiol. Plant. 6*, 333–405.

Jarvis, P. and House, C.R. (1967). The radial exchange of labelled water in maize roots. *J. Exp. Bot. 18*, 695–706.

Laties, G.G. and Budd, K. (1964). The development of differential permeability in isolated steles of corn roots. *Proc. Nat. Acad. Sci. 52*, 462–469.

Morris, D.A., Briant, R.E. and Thompson, P.G. (1969). The transport and metabolism of ^{14}C-labelled indole acetic acid in intact pea seedlings. *Planta (Berl.) 89*, 178–197.

Scott, T.K. (1972). Auxins and Roots. *Ann. Rev. Plant Physiol. 23*, 235–258.

Steward, F.C. and Sutcliffe, J.F. (1959). Plants in relation to inorganic salts. In, Plant Physiology ed by F.C. Steward, Academic Press. Vol. II. 253–465.

Zenk, M.H. (1961). 1-(Indole-3-acetyl)-β-D-Glucose, a new compound in the metabolism of indole-3-acetic acid in plants. *Nature 191*, 493–494.

CHAPTER 16

Phloem Loading in Source Leaves

D.R. Geiger
University of Dayton, Dayton, Ohio 45469, U.S.A.

INTRODUCTION

Phloem loading can be defined as the process in which the major translocate species are transported into and concentrated in the conducting elements of the phloem of a translocation source. Excluded from the definition is the process of passive entry of substances into the phloem without undergoing active uptake. Compounds which undergo phloem loading are taken up and concentrated to a level above their concentration in the apoplast, usually above that in the surrounding tissues. Presumably most of the compounds which are phloem-mobile (Crafts and Crisp 1971) enter passively by diffusion and are carried along the transport stream. Only relatively few undergo phloem loading.

Evidence for phloem loading was observed by Roeckl (1949), who found that the osmotic pressure of the sieve sap was considerably higher than that of the surrounding leaf tissue. Earlier, Curtis and Scofield (1933) concluded that sugar concentration gradient in the translocation system is not favorable for diffusion of the major translocation species into the phloem of the source or out into the sink tissues. The original Münch hypothesis has been refined by a number of workers to include a concentrating step in which compounds to be translocated are actively taken up into the conducting elements of the phloem (Crafts 1951, Wanner 1952, Bauer 1953, Geiger 1975).

The sieve elements and companion cells of minor veins in a source leaf appear to have a much higher osmotic pressure than the surrounding tissues (Geiger et al. 1973). Figure 1 is a micrograph showing the relative solute content of the phloem and of the surrounding tissue. In developing leaves, the osmotic pressure of sieve elements and companion cells of minor veins increases as that particular region of leaf begins

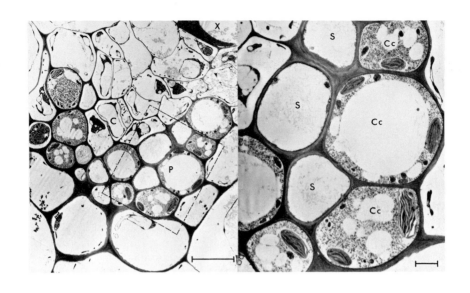

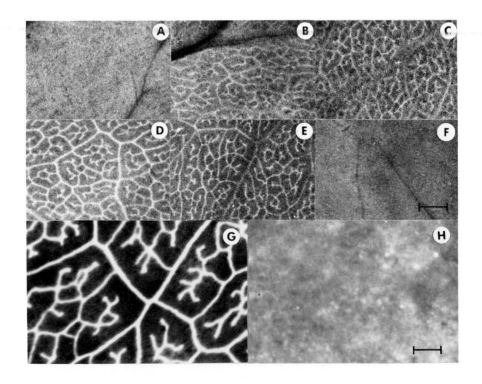

Fig. 1 - Cross section of minor vein from a source leaf of sugar beet showing the high osmotic pressure of the contents of sieve elements and companion cells. Leaf tissue was equilibrated with 1 M mannitol for 30 min, frozen quickly and prepared by freeze-substitution for electron microscopy. X = xylem, C = cambium, P = phloem, S = sieve tube, Cc = companion cell. (Left) Electronmicrograph showing plasmolyzed cells surrounding intact sieve elements and companion cells of the phloem. Marker = 5 μ. (Right) Detail of three se-cc complexes. Marker = 1 μ. Micrograph by R. Giaquinta. Geiger et al. (1973).

Fig. 2 - Autoradiographic evidence of phloem loading. A to F: Autoradiographs of freeze-dried sugar beet leaf showing time course of phloem loading following a 3-min pulse of $^{14}CO_2$. Following labeling, samples were taken after 90 sec (A), 5 min (B), 8 min (C), 12 min (D), 20 min (E) and 100 min (F) of a chase period in air. Marker = 1 cm (Results of quantitative autoradiography shown in Fig. 3). G: Autoradiograph of freeze-dried leaf of sugar beet source leaf following a 180-min period of labeling with a 10-mM ^{14}C-sucrose solution. H: Same leaf following extraction of soluble materials with 80% aqueous ethanol, showing distribution of insoluble material. Marker = 2 cm.

to export assimilate (Fellows and Geiger 1974). Concentrating of solutes by minor veins of leaves photosynthesizing in $^{14}CO_2$ can be shown by autoradiography (Fig. 2A to F). Label can be seen in the veins after 5 min of labeling. Phloem loading of externally applied ^{14}C-sucrose can also be shown by the same technique (Fig. 2G,H). The time course for ^{14}C-incorporation from $^{14}CO_2$ in mesophyll and minor veins, obtained by scanning the grain density of autoradiographs, is given in Figure 3.

FUNCTIONS SERVED BY PHLOEM LOADING

Only a few compounds are translocated in large amounts in the phloem, including several sugars, some amino acids and several inorganic ions (Pate et al. 1974). Restriction of phloem loading to a limited number of compounds limits entry of materials which are best kept in the source region or which might be degraded in transit and therefore transported inefficiently. For instance, certain amino acids and certain ions are presumably best conserved in the source region during

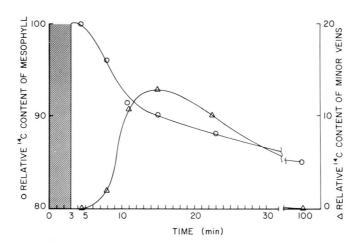

Fig. 3 - Time course of ^{14}C-content of mesophyll and minor veins of a sugar beet source leaf following a 3-min pulse labeling period in $^{14}CO_2$ and a chase period in air. (Autoradiographs in Figure 2A to F).

certain parts of the growing season. Also, sugars such as glucose are thought to be too readily metabolized to be efficient transport sugars. Little is known about the control function of phloem loading in relation to these adaptations nor about the mechanisms for selective entry of compounds.

At least from a theoretical point of view, it is not necessary that all major transport species be loaded as such. Some could arise by conversion from another species which was actively taken up. For instance, sucrose might be loaded and then metabolized to raffinose, stachyose or mannitol. The extent to which phloem loading controls entry of compounds and the identity of those substances which can be actively taken up are topics which need further study. Aspects of selectivity of phloem loading will be discussed in a later section.

Phloem loading is capable of accumulating solutes in minor vein phloem to a level above that in surrounding tissues (Geiger et al. 1973). It is likely that the high osmotic pressure in the minor vein phloem causes water to enter and produces a high hydrostatic pressure in the phloem. Entry of solutes by active transport seems ideally suited to produce a gradient of osmotic pressure in the phloem from source to sink. Because of the pattern of solute distribution in cells of the source and sink regions (Geiger et al. 1973; Roeckl 1949; Curtis and Scofield 1933), it does not seem reasonable to include the mesophyll as the translocation source. Neither does it seem likely that the tissues of a sink region, as a whole, constitute the translocation sink in the sense used by

Münch (Curtis and Scofield 1933). The mesophyll cells of the source leaf and the cells of sinks, especially of storage regions, are probably isolated from the translocation system by membranes. The situation for the sink region seems to be more variable and is not as clear at present. It appears that the Münch translocation system consists of the sieve element-companion cell complexes of the minor veins as the source, with sieve tubes of the path and the phloem of the sink regions constituting the rest of the system. (Fellows and Geiger 1974; Geiger 1975).

While it is not proven beyond doubt that osmotically gen-'
erated mass-flow is responsible for translocation, this model seems to be a fruitful working hypothesis. Recent work with mathematical modeling adds support to the role of pressure flow as the driving force of translocation (Christy and Ferrier 1973; Tyree et al. 1974).

CHARACTERISTICS OF PHLOEM LOADING

1. *Sugars Loaded from Free Space*. - A vital question, from both a theoretical as well as a methodological view, is whether phloem loading occurs from the free space or whether it occurs from the cytoplasm of cells surrounding the conducting tissue. Studies of kinetics and of selectivity, carried out with exogenous sugars, can be interpreted readily only if it is assumed that loading takes place from the free space. Several lines of evidence indicate that phloem loading occurs from extra-
cellular space. Electron microscopy of rapidly frozen, freeze-
substituted leaf tissue, equilibrated with solutions of various osmotic concentration, provides evidence for a sharp difference in solute concentration between the sieve element-companion cell (se-cc) complex and the surrounding tissues (Fig. 1). The pattern observed indicates that the plasma membrane constitutes the barrier across which phloem loading occurs. However, the distribution of osmotic pressure does not actually settle the path for phloem loading. Solute could be loaded by active transport across the plasma membrane or perhaps by active transfer across plasmodesmata.

Several other studies supply data which pertain to the path of solutes during phloem loading. When sucrose is supplied to free space of a mature sugar beet leaf, phloem loading occurs (Fig. 2G). When applied at a concentration of 10 to 20 mM, sucrose is translocated out of the leaf at a rate which is of the same order as that for export of photosynthate (Geiger et al. 1974). These data suggest that exogenous sucrose is processed in the same way as photosynthate and is translocated at the same rate. It can be argued that sucrose is taken up by the mesophyll and then is loaded via plasmodesmata. This latter route cannot be definitely ruled out but data of several types make the path seem unlikely.

    Another source of evidence for involvement of free space in phloem loading comes from isotope trapping experiments with $^{14}CO_2$ labeling. An unlabeled sugar is introduced into the free space to trap the labeled species of this same sugar as it passes through the free space (Geiger et al. 1974). Typical data for an experiment with a sucrose trap are shown in Figures 4A and 4B. As labeled assimilate is exported from the leaf,

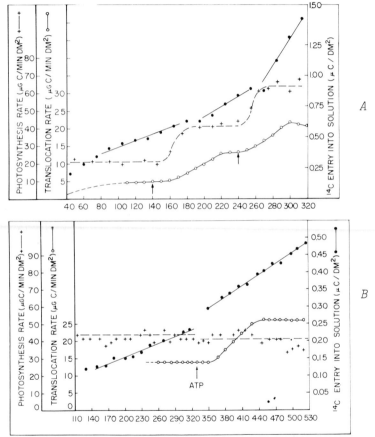

Fig. 4. - Isotope trapping of ^{14}C-sucrose by an unlabeled 10-mM sucrose solution circulating over the abraded upper surface of sugar beet source leaves. Leaves photosynthesizing in $^{14}CO_2$ throughout the experimental period. A: Effect of increasing light intensity on the rate of photosynthesis, translocation, and the trapping of ^{14}C-sucrose. B: Effect of external application of 4-mM ATP on translocation rate and the trapping of ^{14}C-sucrose.

[14]C-sucrose is continuously trapped in the unlabeled sucrose
solution circulating over the abraded surface of the leaf.
The patterns of ^{14}C-sucrose trapping with various treatments
are generally those which would be expected if sucrose is
passing through free space prior to phloem loading. As translocation is increased by increasing light intensity, the rate
of trapping increases accordingly as shown by the increased
slope for sucrose trapped in the external solution (Fig. 4A).
Glucose does not show a rate of trapping which changes with a
change in translocation rate. Treatment of the leaf with ATP,
which serves to increase the rate of translocation, correspondingly causes sucrose to be trapped at a faster rate (Fig.
4B).

A third piece of evidence that the translocate passes
from free space into the minor veins phloem, by way of the
plasma membranes of the sieve element-companion cell (se-cc)
complex, comes from the use of agents which change the rate of
translocation of assimilate when they are applied to the free
space. Application of ATP to the free space both speeds translocation of assimilate and the turnover of sucrose in the free
space (Fig. 4B). Certain chelating agents also increase the
rate of translocation of assimilate without altering the rate
of photosynthesis. (Diane Doman, University of Dayton, private
communication). Reagents such as p-chloromercuribenzoate sul-
fonate (PCMBS), which react with sulfhydryl groups, slow the
translocation of assimilate when applied to the free space.

The sulfhydryl blocking agent, PCMBS, penetrates the
plasma membrane to only a minor extent as judged by its failure
to inhibit photosynthesis or respiration during the treatment
regime used. PCMB and other sulfhydryl reagents which penetrate into the cell interior inhibit photosynthesis and respiration. When a 5-mM solution of PCMBS was applied to the
abraded upper surface of a sugar beet leaf, which was photosynthesizing in [14]CO$_2$, translocation was inhibited noticeably
while photosynethesis rate remained nearly unchanged (Fig. 5).
Prior to treatment with PCMBS, 30 to 33% of the carbon that was
assimilated was being translocated, while after treatment only
24% was being exported. Application of 10-mM cystein failed
to restore translocation following inhibition with PCMBS. Cysteine also produced a decline in photosynthesis. Interpretation
of this latter portion of the experiment must await further
data.

It appears that the sulfhydryl-blocking agent PCMBS inhibits entry of sucrose into minor veins by affecting the outer
surface of the plasma membrane of the sieve elements and companion cells. Data from isotope trapping experiments, from
ATP and PCMBS treatments applied to the free space and from

studies of translocation of exogenous sucrose are consistent
with entry of sucrose into free space prior to phloem loading.
Although none of the evidence is directly conclusive, taken
together it suggests that phloem loading occurs from the apoplast.

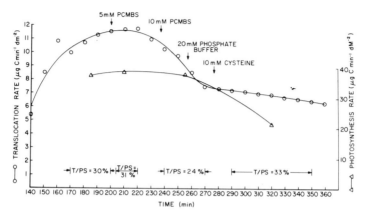

*Fig. 5. - Effect of free-space application of 5-mM PCMBS on
translocation rate of leaves photosynthesizing in
$^{14}CO_2$. The PCMBS solution was removed, the leaf
rinsed and a 5 mM cysteine solution was applied at
the points indicated (D. Doman, private communication).*

2. *Phloem Loading Shows Saturation Kinetics, High Transport Rates and Ability to Concentrate Sugar.* - Loading of
exogenous sucrose into phloem of a sugar beet leaf shows saturation kinetics (Fig. 6). Rates of uptake are higher for
intact leaves than for an equal area of isolated leaf disk,
presumably because the sucrose is able to be translocated out
of intact leaves but not out of the isolated disk. A high
degree of saturation occurs by 100-mM sucrose concentration.
The Kj values found in experiments of this type usually range
from 20 to 40 mM. Rates found for 10 to 20 mM sucrose,
supplied in the light, are similar to the translocation rates
for photosynthesis in air at 2000 ft. c. Rates in the dark
are about 1/3 of those for leaves kept in the light.

With quantitative autoradiography, it is possible to
determine kinetics for both uptake by mesophyll and by minor
veins (Fig. 7). Both minor veins and mesophyll show saturation kinetics of the Michaelis-Menten type, indicating that
transport involves transient and specific interaction with a
limited number of entities in the membrane. Based on the
working hypothesis that sucrose enters minor veins via the
plasma membrane of sieve elements and companion cells, it is

174

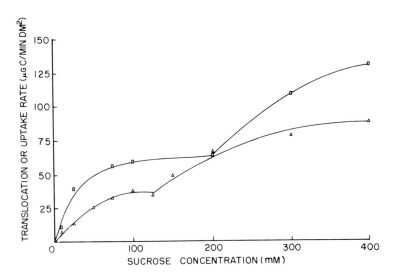

Fig. 6. - *Dependence of sucrose uptake by leaves of* **Beta vulgaris** *on the concentration of sucrose supplied. Solution was applied through the abraded upper surface of source leaves (□) and through the abraded surface of disks from source leaves (△). (Sovonick et al. 1974).*

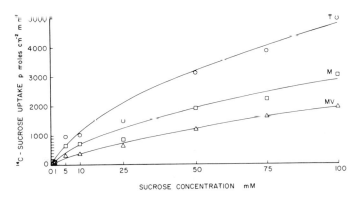

Fig. 7. - *Rate of uptake of sucrose into leaf, (T), into mesophyll (M) and into minor veins (MV). Solutions of various concentrations of* ^{14}C-*sucrose were applied to the abraded upper leaf surface for a 30-min period in darkness.*

possible to estimate the flux of sucrose across a known total area of the se-cc complex plasma membrane (Sovonick, Geiger and Fellows 1974). For a translocation rate of 0.95 μg sucrose

$min^{-1}cm^{-2}$ leaf blade, the flux is calculated to be 3.2 x 10^{-9} mol $min^{-1}cm^{-2}$ plasma membrane, a very high rate (Sovonick et al. 1974). The magnitude of the sucrose transport rate and saturation kinetics indicate that an active transport system is at work during phloem loading.

The rapid turnover of material during phloem loading was visualized by following the time course of movement of ^{14}C-assimilate through mesophyll and minor veins after pulse labeling. A 3-min pulse of $^{14}CO_2$ was supplied to an illuminated sugar beet leaf. Leaf punches, which were removed at intervals, were freeze dried and autoradiographs were prepared. Quantitative grain-density scans were made of mesophyll and of minor veins. For the first several minutes of pulse labeling, radioactivity was found only in mesophyll. Larger veins did not show labeling (dark lines in Fig. 2A) and the level of label in the minor veins did not surpass that in the mesophyll. By 5 min after the beginning of pulse labeling, ^{14}C-assimilate accumulated to a level visibly above that in the surrounding mesophyll (Fig. 2B). Label in the veins reached a maximum approximately 12 min after the end of the pulse (Fig. 2D). Thereafter labeled material was displaced from the veins by unlabeled assimilate from $^{12}CO_2$ (Fig. 2E and F). Quantitative grain density scans (Fig. 3) show that loss of label from mesophyll resulted in a rapid increase in the level present in minor veins. After products of $^{14}CO_2$ assimilation no longer entered the minor veins, there was loss of label from the veins. Extraction of tissue that contains labeled minor veins with 80% ethanol resulted in the loss of label from the veins (Fig. 2H), which indicates that little or no starch was stored in these veins. The time course of movement of the pulse from the mesophyll through the minor veins indicates that there is rapid turnover in mesophyll and minor veins of material destined for export.

3. *Energy Required for Phloem Loading.* - Phloem loading, like other processes in which active membrane transport occurs, is inhibited by treatments which lower energy metabolism. Application of 4 mM 2,4 dinitrophenol (DNP) to a translocating sugar beet leaf causes CO_2 release to be accelerated and lowers the ATP level in the leaf, indicating an uncoupling of respiration (Sovonick et al. 1974). This treatment reduced translocation of exogenously supplied ^{14}C-sucrose out of the leaf to approximately 20% of the control rate. Treating the leaf with 0°C temperature or by excluding oxygen also inhibits phloem loading.

Application of 4 mM ATP stimulates the rate of translocation of exogenous sucrose by 60 to 80% (Sovonick et al. 1974). The proportion of stimulation is approximately the same for both control and DNP-treated leaves.

Similarly, application of 4 mM ATP increases translocation of assimilate derived from photosynthesis by 40% (Geiger et al. 1974). Although the increase in translocation may be a result of an energy-related effect of ATP, it may also be due to some other effect such as chelation of ions.

4. *Phloem Loading of Sugars is Selective.* - Selectivity is a characteristic of active membrane-transport systems. It is known that translocation of organic molecules is limited to relatively few compounds including chiefly sucrose, members of the raffinose family of oligosaccharides, plus mannitol and sorbitol (Zimmermann 1957, 1960). In view of the limited number of compounds that are major components of the translocate, one expects to find specificity of membrane transport to be part of phloem loading. However, the manner in which selective active transport regulates the composition of translocate is not clear.

Higher plants can be arranged into several major groups on the basis of translocate sugars present in the phloem (Zimmermann 1960). Some possible mechanisms for regulating these patterns are summarized in Table 1. The specific pattern encountered may be a result of directed metabolism in the sieve tubes and companion cells of the phloem (Mechanism A). Alternatively, the pattern may be a result of the specific carrier or carriers present in the se-cc membranes; the sugars being translocated would be the only ones capable of being actively loaded (Mechanism B). Finally, the particular spectrum of sugars in the phloem may be a result of the particular sugars available in the leaf, that is, the ones which alone can be synthesized in quantity in the mesophyll (Mechanism C). In summary, the major agent responsible for generating the pattern of translocate sugars in the phloem may be the phloem tissue; the selective plasma membrane of the se-cc complex or the mesophyll with its particular ensemble of enzymes.

A study of specificity of sugar loading was undertaken with sugar beet. Six sugars were presented singly, in solution, to the abraded upper surface of source leaves. Three of the sugars, sucrose, stachyose and mannitol, are commonly translocated by higher plants. Of these, only sucrose is normally found in the phloem of sugar beet (Fife et al. 1962; Geiger et al. 1969). The non-biological sugar, L-glucose, was used as a control to assess entry by passive, non-mediated permeation. Fructose was used because it is a hexose precursor of sucrose, while 3-0-methyl glucose is a non-metabolizable analog of glucose. The uptake of these six sugars into mesophyll and minor veins, when they were presented to the abraded epidermis as 10-mM solutions, was studied (Fondy and

Geiger unpublished). Four of the sugars showed signs of contributing to vein loading: sucrose, fructose, stachyose and
mannitol (Fig. 8). Sucrose entered the metabolic space of the
leaf tissue 3 to 25 times faster than did the other sugars.

TABLE 1. Possible mechanisms which may give rise to patterns
observed with respect to the major sugars that are translocated
in higher plants.

Manner of establishing pattern of major components of translocation stream	Number of carriers present	Type of specificity of the carriers
A. One major sugar, sucrose is loaded. All other sugars arise by directed metabolism of the sucrose	One	Narrow specificity. Recognized only sucrose. May transport other sugars to a slight extent.
B1. The pattern of translocate is determined by the sugars which are actively loaded. Only sugars which are translocated by the species can be loaded.	One	Broad specificity carrier which is characteristic of the species of higher plant in question.
B2. Ditto	Several	One carrier present for each sugar or perhaps for each class of sugar that is translocated.
C1. The pattern of translocate is determined primarily by the sugars produced by the metabolism of the leaf; loading selectivity is secondary	One	Broad specificity carrier capable of transporting any of several sugars that are found to be translocated in higher plants.
C2. Ditto	Several	One carrier present for each sugar or perhaps for each class of sugar that can be translocated by higher plants.

Quantitative autoradiography was used to determine the ^{14}C-content of minor veins after 30 min of labeling with various exogenous sugars. Freeze-dried leaf tissue was

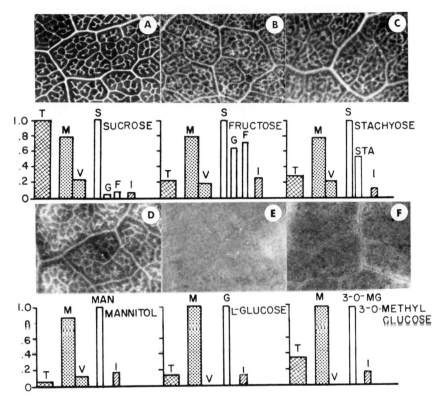

Fig. 8. - *Uptake, metabolism and phloem loading of sugars supplied to the leaves of* Beta vulgaris. *Sugars supplied at 10 mM solutions to abraded upper surface of sugar beet leaf. Sugars supplied are sucrose (A), Fructose (B), Stachyose (C), Mannitol (D), L-Glucose (E) and 3-0-Methyl Glucose (F). Bar graphs depict the results of silver grain scans (T, M and V) and of chromatographic analysis. Total uptake (T), label in mesophyll (M) and minor veins (V) is shown for scans. Relative ^{14}C-content is shown for sucrose (S), glucose (G), fructose (F), stachyose (STA), mannitol (MAN), 3-0-methyl glucose (3-0-MG) and material insoluble in 80% ethanol (I). L-glucose in E is given by (G).*

analyzed by chromatography to find the amount of label present in major sugars (Table 2). Neither ^{14}C-stachyose nor ^{14}C-fructose were present in sufficient quantity to account for label content of the veins. Sufficient mannitol was present

TABLE 2. Comparison of ^{14}C-content of sugar beet minor veins with ^{14}C-content of major sugars present in leaf. Total content of ^{14}C-compounds in the veins, of ^{14}C-sucrose and of the ^{14}C-labeled sugar supplied are compared for a unit area of leaf after 30 min of labeling with exogenous sugars in the dark.

[14]C-sugar supplied	Total ^{14}C-content in minor veins	^{14}C-sucrose	^{14}C-sugar supplied
	ngC cm^{-2}	ngC cm^{-2}	ngC cm^{-2}
Sucrose	857 ± 8	2175 ± 22	2175 ± 22
Fructose	150	117	84
Mannitol	14	0	48
Stachyose	196	290	75

to account for the label content of the minor veins in the leaf supplied with mannitol. However, uptake of this sugar alcohol into the minor veins from the exogenous supply was very low. Of the six sugars presented, sucrose alone gave evidence of rapid uptake and of accumulation in the minor veins to a concentration several-fold higher than that in the external solution (Fondy and Geiger unpublished). It appears that neither stachyose nor fructose can be activity loaded by sugar beet. The data present evidence against Mechanism C, in which sugar availability is the major determinant of the pattern of sugars translocated. Both Mechanism A, in which translocate sugars are all derived from sucrose, or Mechanism B, in which carrier specificity is responsible for the pattern of translocate sugars, are compatible with the data.

To distinguish these mechanisms, further work needs to be done with plants that have a broad spectrum of trans-locate sugars. Preliminary studies with white ash and squash are in progress.

5. *Phloem Loading Produces Osmotic Pressure in Minor Vein Phloem.* - The process of active loading not only contributes to the spectrum of translocate sugars. Phloem loading also gives rise to high osmotic pressure in sieve tubes of the source; the resulting turgor pressure is likely to be res-ponsible for generating mass flow. For data obtained by plasmolysis studies (Geiger et al. 1973) it appears that the osmotic pressure in sieve tubes of the source is approxi-mately 1 M. Similar values have been obtained from the data

obtained by scanning autoradiographs of minor veins, by
sucrose assay and by specific activity measurements for
sugar beet source leaves (Fondy and Geiger unpublished).

There is evidence that buildup of osmotic pressure in se-
cc complexes of the minor veins of developing leaves of
sugar beet is responsible for the onset of export (Fellows
and Geiger 1974). A number of events precede cessation of
import and onset of export from a particular area of a developing sugar beet leaf. Preparatory events appear to include
changes in photosynthetic capacity of the mesophyll, opening
of sieve plates, clearing of sieve tube lumens, and the start
of phloem loading in minor veins. Plasmolysis data comparing
the osmotic pressure of sieve tubes and companion cells in
importing, transition and exporting regions of a typical
developing leaf are shown in Table 3. Because of tissue
shrinkage and continued loading during equilibration, the
measured values are likely to be higher than the values in the
intact plant.

TABLE 3. Osmotic pressure values for estimated 50% plasmolysis
of cells from tip, middle (intermediate), and base (nonload-
ing) from 42% FLL leaf and source leaf from same plant.
(Fellows and Geiger, 1974).

Vein order	Osmotic pressure values						Source leaf	
	42% FLL leaf							
	Tip loading		Middle intermediate		Base nonloading			
	cc	se	cc	se	cc	se	cc	se
					$J kg^{-1}$			
I			2420	2450	1900	1900	2390	2450
II	2550	3250	2550	2580	1900	1900	2320	2450
III	2580	3250	2390	2580	1900	1900	2320	2450
IV	2580	3250	2520	2720	1900	2320	2450	2780
V	2580	3250	2200	2580	2100	2320	2360	2780

The water potential of sieve tubes may not be the result
of osmotic pressure alone but may be the result of matric
potential also. In the plasmolysis experiments, withdrawal
of water during plasmolysis would likely cause the matric
potential to become more negative decreasing water potential.
This occurrence would result in values for osmotic pressure
which are too high if the matric potential is ignored.

However, the decrease in matric potential could well be a function of the p-protein content. Under water-stress conditions, decreased matric potential in the sieve tubes and companion cells could maintain entry of water even as the water potential of the xylem decreases". From the pattern of osmotic pressure values, it appears that when the osmotic pressure in the se-cc complexes of a developing leaf reaches a level greater than that for the source regions, import stops. A further increase in the value brings about the start of export.

In summary, phloem loading in higher plants is an active, energy-requiring process which is capable of loading sucrose and perhaps a few other sugars against a concentration gradient and at a high rate. Besides being responsible, in part, for regulating the spectrum of compounds translocated, phloem loading is also thought to generate an osmotic gradient capable of driving mass-flow.

ACKNOWLEDGEMENTS

The work reported here was supported by NSF grants GB 33808 and BMS 71-1572.

REFERENCES

Bauer, L. (1953). Zur Frage der Stoffbewegungen in der Pflanze mit besonderer Berucksichtigung der Wanderung von Fluorochromen. *Planta (Berl.) 42*, 362-451.

Christy, A.L. and Ferrier, J.M. (1973). A mathematical treatment of Münch's pressure-flow hypothesis of phloem translocation. *Plant Physiol. 52*, 531-538.

Crafts, A.S. (1951). Movement of viruses, auxins and chemical indicators in plants. *Bot. Rev. 17*, 203-284.

Crafts, A.S. and Crisp, C.E. (1971). Phloem Transport in Plants. Freeman and Co: San Francisco.

Curtis, O.F. and Scofield, H.T. (1933). Comparison of osmotic concentrations of supplying and receiving tissue and its bearing on the Münch hypothesis of the translocation mechanism. *Amer. J. Bot. 20*, 502-513.

Fellows, R.J. and Geiger, D.R. (1974). Structural and physiological changes in sugar beet leaves during sink to source conversion. *Plant Physiol. 54*, 877-855.

Fife, J.M. Price, C. and Fife, D.C. (1962). Some properties of phloem exudate collected from root of sugar beet. *Plant Physiol. 37*, 791-792.

Geiger, D.R. (1975). In "Encycl. Plant Physiol. (N.S.) 1. Transport in plants I Phloem Transport" ed. by M.H. Zimmermann and J.A. Milburn, Springer, New York. Ch 17.

Geiger, D.R., Saunders, M.A. and Cataldo, D.A. (1969). Translocation and accumulation of translocate in the sugar beet petiole. *Plant Physiol.* *44*, 1657-1665.

Geiger, D.R., Giaquinta, R.T., Sovonick, S.A. and Fellows, R.J. (1973). Solute distribution in sugar beet leaves in relation to phloem loading and translocation. *Plant Physiol.* *52*, 585-589.

Geiger, D.R., Sovonick, S.A., Shock, T.L. and Fellows, R.J. (1974). Role of free space in translocation in sugar beet. *Plant Physiol.* *54*, 892-898.

Pate, J.S., Sharkey, P.J. and Lewis, O.A.M. (1974). Phloem bleeding from legume fruits - A technique for study of fruit nutrition. *Planta (Ber.)* *120*, 229-243.

Roeckl, B. (1949). Nachweis eines Konzentrationshubs zwischen Palisadenzellen und Siebrohren. *Planta (Berl.)* *36*, 530-550.

Sovonick, S.A., Geiger, D.R. and Fellows, R.J. (1974). Evidence for active phloem loading in the minor veins of sugar beet. *Plant Physiol.* *54*, 886-891.

Tyree, M.T., Christy, A.L. and Ferrier, J.M. (1974). A simpler iterative steady state solution of Münch pressure flow systems applied to long and short translocation paths. *Plant Physiol.* *54*, 589-600.

Wanner, H. (1952). Phosphataseverteilung und Kohlenhydrattransport in der Pflanze. *Planta (Berl.)* *41*, 190-194.

Zimmermann, M.H. (1957). Translocation of organic substances in trees. I. The nature of the sugars in the sieve-tube exudate of trees. *Plant Physiol.* *32*, 288-291.

Zimmermann, M.H. (1960). Transport in the phloem. *Ann. Rev. Plant Physiol.* *11*, 167-190.

CHAPTER 17

Transfer of Sorbitol in Pear Leaf Slices

R.L. Bieleski

Plant Diseases Division, D.S.I.R., Private Bag, Auckland, New Zealand

INTRODUCTION

Sorbitol is the hexitol produced by reduction of glucose or fructose. Although it is generally uncommon in higher plants, it is present in many woody species of the family Rosaceae; and can be the major soluble carbohydrate in the leaf, and the major carbohydrate moving in the phloem (Bieleski 1969; Reid and Bieleski 1974).

In this paper, I want to consider the question, "What is the common factor that causes sucrose to be the major sugar formed in the leaf and translocated in the phloem in most species, but sorbitol to be the sugar both synthesized and translocated in the woody Rosaceae?" In my view, sorbitol is, like sucrose, a "transfer carbohydrate" : that is, it is in a form that can readily be shifted across membranes by transfer processes, and yet is blocked from metabolism. Thus it can be stored in vacuoles of the leaf cells, or loaded into the phloem by the cells at the vein ends. At the destination, transport is followed by metabolism, and sorbitol is converted to other compounds.

METHODS

I have used leaf slices of pear (*Pyrus domestica* Medik.) as a model system for studying transport of sorbitol. Strips 1 cm wide were cut, parallel to the midrib, from the blades of young, fully-expanded leaves, and then chopped into 2 mm segments. The sugar whose uptake was to be studied was generally supplied at 1×10^{-3}M, and was labelled with ^{14}C-sugar at 20 nCi/ml. Uptake was measured in terms of radioactivity appearing in the tissue, and each uptake rate was measured as the slope of the calculated line of best fit for 4 sample times (1, 2, 3 and 4 h).

When the products of uptake were to be studied, the
radioactivity of the sugar solution was 1 µCi/ml at 10^{-3}M,
and extracts were prepared and chromatographed as in Bieleski
(1969). A detailed account of these experiments is being prepared for publication elsewhere.

RESULTS AND DISCUSSION

In preliminary experiments I found that uptake rates as
measured were mot markedly affected by experimental technique,
or the time of year when leaves were taken.

Uptake of sorbitol from 10^{-3}M solutions was continuous
for at least 8 h, and at the end of this time, the concentration of ^{14}C-radioactivity in the tissue was 5 times that in
the external solution (Fig. 1). Typically, 75-85% of the
extracted radioactivity was in the form of sorbitol, so that
accumulation of sorbitol had occurred against a 4-fold gradient, without any reference to endogenous sorbitol (if this
was taken into account, the gradient surmounted was over 100-
fold). There was no change in the rate of uptake when the
concentration in the tissue reached that in the external solution. On the other hand, the presence of metabolic inhibitors decreased uptake by up to 90% (Table 1). Glucose uptake
showed similar features to sorbitol uptake.

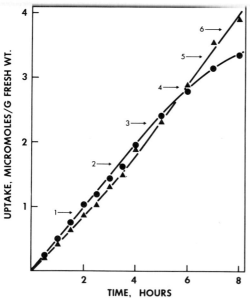

Fig. 1. - Uptake with time
into pear leaf slices of
sorbitol (●) and glucose
(▲) from 1 x 10^{-3}M solutions. Arrows mark the
stages at which the ratio
(internal:external ^{14}C concentration) had reached
the value shown. A ratio
of 5 corresponds to a 33%
depletion of the medium.

The first conclusion, therefore, is that pear leaf slices
are able to accumulate sorbitol by the operation of an active
transfer mechanism.

TABLE 1. Rates of uptake of sorbitol and glucose into pear leaf slices from 10^{-3}M solution in the presence of various compounds

Compound	Sorbitol uptake, % control rate	Glucose uptake, % control rate
10^{-4}M dinitrophenol	48	45
10^{-3}M azide	19	11
10^{-2}M mannitol	92	109
10^{-2}M fructose	53	74
10^{-2}M mannose	76	53

I then studied the effect of various sugars on the rates of sorbitol and glucose uptake (Table 1). ^{14}C-sorbitol or ^{14}C-glucose was supplied to the tissue at 10^{-3}M while the inhibiting sugar was supplied at 10^{-2}M. Uptake of the two sugars was affected to different degrees by the various sugars, but was in general, rather insensitive to the presence of other sugars. Glucose had little effect on sorbitol uptake, and sorbitol little effect on glucose uptake. Two sugar analogues known to inhibit glucose uptake in other tissues, 2-deoxyglucose and 3-0-methylglucose, had no effect on sorbitol uptake but inhibited glucose uptake in pear leaf slices by 30%.

Other treatments also differed in their effects on sorbitol and glucose uptake. Aging the tissue did not affect sorbitol uptake but stimulated glucose uptake by 35%. Darkness inhibited sorbitol uptake significantly less than glucose uptake.

The second conclusion, therefore, is that sorbitol uptake and glucose uptake, though similar, occur through different transfer mechanisms.

In other studies (Vicery and Mercer 1967), the relationship between sugar uptake rate and concentration has often been found to fit a rectangular hyperbola, suggesting the operation of a permease system. Data for sorbitol and glucose uptake in pear leaf slices failed to fit a simple hyperbola, but could be neatly expressed as the sum of two hyperbolas, as if there were two such systems working in parallel for each sugar. The basis for the co-existence of two hypothetical transfer systems is not known.

The third conclusion is that in pear leaf slices, neither sorbitol uptake nor glucose uptake occurs through the operation of a single classical permease system.

The abilities of leaf slices from different species to take up sorbitol were then compared. Those species that, like pear, contain sorbitol as a major sugar all accumulated sorbitol from 10^{-3}M solutions as readily as they did glucose. Those species that do not contain sorbitol mostly accumulated it at about one-third the rate of glucose (Table 2). The lowest value was shown by rose leaf tissue (12%).

TABLE 2. Uptake of sorbitol from 10^{-3}M solutions into leaf slices of various species. Rates are relative, being expressed as a percentage of the glucose uptake rate from 10^{-3}M solutions by that tissue

Group of species	Number of species	Sorbitol uptake % glucose rate
Sorbitol makers	8	105 ± 32
Sucrose makers	10	33 ± 11

The glucose uptake rates used as reference points ranged from 100 to 920 nmols/g fr.wt/h. The relative sorbitol uptake rate in the "sorbitol makers" group was significantly different from that in the "sucrose makers" at p = 0.001.

The fourth conclusion is that the synthetic and transfer activities of the tissue are related: tissues able to synthesize sorbitol are able to accumulate it readily. However, even tissues that contain no detectable sorbitol appear to possess accumulation mechanisms for sorbitol.

The products formed from accumulated sorbitol were compared with those formed from glucose. This was done both for pear leaf tissue (which forms sorbitol as the main product of photosynthesis) and rose leaf tissue (which forms sucrose and does not contain sorbitol). In pear leaf tissue, there was very little metabolism of the accumulated sorbitol so that over 96% of the radioactivity taken up was recovered in the soluble extract after 4 h; accumulated ^{14}C-glucose was partly metabolized to CO_2 (9% of accumulated radioactivity) and insoluble products (14%), so that less than 80% was recovered in the soluble extract. This greater metabolism of glucose was also apparent when the composition of the soluble extracts was studied (Table 3). Accumulated sorbitol remained largely in its original form, while glucose was converted to sorbitol, sucrose and other compounds. In rose leaf tissue, on the other hand, sorbitol and glucose were both metabolized, CO_2 and insoluble products were formed, and much of the ^{14}C in the extract was converted to sucrose, fructose and other compounds (Table 3).

Thus the fifth conclusion may seem paradoxical: the tissue that accumulated sorbitol readily and forms it as a major photosynthetic product does not metabolize sorbitol readily, whereas the tissue that accumulates sorbitol slowly and does not form it is able to metabolize accumulated sorbitol almost as readily as glucose.

TABLE 3. Products formed from sorbitol or glucose taken up for 4 h from 10^{-3}M solutions into pear leaf or rose leaf slices. Amounts in extracts corresponded to the following uptakes (in nmols/g fr.wt/4 h): pear, sorbitol, 1552; pear, glucose, 1608; rose, sorbitol, 392; rose, glucose, 2740

| | % total ^{14}C in alcohol extract | | | |
| | Pear leaf tissue | | Rose leaf tissue | |
Product	Sorbitol supplied	Glucose supplied	Sorbitol supplied	Glucose supplied
Sorbitol	61.2	43.1	5.8	0.4
Glucose	0.6	3.6	16.6	7.7
Fructose	0.5	8.9	33.7	24.5
Sucrose	0.3	18.8	20.3	57.2
Malate	31.6	20.8	<0.5	1.1
Other compounds	5.8	4.8	23.5	9.0

The conclusions I have drawn can be fitted into the following coherent pattern. I believe that in pear leaf tissue, entering sorbitol either fails to reach the metabolic site, or is rapidly removed from thence by active transfer processes and is segregated in a storage space, presumably the vacuole. When glucose is supplied, this reaches the metabolic site. Some is metabolized to CO_2, starch and other products, while most is converted into sorbitol, which is then transferred to the storage space as before (Fig. 2). In rose leaf, on the other hand, sorbitol is not actively transferred into storage: instead it is metabolized to fructose and glucose, so that its pattern of utilization resembles that of glucose. In this tissue, sucrose is the carbohydrate that is transferred to the storage space. I further envisage the sieve elements as being like a special sort of vacuole or storage space, so that they also receive sorbitol in the woody Rosaceae but sucrose in other plants. On this basis, sucrose when supplied to most leaf tissues should behave like sorbitol when supplied to pear - that is, be poorly metabolized. There is however a fundamental difficulty in studying sucrose transfer: active invertases present in tissue slices usually hydrolyse any sucrose supplied, so that glucose is the sugar actually transferred to the cell. For this reason, sorbitol movement may provide a suitable model system for studying sugar accumulation and phloem transport mechanisms.

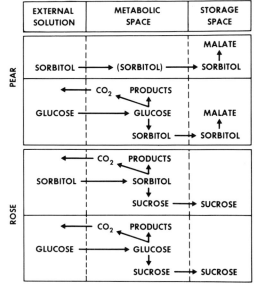

Fig. 2. - *Scheme of sugar interconversion and transfer in pear leaf and rose leaf slices.*

Other questions remain unanswered. Has sorbitol in the Rosaceae arisen during evolution as a completely new compound, or has there been a greatly increased production of a compound normally present at low concentration in most plant tissues? And why has one particular group of plants turned to a sugar other than sucrose for the purposes of accumulation and transport? At the moment I have no ready answer.

ACKNOWLEDGEMENTS

This study has been greatly helped by the excellent technical work of R.J. Redgwell.

REFERENCES

Bieleski, R.L. (1969). Accumulation and translocation of sorbitol in apple phloem. *Aust. J. Biol. Sci. 22*, 611-620.

Reid, M.S. and Bieleski, R.L. (1974). Sugar changes during fruit ripening - whither sorbitol? In: "Mechanisms of Regulation of Plant Growth" ed. R.L. Bieleski, A.R. Gerguson and M.M. Cresswell. *Bulletin 12, The Royal Society of New Zealand,* Wellington, pp. 823-830.

Vickery, R.S. and Mercer, F.V. (1967). The uptake of sucrose by bean leaf tissue. II. Kinetic experiments. *Aust. J. Biol. Sci. 20*, 565-574.

CHAPTER 18

Metabolite Transport in Leaves of C_4 Plants: Specification and Speculation

P.W. Hattersley, L. Watson and C.B. Osmond
Research School of Biological Sciences
Australian National University,
Canberra, A.C.T. 2601 Australia

INTRODUCTION

Two major symplastic transport phenomena may be identified in leaves of C_4 plants. First, *intermediates* of photosynthesis move between two chlorenchymatous cell types, one of which is specialized for carbon assimilation and the other for carbon reduction. Second, *products* of photosynthesis move from the latter, or both cell types, to the phloem. Here, we review the biochemical and anatomical bases of these phenomena at the cellular level, emphasizing major variations of C_4 leaf anatomy, introduce a new histochemical approach to these systems, and discuss the implications of structural variation for transport.

SPECIFICATION

Biochemical bases. - The C_4 pathway of photosynthesis is best defined in terms of radiotracer kinetic experiments, which demonstrate that C_4 dicarboxylic acids are both initial products and primary intermediates of $^{14}CO_2$ assimilation (Hatch 1976). Studies on cellular localization of C_4 photosynthetic enzymes, and on the complementary metabolic capacity of the two cell types isolated from leaves, clearly show that primary carboxylation of phosphoenolpyruvate is restricted to one cell type, with decarboxylation of C_4 acids and secondary carboxylation of ribulose-1,5-bisphosphate (RuP2) restricted to the other (Black 1973; Hatch and Osmond 1976). We propose to denominate these cell types on the basis of function rather than histology. Cells having the capacity for C_4 acid synthesis we call primary carbon-assimilation (PCA) cells, and those with the capacity for carbon reduction via the photosynthetic carbon reduction cycle, PCR cells. Thus mesophyll cells of C_4 plant leaves correspond to PCA cells, and chlor-

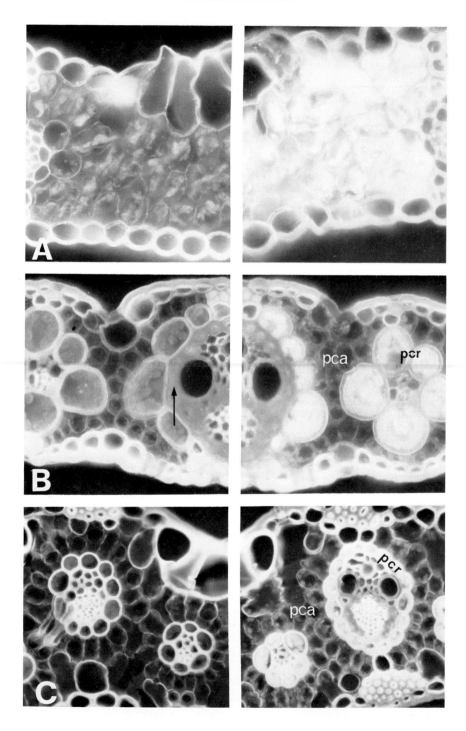

enchymatous bundle sheath, or Kranz, cells correspond to PCR cells.

Sources and sinks of photosynthetic *intermediates* may be specified at the intracellular level; e.g., for C_4 acids, the source is either the chloroplasts or cytosol of PCA cells, depending on whether malate or aspartate is transported to PCR cells. The sink is either the chloroplasts or mitochondria, depending on which of the three principal C_4 acid decarboxylase systems is found in PCR cells (Gutierrez et al. 1974; Hatch et al. 1975).

The source of photosynthetic *products* has not yet been delimited. PCR, and PCA cells, are capable of sucrose synthesis (Downton and Hawker 1973; De Fekete and Vieweg 1973; Bucke and Oliver 1975), but if PCA cells synthesize sucrose *in vivo*, sugar phosphate precursors presumably are derived from PCR cells, increasing the intricacy of intercellular coordination and transport. We assume the phloem to be the sucrose sink.

Structural bases. – An essential starting point in the structural specification of the transport pathway for intermediates is the relationship between PCA and PCR cells, ideally identified on the basis of functional properties. Further, the transport of products demands detailed analysis of the pathways between PCR and vascular tissues. The location of PCR cells is central to both problems, and we have sought to identify these cells by fluorescent antibody labelling techniques (of. Knox 1976/7/7).

We have used rabbit antiserum raised to purified RuP_2 carboxylase (anti-RuP_2C) from wheat or spinach, to locate RuP_2C in hand-cut leaf transections of both C_4 and C_3 species. Fixed material, treated with anti-RuP_2C and fluorescein isothiocyanate (FITC) labelled anti-rabbit immunoglobulin, was examined in an epifluorescence microscope. Figure 1 shows leaf transections of *Triticum aestivum* (C_3), *Panicum maximum* (C_4), and *Pennisetum villosum* (C_4) (test serum, right; normal serum control, left). As expected, conspicuously FITC-

Fig. 1. Hand transections of grass leaf blades showing fluorescent antibody labelling of RuP_2C in chloroplasts; normal serum control, left; anti-RuP_2C test, right; adaxial epidermis uppermost; pca = PCA cells; pcr = PCR cells. A. Triticum aestivum, C_3, fluorescing chloroplasts in all chlorenchymatous cells in test (control, x375; test, x460). B. Panicum maximum, C_4, XyMS+; fluorescing chloroplasts only in PCR cells in test; arrow indicates cells in the XyMS position (both, x275). C. Pennisetum villosum, C_4, XyMS-; fluorescing chloroplasts only in PCR cells in test; note absence of cells in the XyMS position in extreme right vein (both, x210).

Fig. 2. Hand transections of leaf blade of Arundinella
nepalensis *(C_4, Gramineae); adaxial epidermis uppermost; pca =*
PCA *cells; pcr =* PCR *cells; arrows indicate isolated* PCR *cells
of longitudinal interveinal strands.* Top, *normal serum control;* Bottom, *anti-RuP_2C test, showing strong FITC-fluorescence in* PCR *cells only, including those of the isolated
strands. Note weak fluorescence of* PCA *cell chloroplasts cf.
control, and absence of cells in the* XyMS *position cf. Fig.
1B. Both x450.*

---

fluorescing chloroplasts are found in all chlorenchymatous
cells in wheat, but only in a sheath of cells (PCR) around
the veins in the C_4 species. Weak fluorescence of PCA cell
chloroplasts (cf. control), which persists when anti-serum
raised to the large subunit only of RuP_2C is used, may be due
to non-specific binding.

The technique is particularly valuable for leaves not
amenable to mechanical or enzymatic separation of PCA and PCR
cells e.g., those of *Arundinella nepalensis* (Fig. 2). Here,
longitudinal strands of cells are found, resembling sheath
PCR cells, but isolated from the veins (Crookston and Moss
1973). Strand cell chloroplasts, as well as those of sheath
cells, fluoresce markedly when treated with anti-RuP_2C. Table
1 gives PCR cell locations in other C_4 species with unusual
leaf anatomy, as detected by this technique.

Panicum milioides, which exhibits "intermediate" CO_2
compensation point (Krenzer et al. 1975), was originally reported to have C_4 leaf anatomy (Downton 1971). However, recent anatomical assessment suggests it is a C_3 plant
(Hattersley and Watson 1975); a view confirmed by physiological studies (Kanai and Kashiwagi 1975), $\delta^{13}C$ value (W.V.Brown,
pers. comm.), and marked FITC-fluorescence of all chlorenchymatous cell chloroplasts. It will be interesting to apply
fluorescent antibody labelling to other species claimed to
represent intermediates between C_3 and C_4, and to artificially
produced hybrids (Boynton et al. 1971).

It is now possible to pinpoint the PCA/PCR and PCR/vas-
cular tissue interfaces in all types of C_4 leaves. For the
latter interface, a major distinction may be drawn between
two groups of C_4 grasses. The presence (XyMS+; e.g., in
Panicum maximum, Fig. 1B) or absence (XyMS-; e.g., in
Pennisetum villosum, Fig. 1C) of essentially chloroplast-free
cells, in the mestome sheath position between metaxylem
vessels and laterally adjacent PCR cells, seems to be correlated with the type of decarboxylase system in the PCR cells
(Hattersley and Watson 1976).

Concerning PCA/PCR intercellular transport, it has been
shown, in a quantitative assessment of the chlorenchyma of

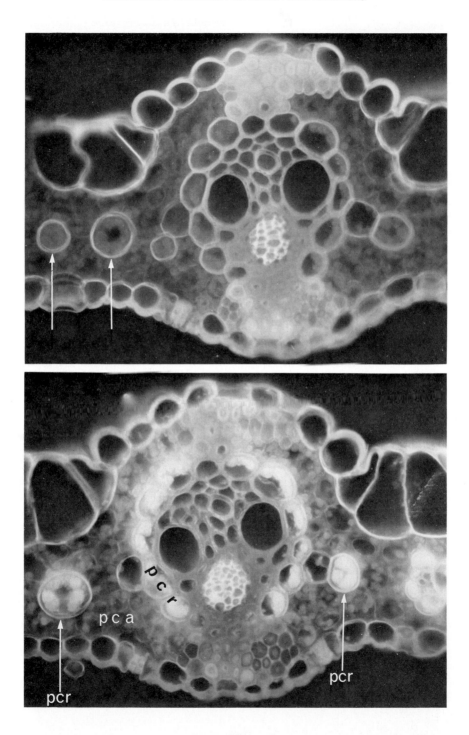

grass leaves, that no typical PCA cell is ever more than one
other PCA cell from its nearest PCR cell (Hattersley and
Watson 1975). Indeed, most PCA cells are in direct contact
with PCR cells. Such relationships have not been examined in
other families, and indeed would be more difficult to assess
in reticulately-veined leaves.

SPECULATION

Transport of photosynthetic intermediates. - Measured
rates of net carbon assimilation in C_4 plants imply rapid bidirectional movement of intermediates between PCA and PCR
cells. The path taken is presumably symplastic, since plasmodesmata provide a lower resistance pathway than two plasma-
membranes and a cell wall (Tyree 1970; Gunning and Robards,
this volume). Water flux, about 300 times greater than the
metabolite exchange rate, is assumed to be apoplastic.

Two simple treatments of PCA/PCR symplastic metabolite
exchange have been proposed. On one hand, plasmodesmata are
regarded as open pores occupying approximately 2% of the PCA/
PCR cell interface (Osmond 1971). This estimate is supported
by measurements of plasmodesmatal frequency and pore radii in
Salsola kali (Oleson 1975) and *Zea mays* (O'Brien, pers. comm.).
The pathlength for C_4 acid diffusion is taken as the distance
between the mean radii of the source (chloroplasts or cytosol
of PCA cells) and sink (chloroplasts or mitochondria of PCR
cells). In simple systems, with adjacent PCA and PCR cells,
the pathlength ranges from about 10 µm in *Zea mays* to about
50 µm in *Amaranthus edulis* or *Atriplex spongiosa* (Osmond 1971;
Hatch and Osmond 1976). Even in a complex system (e.g.
Fimbristylis: Table 1), the general pathlength is only 40 µm,
with a maximum of 65 µm. However, two or more sets of plasmodesmata must be traversed.

On the other hand it is assumed that the transport pathway is restricted to plasmodesmatal desmotubules (which occupy
only 0.1% of the PCA/PCR interface: Oleson 1975), that resistance to diffusion is restricted to the length of plasmodesmata, and that metabolites are perfectly mixed in the sources
and sinks. This pathway would involve loading of a membrane
system contiguous with the desmotubule (endoplasmic
reticulum?); a process currently defying speculation.

Osmond and Smith (1976) calculated that in either model,
gradients of 10 mM between PCA and PCR cells would be sufficient to sustain diffusive metabolite fluxes compatible with
net photosynthetic rates. Such gradients agree with independently derived estimates of the concentrations of C_4 intermediates (Osmond 1971; Hatch and Osmond 1976). Hatch (1971)
estimated that C_4 acid decarboxylation produces "total CO_2"
concentrations of the order of 1 mM in PCR cells. The back

TABLE 1. Site of PCR cells in leaf blades of C_4 species exhibiting unusual C_4 leaf anatomy, as detected by fluorescent antibody labelling of RuP_2C in chloroplasts

Species	Location of PCR cells
Gramineae:	
1). *Alloteropsis semialata*	innermost of the two recognizable sheaths surrounding the veins (cf. Ellis 1974; C_4 form)
2). *Aristida biglandulosa*	both recognizable sheaths surrounding the veins (cf. Brown 1975)
3). *Triodia pungens*	the sheath associated with veins and its extensions (cf. Hattersley and Watson 1976)
Cyperaceae:	
4). *Fimbristylis dichotoma*	the interrupted sheath, which lies between vascular tissue and the two recognizable complete sheaths (cf. Brown 1975)
Chenopodiaceae:	
5). *Salsola kali*	innermost of the two chlorenchymatous cell layers which surround the entire leaf (cf. Olcson 1974; Carolin et al. 1975).

flux of CO_2 to PCA cells may only be 10% of that of the metabolite flux (Hatch and Osmond 1976), and a tight stoichiometry between PCA and PCR cell reactions is probable.

Two implications of this discussion for C_4 plants with more complex leaf anatomy could be suggested. If the more tortuous transport pathway in species of genera such as *Aristida, Alloteropsis,* and especially *Fimbristylis,* limits metabolite transfer, we might predict larger pool sizes of photosynthetic intermediates than found in *Zea mays* for example. Alternatively, such species may photosynthesize at rates lower than normally found in C_4 plants.

Transport of photosynthetic products. - Export of [14]C-

labelled photosynthetic products and estimated total assimilate export are faster in leaves of C_4 plants than in C_3 plants
(Hofstra and Nelson 1969; Lush and Evans 1974; Gallaher et al.
1975). Generally, the average pathlength between source and
sink is shorter in C_4 plants, as reflected by lower interveinal
distances (Chonan 1972), and fewer plasmodesmata must be traversed, as indicated by lower lateral cell counts (Crookston
and Moss 1974; Hattersley and Watson 1975). C_4 plants do not
necessarily have more phloem per leaf cross-section (Lush and
Evans 1974; cf. Gallaher et al. 1975), but the key relationship of phloem area/chlorenchyma area remains to be assessed.
Our speculations on sucrose transport will be confined to
discussing likely intercellular pathways between chlorenchyma
and phloem.

If PCA cells synthesize sucrose *in vivo*, the pathway of
required sugar phosphate precursors, from PCR to PCA cells,
would be the same as for other metabolites. Sucrose from PCA
cells may take a similar path to PCR cells, *en route* to the
phloem, since in all types of C_4 leaf anatomy, PCR cells lie
between PCA cells and the veins. (An exception is the
Arundinella type, where a minor proportion of PCR cells are
embedded in PCA tissue.) Such positioning means that if only
PCR cells synthesize sucrose *in vivo*, the pathlength from
source to sink is shorter than if PCA cells are also involved
and thus still shorter than in C_3 plants.

There are several possible pathways for sucrose transport from PCR tissue to phloem. In leaves of XyMS+ C_4 plants
(e.g., *Panicum maximum*: Fig. 1B), sucrose from adaxial PCR
sheath cells may move radially, through cells in the mestome
sheath (MS) position, to the cells abutting upon the phloem.
Such a pathway is analogous to that suggested in wheat (Kuo
et al. 1974). Two XyMS+ C_4 plants examined by Lush and Evans
(1974) had lower phloem areas per leaf catchment area than
two XyMS- plants also examined, suggesting that cells in the
MS position may compensate for reduced phloem area. However,
the XyMS distinction in C_4 grass leaves refers only to primary
vascular bundles. In smaller bundles, as well as in *all*
bundles of XyMS- plant leaves (Figs. 1C, 2), sucrose from
adaxial PCR cells presumably moves radially through adjacent
PCR cells to those abutting upon the phloem. Plasmodesmata
occur in radial PCR cell walls of *Zea mays* (O'Brien and Carr
1970), but ignorance of plasmodesmatal frequency and distribution in PCR cells is substantial.

Transport of sucrose between PCR cells is also likely in
species where many PCR cells are not associated with vascular
tissue (e.g., *Triodia pungens, Salsola kali*: Table 1). In
leaves where the PCR sheath is interrupted by metaxylem
vessels (e.g., *Fimbristylis dichotoma*: Table 1), direct radial

transport from adaxial to abaxial PCR cells cannot occur, and the path may be through the sheath of essentially chloroplast free cells found external to the PCR sheath in such species. *Alloteropsis semialata* (Table 1) possesses a similar sheath, but the PCR sheath is complete.

Perhaps the greatest distances between PCR cells and phloem are encountered in C_4 leaf anatomy of the *Arundinella* type (Fig. 2). In *A. hirta*, there are no plasmodesmata between adjacent PCR cells of an isolated strand (Crookston and Moss 1973), and sucrose from these cells presumably moves through a symplast as extensive as that in some C_3 plants. In *Garnotia stricta*, there may be six such PCR strands between adjacent veins, with an interveinal distance up to 280 μm.

Comparison of assimilate translocation in different forms of C_4 leaf anatomy may help to specify which structural features influence rates of assimilate transport to phloem loading sites.

ACKNOWLEDGEMENTS

We thank Dr. G.H. Lorimer, Ms. E. Marchant, and Dr. N.E. Stone for help in preparation of wheat RuP_2C, Dr. T. Akazawa for antisera to spinach RuP_2C, and Mr. I. Oliver for material of *Panicum milioides*.

REFERENCES

Boynton, J.E., Nobs, M.A., Björkman, O. and Pearcy, R.W. (1971). Hybrids between *Atriplex* species with and without β-carboxylation photosynthesis. Leaf anatomy and ultrastructure. *Carnegie Inst. Washington, Yearb. 69*, 629–632.

Black, C.C. Jr. (1973). Photosynthetic carbon fixation in relation to net CO_2 uptake. *Ann. Rev. Plant Physiol. 24*, 253–286.

Brown, W.V. (1975). Variations in anatomy, associations, and origins of Kranz tissue. *Amer. J. Bot. 62*, 395–402.

Bucke, C. and Oliver, I.R. (1975). Location of enzymes metabolizing sucrose and starch in the grasses *Pennisetum purpureum* and *Muhlenbergia montana*. *Planta (Berl.) 122*, 45–52.

Carolin, R.C., Jacobs, S.W.L. and Vesk, M. (1975). Leaf structure in Chenopodiaceae. *Bot. Jahrb. Syst. Pflanzengesch. Pflanzengeogr. 95*, 226–255.

Chonan, N. (1972). Differences in mesophyll structures between temperate and tropical grasses. *Proc. Crop Sci. Soc. Japan 41*, 414–419.

Crookston, R.K. and Moss, D.N. (1973). A variation of C_4 leaf anatomy in *Arundinella hirta* (Gramineae). *Plant Physiol.* *52*, 397–402.

Crookston, R.K. and Moss, D.N. (1974). Interveinal distance for carbohydrate transport in leaves of C_3 and C_4 grasses. *Crop Sci. 14*, 123–125.

De Fekete, M.A.R. and Vieweg, G.H. (1973). Synthesis of sucrose in *Zea mays* leaves. *Ber. Deutsch Bot. Ges. 86*, 227–231.

Downton, W.J.S. (1971). Check list of C_4 species. In: "Photosynthesis and Photorespiration" ed. by M.D. Hatch, C.B. Osmond and R.O. Slatyer, Wiley-Interscience, New York pp. 554–558.

Downton, W.J.S. and Hawker, J.S. (1973). Enzymes of starch and sucrose metabolism in *Zea mays* leaves. *Phytochemistry 12*, 1551–1556.

Ellis, R.P. (1974). The significance of the occurrence of both Kranz and non-Kranz leaf anatomy in the grass species *Alloteropsis semialata*. *S. Afr. J. Sci. 70*, 169–173.

Gallaher, R.N., Ashley, D.A. and Brown, R.H. (1975). ^{14}C-photosynthate translocation in C_3 and C_4 plants as related to leaf anatomy. *Crop Sci. 15*, 55–59.

Gutierrez, M., Gracen, V.E. and Edwards, G.E. (1974). Biochemical and cytological relationships in C_4 plants. *Planta (Berl.) 119*, 279–300.

Hatch, M.D. (1971). The C_4 pathway of photosynthesis. Evidence for an intermediate pool of carbon dioxide and the identity of the donor C_4-dicarboxylic acid. *Biochem. J. 125*, 425–432.

Hatch, M.D. (1976). Photosynthesis: the path of carbon. In: "Plant Biochemistry" ed. by J. Bonner and J. Varner, Academic Press, New York (in press).

Hatch, M.D., Kagawa, T. and Craig, S. (1975). Subdivision of C_4 pathway species based on differing C_4 acid decarboxylating systems and ultrastructural features. *Aust. J. Plant Physiol. 2*, 111–128.

Hatch, M.D. and Osmond, C.B. (1976). Compartmentation and transport in C_4 photosynthesis. In: "Intracellular Transport and Interactions among Cell Compartments" ed. by C.R. Stocking and U. Heber, Encyclopedia of Plant Physiology (New Series), Springer, Berlin (in press).

Hattersley, P.W. and Watson, L. (1975). Anatomical parameters for predicting photosynthetic pathways of grass leaves: the 'maximum lateral cell count' and the 'maximum cells distant count'. *Phytomorphology 25*, 325–333.

Hattersley, P.W. and Watson, L. (1976). C_4 grasses: an anatomical criterion for distinguishing between NADP-malic enzyme species and PCK or NAD-malic enzyme species. *Aust. J. Bot. 24*, 297-308.

Hofstra, G. and Nelson, C.D. (1969). A comparative study of translocation of assimilated ^{14}C from leaves of different species. *Planta (Berl.) 88*, 103-112.

Kanai, R. and Kashiwagi, Misako. (1975). *Panicum milioides*, a Gramineae plant having Kranz leaf anatomy without C_4-photosynthesis. *Plant Cell Physiol. 16*, 669-679.

Knox, R.B. (1972/73). Localization of proteins in plant cells by immunofluorescence. *Zeiss Inform. 20*, 52-55.

Krenzer, E.G. Jr., Moss, D.N. and Crookston, R.K. (1975). Carbon dioxide compensation points of flowering plants. *Plant Physiol. 56*, 194-206.

Kuo, J., O'Brien, T.P. and Canny, M.J. (1974). Pit-field distribution, plasmodesmatal frequency, and assimilate flux in the mestome sheath cells of wheat leaves. *Planta (Berl.) 121*, 97-118.

Lush, W.M. and Evans, L.T. (1974). Translocation of photosynthetic assimilate from grass leaves, as influenced by environment and species. *Aust. J. Plant Physiol. 1*, 417-431.

O'Brien, T.P. and Carr, D.J. (1970). A suberized layer in the cell walls of the bundle sheath of grasses. *Aust. J. Biol. Sci. 23*, 275-287.

Oleson, P. (1974). Leaf anatomy and ultrastructure of chloroplasts in *Salsola kali* L. as related to the C_4-pathway of photosynthesis. *Bot. Notiser 127*, 352-363.

Oleson, P. (1975). Plasmodesmata between mesophyll and bundle sheath cells in relation to the exchange of C_4-acids. *Planta (Berl.) 123*, 199-202.

Osmond, C.B. (1971). Metabolite transport in C_4 photosynthesis. *Aust. J. Biol. Sci. 24*, 159-163.

Osmond, C.B. and Smith, F.A. (1976). Symplastic transport of metabolites during C_4-photosynthesis. In: "Intercellular Communication in Plants: Studies on Plasmodesmata" ed. by B.E.S. Gunning and A.W. Robards, Springer, Berlin. pp. 229-241.

Tyree, M.T. (1970). The symplast concept. A general theory of symplastic transport according to the thermodynamics of irreversible processes. *J. Theor. Biol. 26*, 181-214.

CHAPTER 19

Transfer of Ions and Products of Photosynthesis to Guard Cells

K. Raschke

*MSU/ERDA Plant Research Laboratory, Michigan State University.
East Lansing, Michigan 48824, U.S.A.*

TRANSPORT AND STOMATAL ACTION

Stomata have the task of admitting CO_2 to the leaf when conditions favorable for photosynthesis prevail. Simultaneously, stomata have to watch the water balance of the plant; they have to interrupt the connection between the intercellular spaces and the atmosphere when water loss exceeds water uptake. This dual task is accomplished by several feedback loops between the plant body and the guard cells of the stomata. Stomata open and admit CO_2 when the intercellular $[CO_2]$ declines; they close again when a rising $[CO_2]$ indicates satisfaction of the requirement for CO_2. Open stomata close also in response to water stress. This response can either be a purely physical one: a loss of turgor of the guard cells as a consequence of a lowered water potential of the leaf, or it is a metabolic response to abscisic acid (ABA). ABA is rapidly formed in leaf tissue under stress and then swept, presumably by the transpiration stream, to the guard cells where it causes loss of solutes and deflation. In the last case, transport of a phytohormone is involved in the moderation of water loss by the stomata.

In turn, transport in the plant is affected by stomatal action. Admission of CO_2 into the leaves determines the rate of assimilation of CO_2 and, in consequence, the strength of the sources of carbohydrates. The degree of stomatal moderation of the transpiration stream determines magnitude and, in some cases, even the direction of gradients of water potential in the plant. This has consequences for transport in the phloem. The pressure gradient in the sieve tubes depends not only on the distribution of osmotic pressures but also on that of the water potentials in the surrounding tissue. For instance, a low water potential in the source or conducting regions of the phloem will reduce translocation

while a low water potential in the discharge region could
enhance the rate of transport. Stomatal opening may thus have
opposing effects on the export of assimilates from a source
leaf: The increased availability of CO_2 increases source
strength; the reduced water potential lowers the pressure in
the source region of the phloem (if the osmotic pressure in
the phloem stays the same).

After this brief glance at a few interactions between
long distance transport in plants and stomatal action I wish
to concentrate on the mechanism by which stomata moderate the
gas exchange of leaves. The functioning of the stomatal
apparatus is in itself of interest to the participants in this
workshop, because the opening and closing of the stomatal pore
results from a transfer of matter between the guard cells and
the tissue surrounding them. Possibly also the underlying
tissue communicates with the guard cells. Transport over
short distances happens to be the mechanism by which gas
exchange and the transpiration stream through the plant are
moderated, in some cases perhaps even regulated.

I wish to proceed in the following way. First, I shall
estimate the amounts of water transferred into guard cells
during stomatal opening. Then, I intend to identify the
solutes involved in bringing about this water movement.
Thirdly, I shall speculate on how these processes might be
regulated. Finally, I shall turn to the question whether
guard cells are self-sufficient with respect to the supply of
carbon compounds and energy, or whether they have to depend on
imported fuel. Since there is a recent review available on
stomatal action (Raschke 1975a), I shall not try to be complete
in covering the subject nor in giving the references.

TRANSFER OF INORGANIC IONS TO GUARD CELLS

CHANGES IN VOLUME AND SOLUTE CONTENT

Volume changes. - Stomatal opening results from an inflation of the guard cells. The lumina of a pair of guard cells
of a closed stoma of *Vicia faba* have on the average a volume
of 6.3×10^{-12} ℓ; it can vary widely. This volume doubles
when stomata open from an aperture of 1 μm to one of approximately 16 μm. Within the accuracy of the measurements, the
relationship between stomatal aperture and guard cell volume
was found to be linear in *V. faba*. Indirect evidence indicates
that this might be true for other species too. With this
information and knowing the number of stomata per mm^2 as well
as the average thickness of epidermal cells one concludes
that about 4×10^{-10} ℓ of water has to move in 1 mm^2 of epidermis from the epidermal cells to the guard cells in order to

obtain a stomatal opening of 16 μm. This amount is very small; it constitutes approximately 2% of the epidermal volume. In grasses, the relative amounts of water transferred within the epidermis during stomatal movement are probably still smaller.

Changes in solute content. - The increase in guard cell volume is produced by osmosis. During stomatal opening the solute content of the guard cells increases, the guard cells take up water; the turgor rises. Unexpectedly, the relationship between the amount of solute required to produce a unit increase in guard cell volume and stomatal aperture did not increase with stomatal aperture. Sawhney and Zelitch (1969) as well as Turner (1973) found a linear relationship between aperture and solute content. If guard cells behaved like most other plant cells their Höfler diagram should show a concave turgor curve; this was not the case. Determinations of the relationship between turgor and volume of guard cells were made on epidermes of *V. faba* (Raschke, Dickerson and Pierce 1973). It was found that the walls of the guard cells gave i.e. the elastic modulus of the cell walls decreased with increasing cell volume. The increment of turgor needed to produce a unit increment in cell volume decreased with stomatal aperture. The relationship between guard cell turgor, p, volume modulus of elasticity, ε, and the relative increase in guard cell volume, $v=(V-V_o)/V_o$, (with V=guard cell volume; V_o=guard cell volume at zero turgor) could be described by $p=\varepsilon v^n$; n was found to have the value 0.67. An n of 0.67 happens to result in a nearly linear relationship between solute content, s (in osmoles), and volume increase, v, of guard cells (if $0<v<1$, which is normally the case). Linearity would be even better approached with n=0.5. This can be shown by deriving an expression for ds/dv and inserting particular values. The osmotic pressure of guard cells is $\pi=sRT/\{V_o(1+v)\}$, with R= gas constant and T=absolute temperature. If we assume the water potential ψ to be zero (in order to simplify matters) then $p=\pi$ and $ds/dv=\{\varepsilon V_o/RT\}\{nv^{n-1} + (n+1)v^n\}$. With n=0.67, the second parenthesis has the value of 1.72 for v=0.2 and 1.97 for v=0.6; with n=0.5 the values are 1.79 and 1.81 for v=0.2 and v=0.6 respectively. In conclusion, the amount of osmoticum transferred to guard cells or produced in them during stomatal opening is in first approximation proportional to the increase in stomatal aperture. This conclusion receives support from the observation that the rates of stomatal opening and solute increase in guard cells were well correlated in *Zea mays* and *Vicia faba* (Raschke and Fellows 1971; and unpublished). The measured rates of solute increase were as high as 10^{-15} osmoles min^{-1} in a pair of guard cells of *Z. mays* which corresponds to roughly 220 µosmole per g fresh weight per hour. Over short times (2 min) the change in solute content was ten times as fast. The osmotic pressure of inflated guard cells can be as high as 50 atm.

TRANSFER OF POTASSIUM AND CHLORIDE

The solutes involved in stomatal movement are salts of potassium (an exception will be mentioned later). This knowledge is based on the work of Imamura (1943), Fujino (1959, 1967), and Fischer (1968). The quantities of K^+ migrating into guard cells during stomatal opening were estimated to be sufficient to bring about the observed changes in osmotic pressure (Fischer and Hsiao 1968, Sawhney and Zelitch 1969). This contention was proven correct by the simultaneous determinations of osmotic pressure, volume and elemental content (with an electron microprobe analyzer) of guard cells of *Vicia faba* (Humble and Raschke 1971). Potassium was the only cation taken up in large quantity by the guard cells of opening stomata. The final $[K^+]$ was 0.9 eq ℓ^{-1}, when the stomatal aperture was 12 µm. The content of the guard cells of Cl, N, P and S changed only little during stomatal opening. The positive charges of K^+ must therefore have been balanced in the guard cells by organic anions. From a determination of the osmolarities of various K salts of organic acids it was concluded that the counter ion for K^+ probably was malate. The import of K^+ into guard cells, together with an import or production of malate in the guard cells, was found sufficient to account for the observed changes in osmotic pressure and cell volume.

Specificity for K^+. - All plant species tested so far (more than 50) use K^+ for the operation of their stomata under natural conditons (Dayanandan and Kaufman 1975; Willmer and Pallas 1973). However, one exception has been reported: the halophyte *Cakile maritima* does not exclude Na^+ from uptake and its guard cells contain Na^+ (Y. Waisel, personal communication). The specificity for K^+ is less pronounced if epidermal strips are floated on solutions of alkali ions. It does not come as a surprise that stomata opened as widely on solutions containing 1 meq ℓ^{-1} of Rb^+ as they did on solutions of equal concentrations of K^+ (Humble and Hsiao 1969) but guard cells could also use Li^+, Na^+ and Cs^+. The relationship between ion concentration and stomatal aperture was however different in the case of the latter group of ions. In the light, K^+ and Rb^+ produced maximal opening at 1 meq ℓ^{-1}. Of the other alkali ions, 100 meq ℓ^{-1} were needed to enable the stomata to open fully, in the light and also in darkness. Lower concentrations produced proportionally smaller apertures (Humble and Hsiao 1969). The difference between responses to K^+ and Rb^+ on one hand and to Li^+, Na^+ and Cs^+ on the other occurred only in the light; in darkness stomatal preference for K^+ and Rb^+ disappeared. It seems as if two paths exist for the uptake of alkali ions into guard cells. One allows all alkali ions to enter by diffusion (through a Donnan space?), the other one is selective for K^+ and Rb^+ and is combined with a direct or indirect activation of ion uptake by light. The analog of K^+, thallium ion, can also follow the latter path (Pallaghy 1972).

Unfortunately, the findings of Humble and Hsiao (1969) on the preference of guard cells for K^+ and Rb^+ appears not to be of general applicability because under certain conditions, not all of which are known, stomata are able to use Na^+ to the same extent as K^+ and open as widely as with K^+. Willmer and Mansfield (1969) reported this for *Commelina communis* and Pallaghy (1970) for *Vicia faba*. Pallaghy showed that the presence of Ca^{++} reduced the use of Na^+. Microprobe measurements of Raschke and Pierce (1973) confirmed this result: Guard cells of *V. faba* took up as much Na^+ from a 10mM solution of NaCl as they absorbed K^+ from a KCl solution of equal concentration. In the presence of 0.1 mM $CaCl_2$ the uptake of Na^+ was smaller and that of K^+ larger than in the absence of Ca^{++}. In a further (unpublished) experiment, Na^+ and K^+ were offered simultaneously in several ratios of concentration, with the addition of Ca^{++} and without. Guard cells preferred to use K^+, even if Na^+ was offered in excess. In this experiment, Ca^{++} had no effect on the K^+/Na^+ ratio in the guard cells. The role of Ca^{++} remains thus unclear. But obviously, guard cells can be in a state in which they are able to use Na^+ instead of K^+. One should therefore expect to find Na^+ not only in guard cells exposed in the laboratory to solutions containing an excess of Na^+ but also in guard cells sampled in the field from halophytes like *Cakile maritima* which do not exclude Na^+ but allow it to enter.

The source of K^+. - Bernstein's (1971) experiments indicate that the apoplast of leaves from non-saline plants may contain between 1 and 5 meq K^+ ℓ^{-1}. If these values would also apply to the epidermis and the guard cells are able to scavenge all the K^+ from the epidermal apoplast, then the volume of the epidermal cell walls must at least be 100 times larger than the volume of the guard cells. Using data obtained on *Vicia faba* I estimate the apoplast of the epidermis to be too small by a factor between 10 and 20 to store the K^+ needed for stomatal opening. I reach a similar conclusion if I base the estimation on the ion exchange capacity of onion cell walls as determined by Somers (1973). Admittedly, epidermal cell walls may have a larger ion exchange capacity than mesophyll walls (the cell walls of the epidermis stain with ruthenium red, indicating anionic sites); nevertheless it is likely that some of the K^+ needed by the guard cells comes from the cytoplasm and the vacuoles of the epidermal cells. In several species there are epidermal cells which morphologically cannot be distinguished from other epidermal cells but which stand out after staining for K^+, or show strong K_α fluorescence of the element K under the electron beam of the microanalyzer (Raschke and Fellows 1971; Willmer and Pallas 1973). It is not certain, but likely that the K^+ of these "storage cells" is available to the guard cells. The probability that these "ion stores" can serve the guard cells is high where they are in immediate contact with the guard cells, as is the case, for instance, in ferns and some legumes.

The source of K^+ is quite certain in grasses, particularly in *Zea mays*. For this species it has been shown that the K^+ (and Cl^-) taken up during stomatal opening comes from the subsidiary cells and returns to them during stomatal closure; the absolute content in K^+ (and Cl^-) of a stomatal complex does not change during stomatal movement. The transfer of ions is between guard and subsidiary cells only (Raschke and Fellows 1971).

The function of the subsidiary cells of plants not belonging to the family Gramineae is less clear. The variation of ion content of the subsidiary cells of *Commelina communis* indicates that these cells may function as ion channels, and not as stores (Willmer and Pallas 1973). See however W.G. Allaway's estimates in this volume.

Uptake of Cl^-. - Electroneutrality must be maintained during the transfer of K^+ between guard cells and the neighboring tissue. This requirement could be met if anions accompanied the K^+; a good candidate would be Cl^-. The importance of Cl^- as an associate of K^+ varies greatly between species, even within one species. When epidermal strips were taken from leaves of *Vicia faba* and analyzed with the electron microprobe not more than 5% of the K^+ was balanced by Cl^-, the rest by organic anions (Humble and Raschke 1971). In recent (unpublished) experiments with epidermes of the same species floating on solutions up to 73% of the K^+ was neutralized by Cl^-. We suspected that this high participation of Cl^- was caused by the absence of CO_2 in the experiment. But repetition of the experiment with sufficient CO_2 in the atmosphere as a substrate for the production of malate did not change the result. I do not know what determined the degree of participation of Cl^- in the operation of the stomatal mechanism of *Vicia faba*.

The involvement of Cl^- in stomatal movement seems to be

predictable in *Zea mays*. Cl^- shuttled rapidly between subsidiary cells and guard cells. On the average, 40% of the K^+ was accompanied by Cl^-; in a few individual stomata the balance was found to be complete (Raschke and Fellows 1971).

Exchange of K^+ for H^+. - The question now arises whether the organic anions balancing the K^+s in the guard cells which are not associated with Cl^- are imported or are made in the guard cells. Experiments were conducted in which K^+ was offered to guard cells in combination with non-absorbable anions. Stomata opened as widely as when K^+ was offered in combination with Cl^- (Raschke and Humble 1973). The guard cells therefore must have been able to make organic acids while they imported K^+. Since electroneutrality had to be maintained and since it is known that the pH in guard cells rises during stomatal opening H^+ had to be released while K^+ was taken up. It was possible to titrate the amount of H^+ excreted by guard cells; its order of magnitude was equal to that of the K^+ (and Na^+) taken up in exchange (Raschke and Humble 1973; Raschke and Pierce 1973). The fungal toxin fusicoccin is known to stimulate the excretion of H^+ from plant tissue (Marrè et al. 1974). Fusicoccin did enhance stomatal opening and accumulation of K^+ in guard cells (Turner 1973). The accelerated uptake of K^+ was correlated with an increased H^+ expulsion (our unpublished experiments with epidermes of *Vicia faba*, *Commelina communis*, and *Tulipa gesneriana*).

METABOLISM OF ORGANIC ACIDS
Identity of the organic counter ion for K^+. - The determination of the osmotic balance in guard cells of *V. faba* indicated that malate was probably the main anion there (Humble and Raschke 1971). Then Allaway (1973) demonstrated that there was nearly no malate in epidermis of *V. faba* if the stomata were closed but an amount sufficient to balance 50% of the measured K^+ when the stomata were open. In view of the possible presence of Cl^- as second important anion this was a satisfying result. Meanwhile other laboratories have confirmed that epidermal samples with open stomata contain more malate than samples with closed stomata and that the difference in malate content agrees with the estimated or observed differences in K^+ content.

Malate metabolism in epidermis. - Epidermal extracts possess phosphoenolpyruvate (PEP) carboxylase activity in proportion to the number of guard cells present (Willmer, Pallas and Black 1973). The malate in the guard cells therefore arises most probably from the carboxylation of PEP derived from starch. This view received strong support through the experiments of Willmer and Dittrich (1974). They fed ^{14}C-labelled bicarbonate to epidermes of *Commelina communis* and *Tulipa gesneriana*. The label appeared mainly in malate and aspartate. Similar results were obtained when $^{14}CO_2$ was offered; the labelling pattern obtained in the light was virtually identical with that obtained in the dark (Dittrich and Raschke, unpublished). Epidermal malate could also be made radioactive by feeding ^{14}C-labelled sugars to epidermal strips. However, not all of the radioactivity of the sugars appeared in malate. A substantial portion was incorporated into sugars other than the offered one and into starch. Autoradiograms showed that starch was formed almost exclusively in the guard cells. Carbohydrates became radioactive also after floating epidermal strips on a solution of ^{14}C-malate, indicating that gluconeogenesis can take place in the epidermis. It can be concluded, that the malate required to balance the positive charges of K^+ in the vacuole of guard cells is made in the guard cells from carbohydrates, including starch, and CO_2. The guard cells possess the machinery to convert malate into sugars when the content of the vacuole is released. It is not yet clear how much of the vacuolar malate follows this path during stomatal closing.

IGNORANCE ABOUT THE CONTROL OF ION TRANSFER AND METABOLISM

We do not know which mechanism triggers the uptake of K^+ and Cl^- into guard cells and the production of malic acid, whether a modification of an ubiquitous turgor regulator in plant cells is used or not. We do not know which partial process determines the rates of the other processes. The enhancement of potassium import and of stomatal opening by fusicoccin points to proton expulsion or a K^+/H^+ exchange mechanism as the primary mechanism of stomatal opening. But this need not be true. We know that stomatal opening can be induced by high external concentrations of alkali ions (Humble and Hsiao 1969). Does this mean that the import of alkali ions limits the rate of guard cell inflation and that acid production and pumping of malate (and Cl^-) into the vacuole are driving stomatal opening but are curbed by the availability of alkali ions in the cytoplasm? Active uptake of K^+ (and Na^+?) through the plasmalemma would then control the rate of stomatal opening without supplying all the energy for it. If this is the case we could explain the nonspecific enhancement of stomatal opening by high external concentraions of alkali ions (Humble and Hsiao 1969) as a result of an increased diffusion into the guard cells.

Reversal of stomatal opening.- I can only speculate on the series of events leading to stomatal closure. I first have to resolve the problem of how CO_2 causes stomatal closure although it is substrate for the production of malate which is needed for stomatal opening. As described with more detail elsewhere (Raschke 1975a, b) I envisage the concentrations of H^+ and malate in the cytoplasm to control direction and magnitude of the fluxes into the vacuole and out of it; low pH and high malate content of the cytoplasm lead to a release of the vacuolar content. The concentrations of H^+ and malate in the cytoplasm are determined by the balance between formation of malate, deacidification and the removal of malate into the vacuole and of H^+ out of the cell. At high levels of $[CO_2]$ more acid will be formed than removed; this stops and ultimately reverses stomatal opening. Experiments conducted on *Xanthium strumarium* indicate that stomatal closure in response to ABA could perhaps be explained as resulting from the operation of the same mechanism (Raschke 1975b). Abscisic acid possibly inhibits the expulsion of H^+ from guard cells and thereby induces an acidification of the cytoplasm. I must emphasize that there is no direct evidence for this hypothesis.

TRANSFER OF PRODUCTS OF PHOTOSYNTHESIS TO GUARD CELLS

Willmer and Dittrich (1974) found that the main product of CO_2 fixation by guard cells of *Commelina communis* and *Tulipa gesneriana* was malic acid, in the light as well as in darkness. This work was continued by Dittrich and Raschke (unpublished). Epidermal samples were inspected with a microscope before use and those with adhering mesophyll chloroplasts were excluded from the experiments. After exposure to $^{14}CO_2$, such samples did not contain phosphoglyceric acid or phosphorylated sugars, irrespective of whether the experiments were conducted in darkness or in the light. Guard cells of *C. communis*, and very likely also those of *T. gesneriana* were unable to reduce CO_2 by photosynthesis, in spite of the presence of chloroplasts. The earlier suggestion (Raschke 1975a) that guard-cell chloroplasts may serve as CO_2 scrubbers in the light needs to be corrected. We concluded that guard cells must depend on import in order to obtain carbohydrates and energy. We tried to demonstrate this dependence in a series of additional experiments with leaves and epidermes of *C. communis* and *T. gesneriana*.

Epidermis was separated from leaves of *C. communis* and exposed to $^{14}CO_2$ for 10 min while resting on a film of water. Detached leaves were exposed along with the epidermal samples. After exposure, epidermal samples were pulled off the leaves and, like the initially isolated epidermal tissue, were washed, killed and extracted. The epidermis in contact with

the mesophyll during the exposure to $^{14}CO_2$ contained 17 times
more labelled material than the epidermis exposed in isolation. Apparently, the epidermis on the leaf had imported
products of photosynthesis.

In a second experiment, epidermis was stripped from leaves and transferred to water and kept there while the leaves
deprived of the epidermis were exposed to $^{14}CO_2$ for 10 min.
Then the leaves were washed; the epidermal strips were placed
back on them and kept there for 3 hours. Then leaves and
epidermal samples were extracted and chromatographed. The
labeling patterns of leaf and epidermis were similar; they
showed sugars, sugar phosphates and organic acids (Table 1).
Major differences occurred only in the relative predominance
of sucrose in the leaf extract (60% of the total radioactivity
against 40% in the epidermal extract) and in the relative
increase in malate content in the epidermis (22%, versus 10%
in the mesophyll). Since isolated epidermis of *C. communis*
could not produce sugars or sugar phosphates by photosynthesis
we consider the result of the experiment to be further evidence for the transfer of products of photosynthesis from
the mesophyll to the epidermis.

TABLE 1. Transfer of products of photosynthesis from the
mesophyll to the epidermis of *Commelina communis* after the
mesophyll was labelled with $^{14}CO_2$ for 10 min. Duration of
contact between mesophyll and the epidermis: 3 h after the
labelling period.

Labelled assimilate	Percent of total radioactivity in sample		
	mesophyll	epidermis	epidermis labelled directly[*]
Sucrose	60	40	0
Malic acid	10	22	49
Aspartic acid	1	1	22
Fructose	7	4	0
Glucose	6	12	0
Sugar phosphates	1	3	0
Citric acid	3	2	5
Succinic and fumaric acids	1	2	1

[*]after 90 min exposure to $^{14}CO_2$

In the third experiment, we tested the ability of epidermis to take up and metabolize carbohydrates from solutions.

Epidermal strips from leaves of *C. communis* and *T. gesneriana* were floated on solutions of ^{14}C-sugars and ^{14}C-glucose -1-phosphate. All these compounds were taken up unaltered and then metabolized. A large fraction went into starch (highest from glucose and glucose-1-phosphate, lowest from maltose). Another fraction of the label appeared in the organic acids, particularly malic. Malate formation was very pronounced after feeding glucose-1-phosphate to epidermes of tulip. Autoradiography of epidermal sections showed that almost all of the water-insoluble radioactive material was in the starch grains of the guard cells.

There are several earlier observations (Raschke 1975a) indicating that stomatal responses to light cannot be ascribed to photosynthesis in guard cells. It is therefore likely, but not proven, that the dependence of guard cells on the import of carbohydrates is general.

ACKNOWLEDGMENT

My research was supported by the U.S. Energy Research and Development Administration (formerly A.E.C.) under Contract E(11-1)-1338 as well as by the Deutsche Forschungsgemeinschaft.

REFERENCES

Allaway, W.G. (1973). Accumulation of malate in guard cells of *Vicia faba* during stomatal opening. *Planta (Berl.) 110*, 63-70.

Bernstein, L. (1971). Method for determining solutes in the cell walls of leaves. *Plant Physiol. 47*, 361-365.

Dayanandan, P. and Kaufman, P.B. (1975). Stomatal movements associated with potassium fluxes. *Amer. J. Bot. 62*, 221-231.

Fischer, R.A. (1968). Stomatal opening: role of potassium uptake by guard cells. *Science 160*, 784-785.

Fischer, R.A. and Hsiao, T.C. (1968). Stomatal opening in isolated epidermal strips of *Vicia faba*. II. Responses to KCl concentration and the role of potassium absorption. *Plant Physiol. 43*, 1953-1958.

Fujino, M. (1959). Stomatal movement and active migration of potassium (in Japanese) *Kagaku 29*, 660-661.

Fujino, M. (1967). Role of adenosine triphosphate and adenosine triphosphatase in stomatal movement. *Science Bull., Fac. Educ., Nagasaki Univ. 18*, 1-47.

Humble, G.D. and Hsiao, T.C. (1969). Specific requirement of potassium for light-activated opening of stomata on epidermal strips. *Plant Physiol. 44*, 230-234.

Humble, G.D. and Raschke, K. (1971). Stomatal opening quantitatively related to potassium transport. Evidence from electron probe analysis. *Plant Physiol. 48*, 447-453.

Imamura, S. (1943). Untersuchungen über den Mechanismus der Turgorschwankung der Spaltöffnungsschliesszellen. *Jap. J. Bot. 12*, 251-346.

Marrè, E., Lado, P., Rasi-Caldogno, F., Colombo, R. and deMichelis, M.I. (1974). Evidence for the coupling of proton extrusion to K^+ uptake in pea internode segments treated with fusicoccin or auxin. *Plant Sci. Letters 3*, 365-379.

Pallaghy, C.K. (1970). The effect of Ca^{++} on the ion specificity of stomatal opening in epidermal strips of *Vicia faba*. *Z. Pflanzenphysiol. 62*, 58-62.

Pallaghy, C.K. (1972). Localization of thallium in stomata is independent of transpiration. *Aust. J. Biol. Sci. 25*, 415-417.

Raschke, K. (1975a). Stomatal action. *Ann. Rev. Plant Physiol. 26*, 309-340.

Raschke, K. (1975b). Simultaneous requirement of carbon dioxide and abscisic acid for stomatal closing in *Xanthium strumarium*. *Planta (Berl.) 125*, 243-259.

Raschke, K., Dickerson, M. and Pierce, M. (1973). Osmotic pressures in guard cells of *Vicia faba* redetermined: evidence for an intercellular "hydroactive" mechanism in guard cells obtained; solute requirement for stomatal opening reassessed. Plant Res. '72, MSU/AEC Plant Res. Lab., Mich. State Univ., 149-153.

Raschke, K. and Fellows, M.P. (1971). Stomatal movement in *Zea mays*: shuttle of potassium and chloride between guard cells and subsidiary cells. *Planta (Berl.) 101*, 296-316.

Raschke, K. and Humble, G.D. (1973). No uptake of anions required by opening stomata of *Vicia faba*: guard cells release hydrogen ions. *Planta (Berl.) 115*, 47-57.

Raschke, K. and Pierce, M. (1973). Uptake of sodium and chloride by guard cells of *Vicia faba*. Plant Res. '72, MSU/AEC Plant Res. Lab., Mich. State Univ. 146-149.

Sawhney, B.L. and Zelitch, I. (1969). Direct determination of potassium accumulation in guard cells in relation to stomatal opening in light. *Plant Physiol. 44*, 1350-1354.

Somers, G.F. (1973). The affinity of onion cell walls for calcium ions. *Amer. J. Bot. 60*, 987-990.

Turner, N.C. (1973). Action of fusicoccin on the potassium balance of guard cells of *Phaseolus vulgaris*. *Amer. J. Bot. 60*, 717-725.

Willmer, C.M. and Dittrich, P. (1974). Carbon dioxide fixation
     by epidermal and mesophyll tissues of Tulipa and
     Commelina. *Planta (Berl.) 117*, 123-132.
Willmer, C.M. and Mansfield, T.A. (1969). Active cation
     transport and stomatal opening: a possible physio-
     logical role of sodium ions. *Z. Pflanzenphysiol. 61*, 398-
     400.
Willmer, C.M. and Pallas, J.E. Jr. (1973). A survey of stomatal
     movements and associated potassium fluxes in the plant
     kingdom. *Can. J. Bot. 51*, 37-42.
Willmer, C.M. Pallas, J.E. Jr. and Black, C.C. Jr. (1973).
     Carbon dioxide metabolism in leaf epidermal tissue.
     Plant Physiol. 52, 448-452.

SHORT DISTANCE TRANSFER - SUMMARY AND DISCUSSION

Adapted from presentations given by the Session Chairmen :
R.L. Bieleski, R.N. Robertson and F.L. Milthorpe

The overall impression from these sessions was that as
our knowledge of the structure of plant cells and tissues is
increasing so is our grasp of the transfer processes that are
going on in the plant. There are however many instances
where structural and functional aspects are so intertwined
that not just one, but a series of measurements are necessary
to obtain a correct understanding of how a particular physiological process works. We have an example of this in the
work of N.A. Walker, where the speed and pattern of cytoplasm
movement, measurements on ion transfer and measurements on
structural features needed to be handled as one unit, in order
to decide where the limits are to solute transfer between two
Chara internode cells. Thus it would appear on close analysis
that the passage of ions from one cell to another could be
limited by either cytoplasmic streaming, or the rate of
diffusion through the plasmodesmata and not solely the latter
as suggested by Tyree. Passage through a cell from the zone
of absorption to the zone of export is another factor in
transport. This movement is aided by protoplasmic streaming,
but it was also pointed out that there must be considerable
control of ion and metabolite distribution with the cells due
to compartmentalization within the cytoplasm, an area receiving little emphasis during the conference. In giant algal
cells for example, one zone can be excreting acid and an adjoining zone excreting alkali.

There is a strong case for functional plasmodesmatal connections between cells, based on the electrical coupling between cells, structural evidence for continuous endoplasmic
reticulum, calculated resistances to flow and diffusion and
more directly the localization of silver chloride precipitates within the plasmodesmata. In discussion the question
was asked what happens to plasmodesmatal connections when
cells are plasmolyzed? For example, following drought stress
there is often a residual effect on growth and retarded ion
transport and it would be valuable to know if plasmolysis does
occur under stress, how this affects the plasmodesmata and if
they are broken can they be reformed.

As well as plasmodesmata, transfer cells can be interpreted as agents of transfer - they both have the right
structures, the right qualities and occur in the right place,
and consideration was given to the need for two distinct sorts
of structures. The plasmodesmatal path appears to have the

217

characteristics of a low resistance, high capacity path that
may have rather low selectivity, even though it is able to
discriminate against the larger molecules, thus preserving
the genetic integrity of the cell. These characteristics are
clearly advantageous where a number of things have to be moved
quickly. The transfer cell on the other hand offers an enlarged surface with a relatively high resistance, but a high
selectivity. These distinctions in behaviour are of course
still speculative and require confirmation.

There is little argument that the entry of nutrients
into roots is across wet cellulose cell walls and then through
the outer cell membrane and that this process is repeated
following transfer of nutrients and metabolites to the leaf
through the xylem. However one area that did not receive
much consideration in the work presented was the structure
and mechanism of transfer of ions through membranes. Ions
having entered the symplasm of the outer root cells then
appear to travel from cell to cell through the symplasm and
plasmodesmata until they are inside the endodermis and ready
to be unloaded into the xylem. In discussing ion uptake new
developments in the area of the relation between proton extrusion and potassium uptake, in particular, at cell surfaces
were emphasized. It was also pointed out that the exchange is
more than simply the result of hydrogen ion extrusion, and
could also involve a suitable ATPase in the plasmalemma. In
fact the presence of a special ATPase could help to explain
the discrimination in uptake between sodium and potassium.

An interesting hypothesis on the control of potassium
uptake was suggested by A.D.M. Glass, where the potassium
carrier may have four allosteric sites on the inside, such
that when the four sites are loaded on the inside the carrier
is incapable of picking up another potassium from the outside. This is an interesting use of kinetics, but questions
were directed in the discussion to the standing difficulty
of relating enzyme kinetic interpretations to whole plant
problems.

With the unloading of ions into the xylem of the root we
have the interesting observation that the amino acid inhibitor
parafluorophenylanaline seems to inhibit specifically the release of ions from the symplasm. The central question here
is whether this release therefore depends on a specific protein of short half life which through its action controls the
rate of movement of, for example, potassium across the root.

The passage of ions and metabolites across shoot tissues
are not necessarily analagous to that found in the root, as
can be seen by the series of processes that come into play
in the wheat grain as described by C.F. Jenner. It would

appear that much more detailed consideration could be given
to the question of how various materials in the xylem stream
move into the tissues at their destination. Thus when amino
acids and ions such as K^+ are removed from the xylem, after
entering the leaf, are they first taken up by the nearest
mesophyll cells and subsequently passed from cell to cell
through the symplasm, or does each cell obtain its supply
directly from the apopolasm?

With stomata the major factor leading to changes in the
deformation of the guard cells is almost certainly an export
of protons from the cell and uptake of potassium, but the
mechanism which is involved is still not resolved and may involve either a proton or potassium pump. In this case there
is no obvious reason why the mechanism should differ from
that found in root cells or other tissues. It is well documented that the hydrogen ions produced depend on malate production and there is a close association between stomatal
opening, potassium uptake and malate concentration. However
this association is not complete and it would seem that it
is not malate production which is governing the entry of
potassium into the cell and other mechanisms including the
production of ATP are required. The next stage in the eluc-
idation of stomatal mechanisms is possibly to try and understand something of the CO_2-malate pathway, the carbon metabolism of guard cells and the production of energy.

Another problem of transfer is that between the mesophyll
generally and the epidermis. The present evidence would
suggest that all fluxes, even of water, are probably slow.
It is apparent that there is now a need to concentrate on
quantitative measurements of fluxes in many of the leaf
systems.

A major theme in relation to the leaf was the transfer
of carbon from the chloroplasts to the sieve tubes. There
was some agreement that movement of sucrose through the cells
of C_3 species occurs by the symplastic pathway, at least until
the junction between the inner parenchymatous tissue and the
companion cell, sieve tube complex. At this junction there
is a big build in sugar concentration. The point was made
that the high concentration in the companion cell does not
necessarily mean that the sugar flux passes directly through
the companion cell, although the anatomical connections
would suggest that this is likely to be the case. In C_4-
species the transfer of photosynthate to the sieve tubes is
more complex with a possible malate-pyruvate shuttle between
mesophyll and bundle sheath cells. This bidirectional transfer is presumably accommodated by the plasmodesmata and has
a possible parallel in the opposing movement of carbohydrate
and nutrients across the root endodermis.

LONG DISTANCE TRANSPORT

CHAPTER 20

The Study of Vascular Patterns in Higher Plants

M.H. Zimmermann
Harvard Forest, Petersham, Mass. U.S.A.

INTRODUCTION

The complexity of the vascular system of plants is rarely appreciated by investigators of translocation. This paper describes methods which facilitate the analysis of complex three-dimensional structures. They have been developed over the past twelve years to investigate the vascular system of arborescent monocotyledons (such as palms) where we deal with complexities that are entirely inaccessible by conventional methods. As my primary interest has been translocation for many years, it was clear all along that these methods could play a very useful role in certain phases of translocation research. However, the vascular patterns of arborescent monocotyledons and the patterns of differentiation proved so interesting and new that I spent a far greater time investigating it than I had originally intended. This paper is a brief account of the methods of motion-picture analysis and a description of various aspects of translocation research in which they may play a useful role in the future.

MOTION-PICTURE ANALYSIS OF THREE-DIMENSIONAL STRUCTURES

History. - The use of motion-pictures for three-dimensional structural analysis was reported for the first time in 1907 (Reicher 1907). This early film does not seem to exist any more. A number of publications appeared since then, some of the techniques involving quite elaborate equipment (e.g. Postlethwait 1962). When P.B. Tomlinson and I began to investigate the vascular system of palms, we were unaware of these earlier attempts and designed our own procedures. The reason why the technique was previously largely unknown in plant anatomy was perhaps the fact that it was of limited usefulness. Photography was usually done with paraffin-embedded specimens on a rotary microtome. This meant that only small specimens could be looked at. The fact that the camera "looked" at the cut surface also limited optical resolution, even though surprisingly good results were obtained with surface staining. Our own techniques overcame these difficulties in two ways. A specially designed microtome advances the specimen itself rather than the specimen holder. This enables us to analyze stems of a length up to one meter or more. The second difficulty - lack of resolution - was overcome with methods designed to photograph serial sections through the microscope (Fig. 1).

Surface method. - The motion-picture camera "looks" at the cut surface of a specimen which is held in the microtome clamp. The camera lens is focussed with the help of extension tubes. A longer-than-normal focal length is usually necessary (e.g. 50 mm for a 16 mm camera) so that there is enough working distance between lens and specimen. The specimen is advanced along its axis through the clamp in such a way that after a section has been cut off its end, the newly exposed surface is again precisely in focus (Fig. 1a). Any 16 mm movie camera with ground-glass focussing and single-frame shutter control is suitable. We use a Bolex H 16 REX. As a film, we use Kodachrome Professional Type A or, if the possibility of publication is anticipated or the need for more than one copy, Eastman Color Negative Film 7254.

For many years we used two strong microscope lamps for illumination. The light level was always marginal, especially with dark specimens, we therefore acquired a specially designed electronic flash, giving enough light so that the camera lens can be stopped down for more depth of field. Maximum depth of field is desirable, because it is extremely difficult to focus on the low-contrast surface of the specimen. We have now so much light that it is easy to stop down the lens too much, thereby decreasing the numerical aperture of the optical system to the point where theoretical resolution is less than the resolution of the film. These limits (caused by the extension tubes) can easily be calculated. The electronic flash must have several features which are not normally available. First of all, it must have a short recyling time, because much time would be wasted if one had to wait six seconds between successive photographs. Secondly, the flash must be capable of firing hundreds of thousands of times without light loss, because a single 100-foot roll of film has 4000 pictures. This is accomplished by (1) a special flash tube with long life expectancy, and (2) paper-insulated storage condensers, because the usual electrolyte condensers do not survive more than a few thousand rapid discharges. This makes the power pack very bulky. Thirdly, the

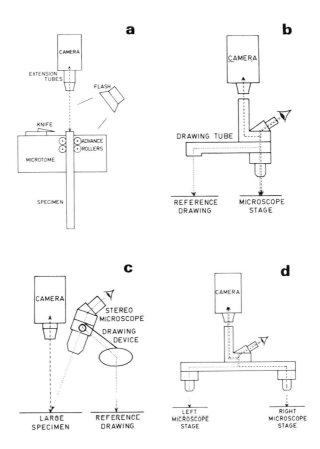

Fig. 1. - Methods of motion-picture analysis. (**a**) *The surface method. Successive surfaces of the specimen are photographed with a motion-picture camera, frame by frame, on the microtome.* (**b**) *Drawing method. Serial microtome sections are photographed, frame by frame, through the microscope (light path shown dashed). Successive sections are lined up with the aid of a reference drawing (dotted light path).* (**c**) *Large sections, such as paleontological peels, can be photographed with the camera directly (dashed light path). A stereomicroscope, fitted with drawing apparatus, is aimed at lowest magnification and at a slight angle. Successive sections are again lined up with the aid of a reference drawing (dotted light path).* (**d**) *The shuttle microscope. Serial sections are photographed frame by frame through the left and the right microscope alternately. Each section is lined up with the previous one by switching on the left and right light source in rapid succession.*

223

unit has a low-voltage triggering circuit, because a high
voltage would deteriorate the camera switch within relatively
short time. A flash synchronization switch has to be built
into the camera, a feature not normally available.

The analysis of very bulky specimens, such as the stem
of larger palms, has to be done in separate sections (Fig. 1
in Zimmermann and Tomlinson 1974).

The photography of a specimen on the microtome is essentially "blind", i.e. one does not see much while the film is
taken. Films are analyzed with a Data Analyzer as described
below.

Drawing methods. - When we began this work some 12 years
ago, we used the drawing methods to take frame-by-frame movies
through the microscope of serial sections. The requirements
for this are a microscope with trinocular outlet for viewing
and photography, into the light path of which a camera lucida
can be inserted. A rotating mechanical stage is essential
for the alignment of sections. We use the Wild M 20, with
Camera Tube H and Drawing Tube (Fig. 1b). Serial sections
are photographed frame by frame. Each section is lined up
with a simple drawing (on paper) of a previous one. The drawing has to be renewed every three to five sections, as the
alignment deteriorates. This procedure is time consuming.
With certain specimens, such as monocotyledonous stems at very
low magnification, it is difficult to use. On the other hand,
the drawing method has the advantage that many serial sections
can be mounted on a single slide - obviously practical if one
deals with thousands of sections. We rarely use this method
today, because the optical shuttle is a very much faster procedure.

Photography of very large sections. - Occasionally, one
encounters the problem of large sections such as paleontological peels, which cannot be accommodated in an ordinary
microscope. There are various ways in dealing with this
situation. The Wild M 5 dissecting microscope can be used in
conjunction with a camera tube and a drawing apparatus. This
admits quite large specimens with either reflected or transmitted light. For still larger specimens, one can mount the
camera directly above the specimen and aim the M 5 with drawing apparatus at a slight angle near the light path of photography. In this case, one again makes outline drawings for
alignment (Fig. 1c).

It would obviously be very useful to analyze motion
pictures of serial electron micrographs. For example, the
distribution and three-dimensional arrangement of P-protein
filaments in a sieve element could be studied. It is usually
said that serial sections are too difficult to make. However,

a film has been produced from electron micrographs by
Dr. Randle Ware of Cal-Tech and is reportedly being published
in Göttingen. Electron micrographs can be photographed with
the above-described method. Simpler, micrographs can be enlarged on Kodak Translite Film rather than on paper, and these
transparencies can be photographed with the movie camera on a
light box whereby they are lined up by superimposing each with
the previous one.

Shuttle microscope. - An elegant device for photograph-
ing serial or sequential microtome sections is the optical
shuttle or shuttle microscope. It consists of two microscope
bases with separate illumination and separate objectives, but
with a single trinocular outlet for viewing and the camera
(Fig. 1d). The two microscopes are connected with an optical
bridge. We use Wild M 20 microscopes with KdGS rotating mechanical stages (Zimmermann and Tomlinson 1966). The connecting
bridge is actually a so-called "Discussion Tube" designed to
be used on a single microscope base with two binoculars so
that two persons can look into the microscope at the same time.
We usé the Discussion Tube upside-down as a bridge, mounted
with special adapters which are available from the manufactur-
er. As a result of this arrangement, the light path above the
objectives is somewhat lengthened and the objectives have to
be used in a slightly lower position. They are therefore not
exactly parfocal, and one of them (the 3x Plan Fluotar) requires an extension ring. A few years ago, Leitz brought out
a "Comparison Bridge" which eliminates this disadvantage.
Comparison microscopes normally show half a field of view of
each microscope, but this bridge has built-in flaps that can
be opened to show the entire fields. However, the Leitz microscopes have some other disadvantages which are more serious
than the lack of parfocal objectives. As the two firms have
recently gone through a partial merger, it is hoped that the
ideal solution will eventually be found.

The illuminators of the two microscope halves are wired
to a foot switch with which one directs the light either to
the left or the right. Ideally, each section is on a separate
microslide. Section No. 1 is placed onto the stage of one of
the microscopes, a position selected and a photograph taken.
Number 2 is placed on the other stage and the light switched
to that side. The same area of the section is brought into
field of view. The light is then rapidly switched back and
forth and the apparent jumping motion of the two slides is
eliminated by moving and turning the stage containing section
No. 2. When alignment is achieved, No. 2 is photographed,
No. 1 replaced by No. 3 and the process repeated.

It is not always necessary to work with every section. It
is often sufficient to use every second, fifth or tenth. As

One begins sectioning, the first ten or so are tentatively
stained and mounted and 1 and 2, 1 and 3, 1 and 5, etc. tried
in the shuttle to find the ideal spacing. Even if the sections are spaced, one often ends up with several thousand
sections and consequently with large numbers of microslides.
To cut down this bulk (and expense!) one can mount more than
one section on a slide, but this makes the process of lining
up somewhat more complicated. The procedure has been described in more detail by Zimmermann and Tomlinson (1969).

Preparation of sections along tracer tracks. - Motion-
picture analysis of dye tracks in stems after selective dye
injection can reveal interesting information about xylem
transport. Movement in the phloem can be studied by motion-
picture analysis of autoradiographic tracks. Here, we have
to distinguish between tracks of insoluble material and of
soluble translocated substances. Let me deal with insoluble
material first. We introduced ^{14}C glucose into a single leaf
vein of the small palm *Rhapis excelsa* for a period of two
days. The stem was then embedded in paraffin and cut into
10,000 sections. These were covered with a drop of liquid
photographic emulsion, exposed, developed and stained in the
regular manner. This produced a very clear track entering
the stem via a single vascular bundle and showing passage of
material through bridge connections in the stem that had been
found before by anatomical analysis (Zimmermann 1973). This
procedure gave us a good deal of information, but it does not
show the movable, soluble substances.

Autoradiography of translocated isotopes in the phloem
has been steadily refined during recent years and numerous
reports have appeared in the literature, showing sugar movement in individual sieve tubes (e.g. Trip and Gorham 1968,
Schmitz and Willenbrink 1968, see also the paper by
D.B. Fisher in this volume). Such autoradiographs always concerned very small areas. Unfortunately, freeze-drying of
large plant parts is not only very slow, but usually of very
poor quality. We are presently working on this problem and
also hope that freeze substitution with liquids of low water
solubility will give us usable results.

Analysis of films. - In following a radioactive track,
or provascular strands in a monocotyledonous bud, one often
would not really have to make a movie because positional
measurements are taken in the shuttle microscope and a plot
is made as one goes along. However, there are two good
reasons for always taking pictures even if they are not
essential. The procedure provides the discipline to line up
subsequent sections precisely, errors of identification are
thus avoided. The resulting motion picture is always useful

for the demonstration of information that has been gathered
during many hours of work.

Surface motion pictures, however, are taken "blindly"
and the results can only be obtained by projection of the
film. The ideal device is a Data Analyzer, a projector which
has a built-in magnetic advance mechanism to move the film
forward or backward at any desired speed or manually frame by
frame. The shutter of the projector remains running at regular speed even at slow film movements; flicker of the pro-
jected image is thus prevented. For convenience, we project
via a mirror onto a white sheet of paper on a table. A scale
(stage micrometer), photographed at the beginning of each
sequence, gives the magnification on the paper and the frame
counter in conjunction with the information about section
thickness and spacing, gives the axial scale. It is thus
quite easy to reconstruct three-dimensional structures from
a film.

PATH OF PRIMARY VASCULAR BUNDLES

Arborescent monocotyledons. - By far the largest amount
of work on vascular patterns has been done with arborescent
monocotyledons, largely in collaboration with P.B. Tomlinson.
Readers who are interested are referred to the published
papers (see Zimmermann and Tomlinson 1974, and the literature
cited therein). Most monocotyledonous stems contain numerous
scattered vascular bundles. Leaves are usually large, borne
in a terminal tuft. New ones are formed from the apical meristem and old ones fall from the lower edge of the crown.
Every leaf is supplied by hundreds or thousands of vascular
bundles. One of the main questions was the means by which
vascular continuity is provided. In other words, if a certain
number of vascular bundles leads into each leaf, why does
the stem not "run out of" bundles? The answer to this question seems to be the upward branching of the leaf trace at the
point of its departure from the stem. Although the patterns
of vascular bundles in monocotyledons show appreciable variation from one taxonomic group to another (even within the
palm family), the upward-branching leaf trace seems to be a
very universal feature.

Patterns of differentiation. - One of the most fascinat-
ing problems is the pattern of differentiation of vascular
bundles in large monocotyledons. It is certainly not easy to
understand how a large primary stem is formed from a single
bud. The problems of transport into such a massive meristem
are most interesting. This can best be visualized by comparing the situation with that of a more familiar dicotyledonous
or coniferous tree. There, we have thousands of tiny buds,

227

all supplied with vascular tissue from the twigs running right up to the meristematic tissue. The cambium is a sheet of tissue enveloping the stem, and here mature vascular tissue runs all along it, within a few microns distance. Nutrient supply is no problem because the meristematic tissue and the vascular tissue feeding it are laid out in such a way that the extrafascicular path of transport is minimal. The apical meristematic tissue of a tree like a palm is very large, the whole area has to be fed, has to contain feeding vascular bundles (protoxylem and protophloem) and has to grow into a mature stem with mature vascular tissue all at the same time.

An understanding of this vascular pattern is the key to many practical problems. For example, when palms are transplanted, most mature leaves are usually cut off. A palm can afford to lose all its leaves and recover. The reason for this is that there are about as many leaf primordia hidden in the bud as there are leaves visible in the crown. Enough storage material is present in the stem to enable the plant to unfold these and produce new primordia.

Lethal yellowing of coconut palms, a terrible disease which has wiped out hundreds of thousands of coconut palms during the past few years, is probably caused by a mycoplasm multiplying in the sieve tubes (Parthasarathy 1973). The disease is now known to affect 14 palm species (Popenoe 1975). Some of them can recover after tetracycline injection (McCoy 1973). However, this is only possible as long as the apical meristem, which produces the new leaf primordia, is still intact. Once the apical meristem is destroyed, the palm is doomed even though the visible leaves and a few additional ones expanding can give it the outward appearance of health.

Many aspects of lethal yellowing are problems of translocation. Transmission is suspected to be caused by leaf hoppers. There is a long incubation time until the first symptom appears. The succession of symptoms indicates that the mycoplasm is translocated from the infected mature leaf into the stem via phloem, then up into younger leaves, to the roots, and finally to the apical meristem. The mechanism of transmission, the spread of the microorganism and its effect, are all translocation problems. In addition, any injection into the stem for the purpose of treatment can be done most effectively only if the vascular pattern and the translocation mechanisms are known.

Dicotyledons. - Our experience with the primary vascular patterns in dicotyledons is still very limited. A good deal of information has been obtained earlier with classical methods (e.g. Esau 1965). However, there is still much to be learned about the more complicated structures. We have recently filmed the nodal area of a *Cucurbita* stem, it is remarkably complex, a careful analysis will be very time consuming. Several years ago, we had a look at the stem of sugar beet in which there are concentric rings of vascular bundles and a curious - but logical - crossing over of vascular bundles from older, outer leaves to older, inner rings in the stem, whereby leaves always connect with more than one concentric circle of vascular bundles.

PATH OF XYLEM VESSELS

Ricinus *petiole*. - When structures such as leaves, petioles or isolated vascular bundles are subjected to increasing water stress, the water columns in individual vessels break. Each time this happens, a small shock wave is emitted which can be detected by acoustical means (Milburn 1973). The number of clicks per organ (isolated piece of vascular bundle, petiole, etc.) is surprisingly high, and it seemed difficult to believe that there are so many individual vessels in the small organs investigated. When we looked at an 11 cm long *Ricinus* petiole, we found 12 vascular bundles with approximately 12 vessels each. When the vessels of a single bundle were followed along the length of the petiole and mapped, they proved to be of variable length, all of them were shorter than the petiole. There were several hundred vessels in the petiole, far more than we would have expected. Direct observations with a dissecting microscope of similar petioles, into which Indian ink had been injected, indicated that such observations can give only the very crudest kind of information. From a functional point of view it can be stated that such a complex vessel structure constitutes a considerable safety factor in the water conduction of a plant.

Vessel length in palms. - We have thousands of feet of film along stems and these contain a wealth of information about vessel length in palms. Unfortunately, we have never had the time to investigate the problem of vessel length thoroughly, we can only make some general statements from observations that were made more or less accidentally along the way. In the small one-vessel palm *Rhapis* (one metaxylem vessel per vascular bundle), a major leaf contact contained at least three separate vessels of which the middle one was *ca.* 10 cm long of which 2 cm at each end overlapped with the next vessel. The vessel structure at the point of leaf-trace departure is rather complex. Bridges, connecting the leaf trace with neighbouring axial bundles all contain individual vessels, and the leaf itself is connected to the stem either

by very short and narrow vessels or by tracheids (see Fig. 2
in Zimmermann 1973). This is of functional importance, because old leaves drop from the stem and the resulting embolism
in the broken vessels must remain confined. Furthermore, the
stem of palms is entirely a primary one. This means that
vascular tissue formed by the bud is not renewed and must remain functional throughout the lifespan of the palm. Therefore, the stem must be provided with over-efficiency and
built-in safety devices such as many short vessels and ample
cross connections (the bridges) to provide alternate pathways
in case of stem injury.

Vessels in dicotyledonous wood. - Skene and Balodis
(1968) introduced an amazingly simple method to obtain a
measure for the distribution of vessel lengths in a given
piece of wood. Previous to their publication, the information
on vessel length had been very scattered and incomplete. Their
mathematical interpretation indicated a continuous range of
lengths. This was questioned by Milburn and Covey-Crump
(1971) who presented convincing arguments that vessel popula-
tions of as few as three lengths (0.4, 1.8 and 4.0 meters
respectively for *Eucalyptus obliqua*) could explain Skene and
Balodis' results equally well. Both of these papers have
their own merits from which the controversy should not dis-
tract. Unpublished film analyses at the Harvard Forest by
D.S. Skene seem to indicate that the continuous-range concept
of vessel-length distribution is probably more correct.

Cinematographic analysis enabled us for the first time
to *see* vessel ends that hitherto were mysteriously hidden
within the wood. Vessels of trees always end in pairs or
groups (Zimmermann 1971). This makes sense functionally, as
water ascending in the xylem must pass from vessel to vessel
via pits. The longer the overlapping ends, the better the
water transfer (Fig. 2).

Large vessels are obviously an enormous evolutionary
advantage in terms of efficiency of water conduction: the
volume of water moved by a given pressure gradient is proportional to the fourth power of the capillary radius. Expressed
in terms of transverse-sectional vessel area, four times as
much water can move through the same vessel cross sectional
area if vessels are of twice the diameter. However, there is
a very significant disadvantage: larger vessel volume carries
a greater risk of embolism, and it is obvious that there is
an upper practical limit of about 0.5 mm vessel diameter. If
evolution ever "tried out" larger vessels, they became ex-
tinct. The evolutionary advantage of small vessels lies in
the greater operational safety (Zimmermann and Brown 1971).
Both the safety and the efficiency "strategy" can be seen in

the genus *Quercus*: oaks of temperate regions have very large vessels, those of dry regions (usually evergreen "live" oaks) have very small vessels.

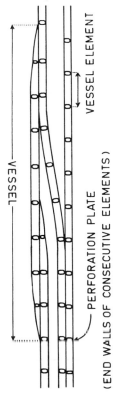

Fig. 2. - Diagrammatic representation of xylem vessels. An entire vessel is shown on the left, the end of another one on the lower left. - It is not known if sieve tubes occur in defined units like this.

Another interesting feature, only fully recognized by cinematographic analysis, is the vessel network. Vessels are almost never entirely straight along the axis of the stem, their paths in successive tangential layers are always more or less at an angle to each other. This results in an intricately-woven vessel network which - as in the case of palms discussed above - provides alternate paths in the case of injury. It is the basis for the spreading of dye within a growth ring when dye is applied via a radial bore-hole (see Figs. IV-5 and IV-6 in Zimmermann and Brown 1971, and the 16 mm film by Zimmermann 1971).

PATH OF SIEVE TUBES

The sieve tube concept. - Vessel elements and sieve-tube elements are derivatives of the same cambial initials and there are many structural similarities between the two. However, there are a number of functional differences. Operational pressures in vessels are usually very low, walls are therefore rigid to prevent collapse, and of course vessel elements are dead and devoid of cell content at functional maturity. Closely related to the functional pressure is the sealing mechanism. With less-than-atmospheric pressure in the vessels, air is drawn into an injured vessel and water is withdrawn to the end of the vessel where the air-water interface is held at vessel-to-vessel pits by capillary forces. With positive operating pressures, as they are commonly found in sieve tubes, a different sealing mechanism must exist, everyone is familiar with the plugging function of P-protein and callose (Eschrich 1975a). It is therefore obvious that the respective sealing mechanisms of vessels and sieve tubes are in different locations. They are all along the whole vessel wherever there is vessel-to-vessel pitting. In sieve tubes the sealing mechanism is situated at the sieve plates which correspond to the perforation plates of vessels. The sealing mechanism is functionally so important that sieve plates did not disappear during evolution.

We know that the sieve tubes form a network in the phloem similar to that of the vessels in the xylem, because there is a tangential spread of axial translocation in the phloem which, in ash for example, has been found with partial defoliation experiments to be not quite 1° (Zimmermann 1961). The present definition of a sieve tube is a chain of sieve elements extending the length of the plant. But we do not know if there are sieve-tube sections homologous in structure to vessels (Fig. 2). If they do exist, they are functionally of lesser importance and primarily of theoretical interest.

Isotope tracks in palms. - ^{14}C sugar introduced into a small leaf flap produces a very clear track along the phloem of a leaf trace into the stem and from this supply bundle via bridges into other leaf traces further down. At the bottom of the analyzed stem a number of vascular bundles are distinctly labeled. At the base of the stem much of the isotope had moved into areas other than phloem, silver grains indicated labeled cell walls of the bundle fibers, labeled starch in parenchyma, etc. (Figs. 3-7 in Zimmermann 1973). One of our major interests is the translocation into the apical region and the quest for the path of transport in meristematic tissues. Are procambial cells able to translocate sugars in a preferential axial direction? Experiments with tritiated thymidine remained inconclusive and we hope to get better results with autoradiography of soluble (i.e. moving) sugars.

Bidirectional transport in dicotyledons. - We do know that in a primary stem with newly-matured leaves there is bidirectional transport in certain internodes (See Fig. V-12 in Zimmermann and Brown 1971). But previous analyses of this area have remained relatively inconclusive. The same is true for the patterns of simultaneous export and import in maturing leaves. These matters have recently been reviewed by Eschrich (1975b). It is obvious that cinematographic analysis of isotope tracks would give us far more precise information than was hitherto available.

Longevity of sieve tubes in Tilia. - One of the problems we are working on presently is an analysis of *Tilia* phloem. *Tilia* bark is relatively easy to section and we have just completed over 200 feet of film containing some 10,000 transverse-sectional views. The film, when analyzed, should show if there are sieve-tube units of a defined length corresponding to vessels in the xylem. If *Tilia* phloem remains functional for more than one growing season, then sieve-tube connections across the growth-ring border must be visible. There are indications that the answer to both these questions is yes, a detailed film analysis is under way and should tell us whether these notions are correct. Future tracer-track analyses will have to confirm the anatomical data.

CONCLUSIONS

It is quite obvious that cinematographic analysis will be a very powerful tool not only in plant anatomy but also in translocation research whenever we are dealing with structural questions concerning the path of translocation. It is obvious, for example, that one reason why the cohesion theory of sap ascent was questioned for many years was the disregard of the intricate and complex three-dimensional vessel arrangement in wood. There are very many short vessels even in woods in which some vessels are long. These and the intricate network that they form make sap ascent safe enough even though the water itself is in a metastable condition. Many phases of phloem transport will benefit from cinematographic analysis, such as the question of bidirectional transport and translocation of nutrients into the massive meristematic tissue of palm buds. This latter problem is almost entirely unexplored, indeed until recently we were absolutely ignorant even about the vascular structure of this area. In addition, cinematographic analysis may play a role in applied areas of research such as the mechanism of xylem failure which is the fatal outcome of the Dutch elm disease, or the lethal yellowing disease of coconut palms. I believe that an investment in research into these problems would not only be of academic interest but of very practical value as well.

REFERENCES

Esau, K. (1965). Vascular Differentiation in Plants. Holt, Rinehart and Winston, New York.

Eschrich, W. (1975a). Sealing systems in phloem. In: Transport in Plants I, Phloem Transport. "Encyclopedia of Plant Physiology, NS, Vol. 1" M.H. Zimmermann and J.A. Milburn eds., Springer-Verlag, New York-Berlin-Heidelberg pp. 39-56.

Eschrich, W. (1975b). Bidirectional transport. In: Transport in Plants I, Phloem Transport. "Encyclopedia of Plant Physiology, NS, Vol. 1" M.H. Zimmermann and J.A. Milburn eds., Springer-Verlag, New York-Berlin-Heidelberg pp. 245-255.

McCoy, R.E. (1973). Antibiotic treatment of lethal yellowing. *Principes 17*, 157-158.

Milburn, J.A. (1973). Cavitation studies on whole *Ricinus* plants by acoustic detection. *Planta (Berl.) 112*, 333-342.

Milburn, J.A. and Covey-Crump, P.A.K. (1971). A simple method for the determination of conduit length and distribution in stems. *New Phytol. 70*, 427-434.

Parthasarathy, M.V. (1973). Ultrastructural studies on palms affected by the lethal yellowing disease. *Principes 17*, 154.

Popenoe, J. (1975). Lethal yellowing of palms. *Bulletin of the Fairchild Tropical Garden 30(2)*, 13-14.

Postlethwait, S.N. (1962). Cinematography with serial sections. *Turtox News 40*, 98-100.

Reicher, K. (1907). Die Kinematographie in der Neurologie. *Neurol. Zbl. 26*, 496.

Schmitz, K. and Willenbrink, J. (1968). Zum Nachweis tritiierter Assimilate in den Siebröhren von *Cucurbita*. *Planta (Berl.) 85*, 111-114.

Skene, D.S. and Balodis, V. (1968). A study of vessel length in *Eucalyptus obliqua* L'Hérit. *J. Exp. Bot. 19*, 825-830.

Trip, P. and Gorham, P.R. (1968). Translocation of radioactive sugars in vascular tissues of soybean plants. *Can. J. Bot. 46*, 1129-1133.

Zimmermann, M.H. (1961). Movement of organic substances in trees. *Science 133*, 73-79.

Zimmermann, M.H. (1971). Dicotyledonous wood structure made apparent by sequential sections, Film E 1735. (Film data and summary available as a reprint). Institut für den wissenschaftlichen Film, Göttingen, Germany.

Zimmermann, M.H. (1973). Transport problems in arborescent monocotyledons. *Quart. Rev. Biol. 48*, 314-321.

Zimmermann, M.H. and Brown, C.L. (1971). Trees: Structure and Function. Springer-Verlag, New York-Berlin-Heidelberg (2nd printing 1974).

Zimmermann, M.H. and Tomlinson, P.B. (1966). Analysis of complex vascular systems in plants: optical shuttle method. *Science 152*, 72-73.

Zimmermann, M.H. and Tomlinson, P.B. (1969). The vascular system of *Dracaena fragrans* (Agavaceae). I. Distribution and development of primary strands. *J. Arnold Arb. 50*, 370-383.

Zimmermann, M.H. and Tomlinson, P.B. (1974). Vascular patterns in palm stems: Variations of the *Rhapis* principle. *J. Arnold Arb. 55*, 402-424.

CHAPTER 21

Histochemical Approaches to Water-Soluble Compounds and their Application to Problems in Translocation

D. B. Fisher

Department of Botany, University of Georgia, Athens, Georgia, 30602, U.S.A.

INTRODUCTION

One of the most difficult experimental problems with working on intercellular transport is that the answers often rely on not only structural observations, but on the distribution and movement, preferably in quantitative terms, of water-soluble compounds within those structures. This is a very difficult problem, and renders useless most tissue preparation procedures, since they almost all rely on aqueous reagents. The few which remain are usually not those which a histologist would otherwise use by choice. In the material that follows, I would like to describe our experiences with some histochemical approaches to water-soluble compounds and to indicate how we have applied them to some problems in translocation. Except for a few comments, I will be concerned almost entirely with freeze-substitution as a means of tissue dehydration. The embedment of freeze-dried tissues in epoxy resin has been discussed thoroughly elsewhere (Eschrich and Fritz 1972) and I have only limited experience with that procedure. For similar reasons, only brief mention will be made of inorganic compounds (see Lüttge 1972).

FREEZE-SUBSTITUTION

Figure 1 summarizes some procedural variations which have been useful with frozen tissue. For observations which require only structural information, it is desirable to remove water-soluble compounds, since the tissue sections much more easily. This is readily accomplished by using methanol as the substituting solvent and warming the tissue to room temperature in methanol.

There are two principal advantages in using freeze-substitution for dehydration, as compared to freeze-drying.

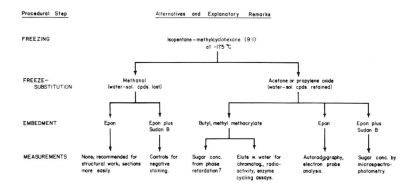

Fig.1.- Summary of the procedural variations used in the localization and quantification of water-soluble compounds in plastic-embedded plant tissues.

(An important disadvantage related to autoradiography is discussed later.) First, it is a simpler procedure, requiring for the dehydration process only a deep-freeze capable of maintaining about -65°C. Second, the very strong tendency of the water-soluble contents of the cells to shrink drastically (Fisher 1972) is more easily avoided during freeze-substitution. At least with the variations I have tried, I have not been able to avoid this problem during freeze-drying, and it almost certainly accounts for the very patchy distribution of silver grains in almost all high resolution autoradiographs of sieve tubes, with the exceptions of those prepared by Fritz and Eschrich's (1970) procedures. The shrinkage is caused by the presence of minute amounts of water in the air and in solvents. It can be eliminated by drying all solvents and resin monomers over molecular sieves before they contact the tissue. In addition, it is necessary to carry out all manipulations of the tissue in a dry box.

METHACRYLATE EMBEDMENT

Phase Retardation. - Despite some of the well-known disadvantages of butyl methacrylate as an embedding medium, we have found it to be very useful for some purposes where water-soluble compounds are concerned. In the first place, when sugars are embedded in epon, their presence cannot be detected by phase contrast microscopy. They are easily visible in methacrylate, and I have attempted to use phase retardation measurements as an indication of sugar concentration (Fisher 1975a). Pith blocks infiltrated with sucrose solutions served as standards. However, the phase retardation measurements can be quite variable, partly because of clumping by the sugar and partly because the presence of sugar apparently affects the polymerization of the methacrylate. I think that some modification of this procedure may reduce its variability.

Quantitative Elution with Water. - Methacrylate embedment has been most useful for work in which we wished to recover quantitatively water-soluble compounds from sections for chromatography and quantitative analysis. In particular, we were interested in the compartmentation of sucrose in leaves, which some of my earlier work suggested could be a significant factor in the interpretation of translocation kinetics. Our approach to this (Outlaw and Fisher 1975) was to pulse label *Vicia faba* leaves and cut (dry) paradermal sections of methacrylate-embedded tissue to obtain, in succession, samples of the upper epidermis, palisade parenchyma, veins plus spongy parenchyma, spongy parenchyma, and lower epidermis. In several experiments we followed only the total soluble and total insoluble ^{14}C in each tissue sample. There was very little insoluble activity, and it showed virtually no change with time. The palisade parenchyma initially contained more ^{14}C than the spongy parenchyma, and there was a continual decline after the labelling period. Paper chromatography of extracts showed that the decline was due almost entirely to a decrease in their ^{14}C-sucrose content. In the spongy parenchyma, there was typically an increase in activity after labeling, and then a decline. The data for veins were quite erratic because of sample variation owing to the variable position of veins in the spongy parenchyma. Nevertheless, the veins clearly contained as much or more ^{14}C as the spongy or palisade parenchyma, in spite of their much smaller volume. Also, their ^{14}C-content was generally high soon after labeling and showed a general decline with time.

These results by themselves demonstrated the compartmentation of sucrose in *Vicia* leaves, but we also wanted to be able to follow the absolute values of sucrose specific activity in the different leaf tissues (Outlaw et al. 1975). The difficulty in doing this lies, of course, in the necessity of assaying very small amounts of sucrose. The procedure we used was an enzyme cycling procedure developed by Lowry (Lowry and Passonneau 1972), and our particular version of it had a range of usefulness of 10^{-12} to 10^{-11} moles, although the sensitivity can be increased greatly. Since there are cycling assays for TPN, TPNH, DPN, DPNH, ADP and ATP, almost any compound of biological interest can be determined by including a preliminary reaction to generate one of those compounds. The sensitivity of the assay can be illustrated by the fact that there is enough sucrose in a single paradermal section one millimeter square and 4 microns thick to run the assay

20 to 30 times, including controls.

The kinetics of sucrose specific activity in separate leaf tissues (Fig. 2) showed that even though the mesophyll tissues are the site of sucrose synthesis in the leaf, its specific activity was the highest in the conducting tissue. Roughly speaking, then, there must be two sucrose pools in mesophyll cells; a storage pool (presumably the vacuoles) and a transport pool (cytoplasm or cell walls) which receives the newly synthesized sucrose.

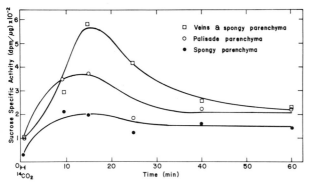

Fig.2.- The kinetics of ^{14}C-sucrose specific activity in various histological samples after pulse-labelling a Vicia faba *leaflet with* $^{14}CO_2$.

EPON EMBEDMENT

Electron Probe Analysis. - Although we have had no direct experience ourselves with inorganic compounds, freeze-substitution in propylene oxide and epon embedment have been used successfully by Pallaghy (1973) to prepare leaf tissue for the intracellular localizaion and measurement of potassium and chloride by electron probe analysis. However, the use of freeze-substitution to retain inorganic compounds substantially predates our own use of it to retain organic compounds (e.g. Lüttge and Weigl 1965; Lüttge 1972).

Microautoradiography. - Certainly one of the main advantages of epon embedment is the retention of sugars in the wetted plastic. Sugars in freeze-substituted tissue, however, are more susceptible to leaching than in freeze-dried preparations, which retain more than 90% of their sucrose on exposure of the sections to water (Eschrich, personal communication). In our material, we have seen retention which varied from 25 to 75%, although it was usually closer to 40% and was fairly consistent in any given experiment. In spite of considerable effort, we have not been able to get consistently high retention. The reason for any retention at all is not clear, but it is probably not simply by physical entrapment. Aqueous stains penetrate the sections fairly readily, and the sucrose which is retained is apparently unreactive.

Aside from the problem of variable retention, we have
worked out satisfactory procedures for the quantitative high-
resolution autoradiography of ^{14}C (Housley and Fisher 1975).
We prepared our own stripping film by dipping parlodion-covered
slides in Ilford L4 emulsion, giving a uniform emulsion layer
about 0.2 micron thick. After drying, the film was stripped
onto a water surface and a slide with sections of measured
thickness was coated with the film. Detection efficiency was
determined from ^{14}C-methacrylate sections of known activity.

These procedures have been used to follow the source
pool kinetics for translocated sucrose in soybean leaves and
in morning glory leaves (Fisher 1975a; Fisher, Housley and
Christy, unpublished). In both of these plants, the translocation profile moves along the path without much change in
shape (Fisher 1975a). The profile shape must, therefore, be
determined mostly by the rate at which ^{14}C-sucrose enters the
translocation stream from the source pool. On anatomical and
other grounds, the sucrose source pool will be contained in
the companion cells in the minor veins. Figure 3 illustrates

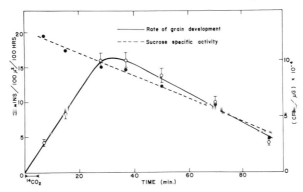

*Fig.3.- The kinetics of ^{14}C in the companion cells of a
soybean leaf, as determined by quantitative
microautoradiography, and of total ^{14}C-sucrose
specific activity after pulse-labeling with
$^{14}CO_2$.*

the kinetics of ^{14}C in the companion cells and of average sucrose specific activity in a soybean leaf after pulse labeling with $^{14}CO_2$. The kinetics for the two were quite different
at early times. The kinetics in the companion cells strongly
resembled both the kinetics of ^{14}C export from a pulse-labeled
soybean leaf, and the arrival kinetics at a sink leaf. Similar
results were obtained in a second experiment. Two experiments
were also run with morning glory with results that, again,
were quite similar to those expected on the basis of the

kinetics in the stem. In morning glory, however, the kinetics
of sucrose specific activity paralleled more closely those for
the companion cells, particularly just after labeling. In
spite of our problems with sucrose retention, we feel that the
reproducibility of the experiments and the smoothness of the
data justify our conclusion that the kinetic behavior of ^{14}C
in the source pool is the most important single factor in
determining the kinetic behavior of ^{14}C along the translocation
pathway.

One of our fondest hopes for our autoradiographic work was
that it might provide us with a means for evaluating the suit-
ability of Münch's osmotically-generated pressure flow
mechanism for phloem transport. As MacRobbie (1975) has
pointed out, this type of movement must inevitably contribute
to transport along sieve tubes; the real question is whether
it, by itself, is sufficient to cause movement at the observed
rates or whether it must be supplemented by some other force.
In order to evaluate this possibility, it is necessary to
measure simultaneously the pressure difference between source
and sink, the path resistance and the velocity of movement.
In soybean, the sieve plate pores are open (Fisher 1975b), so
a calculated value for resistance can be obtained from the
sieve element dimensions. The translocation velocity can be
measured by labeling with $^{14}CO_2$. The real problem lies in
measuring sucrose concentrations in sieve elements, since that
value, in combination with water potential measurements, would
allow calculations of turgor pressure. Our approach to this
(Housley 1974) was to label the entire shoot of a soybean
plant for several hours with constant specific activity $^{14}CO_2$
to raise the specific activity of sucrose along the entire
translocation stream to the same value. By determining the
number of disintegrations per unit volume in sieve elements,
we could calculate the sucrose concentration. Except for our
problems with sugar retention during autoradiography, this
approach worked reasonably well. The calculated sucrose
concentrations in petiolar sieve tubes were always greater
than those in root sieve tubes; the average figures were
4.6% for the petioles versus 2.4% for the roots, for 6
experiments. In most of the experiments, however, the
calculated sucrose concentrations would not have been sufficient even to maintain positive turgor in the sieve elements.
The results were, therefore, generally favorable to the
Münch hypothesis, but too much sucrose had been lost during
autoradiography to make that claim convincing.

Negative Staining.- The most promising approach to measuring sugar concentration seems to be by negative staining.
For this approach, the tissue is embedded in Spurr's epon

containing 6% (w/v) Sudan B, and the per cent transmittance of some area without solutes (usually xylem vessels) relative to the sieve tube contents is used as a measure of sugar concentration. These measurements were made at 610nm with a Leitz MPV II microspectrophotometer. Figure 4 shows the appearance of

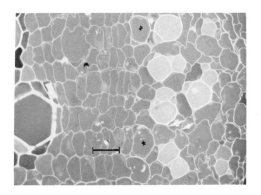

Fig.4.- A 1.5 micron section of bean (Phaseolus) *vascular tissue freeze-substituted with propylene oxide and embedded in epon containing 6% Sudan B. Asterisks (*) indicate immature sieve elements. The bar indicates 20 microns.*

a 1.5 micron section of bean phloem prepared in this way. Negative staining is quite marked, but uniform, in both sieve elements and companion cells. Two differentiating sieve element-companion cell pairs are visible (*), but do not stain differently from other parenchyma cells.

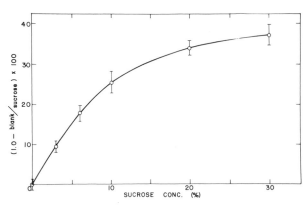

Fig.5.- Standard curve relating staining intensity to sucrose concentration in 1.5 micron sections of sucrose - containing pith blocks which were freeze-substituted in propylene oxide and embedded in epon containing 6% Sudan B.

Figure 5 illustrates a standard curve prepared from 1.5 micron sections of known sucrose concentrations in pith blocks. The relationship is not a linear one, even in a semilog plot, and the decrease in staining is greater than would be expected by a simply volume exclusion mechanism. Nevertheless, it seems to be quite reproducible at sucrose concentrations below about 15%. Using this standard curve, the sucrose concentrations in the sieve elements shown in Figure 4 range from 8.8 to 18.1%, with an average concentration of 11.2±2.1%.

Table 1 shows the differences in sucrose concentrations

TABLE 1. Parameters for the evaluation of Münch's osmotically-generated pressure flow hypothesis in soybean

A. Apparent osmotic, water and turgor potentials in soybean sieve elements. (Average values for 3 experiments.)

Location	Sucrose conc. %	Ψ_s (bars)	Ψ_w (bars)	Ψ_p (bars)	$\Delta\Psi_p$ (bars)
Petiolule	11.5±1.7	-9.7	-3.7[1]	-6.0	
					4.1
Root	5.3±0.8	-3.9	-2.0[2]	-1.9	

B. Path characteristics (ranges for 3 experiments)

Path length (cm)	Velocity (cm min^{-1})	$\Delta\Psi_p$ required with pores 100% open	70% open[3]
70-160	0.8-0.9	1.2-2.0	2.7-4.6

[1] Measured in the leaf by thermocouple psychrometry.
[2] Estimated; nutrient Ψ_w was -0.6 bar. [3] i.e., with the pore diameter equal to 0.7 times its measured value.

between soybean source (petiolule) and sink (young root) sieve elements, as measured by negative staining. The relative values agreed reasonably well with those found by auto-radiography, but the absolute values were much more reasonable, since they indicated the presence of appreciable turgor. The turgor differences between source and sink were about the same for all experiments, even though the path lengths were substantially different. However, the gradient seemed sufficient in all cases to drive an osmotically-generated pressure flow mechanism at the observed velocities.

CONCLUDING REMARKS

In conclusion, I would like to make the observation that most histochemical approaches to water-soluble organic compounds have been directed at their autoradiographic localization. Much of our effort has been directed toward that same goal. This need not be the only approach, however, and I have tried to illustrate that other approaches are not only feasible, but can provide quantitative results that have important implications for our understanding of transport in plants.

ACKNOWLEDGEMENTS

The author's work has been supported by National Science Foundation Grants GB14719 and GB33903.

REFERENCES

Eschrich, W. and Fritz, E. (1972). Microautoradiography of water-soluble organic compounds. In "Microautoradiography and Electron Probe Analysis. Their Application to Plant Physiology," ed U. Lüttge. Springer-Verlag, N.Y. pp. 99-122.

Fisher, D.B. (1972). Artifacts in the embedment of water-soluble compounds for light microscopy. *Plant Physiol.* *49*, 161-165.

Fisher, D.B. (1975a). Translocation kinetics of photosynthates In "Phloem Transport", ed. by S. Aronoff, J. Dainty, P.R. Gorham, L.M. Srivastava and C.A. Swanson. Plenum Press, N.Y. pp. 327-358.

Fisher, D.B. (1975b). The structure of functional soybean sieve elements. *Plant Physiol. 56*, 555-569.

Fritz, E. and Eschrich, W. (1970) ^{14}C - Mikroautoradiographie wasserlöslicher Substanzen im Phloem. *Planta (Berl.) 92*, 267-281.

Housley, T.L. (1974). An evaluation of the pressure flow hypothesis of phloem transport. Ph.D. thesis, Univ of Georgia, Athens.

Housley, T.L. and Fisher, D.B. (1975). The efficiency of ^{14}C-detection in autoradiographs of semithin plastic sections. *J. Histochem. Cytochem. 23*, 678-680.

Lowry, O.H. and Passonneau, J.V. (1972). "A Flexible System of Enzymatic Analysis." Academic Press, N.Y.

Lüttge, U., ed. (1972). "Microautoradiography and Electron Probe Analysis. Their Application to Plant Physiology". Springer-Verlag, N.Y.

Lüttge, U. and Weigl, J. (1965). Zur Mikroautoradiographie wasserlöslicher Substanzen. *Planta (Berl.) 64*, 28-36.

MacRobbie, E.A.C. (1975). Activated mass flow: surface flow. In "Phloem Transport", ed. by S. Aronoff, J. Dainty, P.R. Gorham, L.M. Srivastava and C.A. Swanson. Plenum Press, N.Y. pp. 585-600.

Outlaw, W.H. and Fisher, D.B. (1975). Compartmentation in *Vicia faba* leaves. I. Kinetics of ^{14}C in the tissues

following pulse labeling. *Plant Physiol.* *55*, 699–703.
Outlaw, W.H., Fisher D.B. and Christy, A.L. (1975).
Compartmentation in *Vicia faba* leaves. II. Kinetics of
^{14}C-sucrose redistribution among individual tissues
following pulse labeling. *Plant Physiol.* *55*, 704–711.
Pallaghy, C.K. (1973). Electron probe microanalysis of
potassium and chloride in freeze-substituted leaf sections
of *Zea mays.* *Aust. J. Biol. Sci.* *26*, 1015–1034.

CHAPTER 22

The Distribution of P-protein in Mature Sieve Tube Elements

[1]G.P. Dempsey[1], S. Bullivant[2], and R.L. Bieleski[1]
[1]*Plant Diseases Division, DSIR;* [2]*Department of Cell Biology, University of Auckland, Auckland, New Zealand*

INTRODUCTION

Electron microscopical investigations of sieve element structure are plagued by the extreme sensitivity of the sieve element and its contents to cutting, chemical fixation, and conventional embedding procedures. Consequently controversy still exists over the *in vivo* arrangement of the sieve element contents, particularly the degree of blocking of the sieve plate pores by the P-protein filaments.

Promising alternatives to conventional tissue preparation methods are those involving stabilization by freezing. They have the particular advantage that chemical fixation is not required. Rapid freezing, in association with freeze-substitution, would appear to stabilize the tissue and provide an image of the sieve element as little removed as possible from the *in vivo* condition. Unless special techniques are used, it is necessary to pre-treat the material with cryoprotectants (such as glycerol) in order to avoid structural disruption by ice crystals. However the unphysiological step of cryoprotectant treatment can be eliminated if the freezing rate is so rapid that the ice crystals formed are too small to interfere with the image seen in the electron microscope.

In a series of experiments using the petiole of celery (*Apium graveolens*) we used electron microscopy of thin sectioned and freeze substituted tissue to obtain information on the *in vivo* distribution of P-protein in the lumen and pores of functioning sieve elements (Dempsey et al. 1975).

FIXATION ARTEFACTS

The effect of turgor release. - The effect of turgor release was investigated by conventional fixation and processing of a phloem strand 3.5 cm long. After embedding, the

phloem strand was cut into seven 0.5 cm lengths and thin-
sections from each segment were examined.

Sieve elements within 1 cm of each end displayed sieve
plate pores largely blocked with P-protein. Consistently
there was an accumulation of P-protein on the side of the
sieve plate away from the adjacent cut end (upstream), and a
"trailing" condition on the side towards it (downstream).
Within some sieve plate pores the P-protein was so closely
compacted that individual P-protein filaments could not be
distinguished. P-protein within the lumen of these sieve
elements was largely unoriented and fragmented.

Sieve elements in sections taken from the central 1 cm
of phloem showed less compaction of the P-protein within the
sieve plate pores, such that clear areas existed in the pores
between the P-protein filaments. In general there was no
accumulation of P-protein upstream or downstream. The degree
of orientation of the P-protein filaments within the pores
and close to the pores was much more marked in this region
of the strand.

Chemical fixation and ice crystal damage. - When strands
of celery phloem were glutaraldehyde fixed, frozen in Freon 12
and then freeze-substituted, sieve plate pores were invariably
blocked with P-protein. The P-protein was compacted into a
structureless mass blocking the sieve plate pores and not
extending any great distance into the lumen of the sieve element. However if the tissue was frozen directly without prior
chemical fixation many of the sieve plates displayed partially
open pores.

It should be noted that tissue frozen directly in Freon
12, without the use of a cryoprotectant, always displayed ice
crystal damage. Ice crystals were usually evident in the
cell wall and between the P-protein filaments.

THE USE OF RAPID FREEZING

To maximize the rate of freezing and thus avoid the
formation of damaging ice crystals, use can be made of the
excellent heat conducting properties of copper. By this
freezing method, a surface layer of tissue approximately 12
μm deep can be obtained which is free from electron micro-
scopically-visible ice crystals (Dempsey and Bullivant 1976a,
1976b). Within this layer, the P-protein can be seen to exist
as discrete bundles of parallel filaments, approximately
0.5 μm in diameter, oriented longitudinally within the sieve
element, (Fig.1a), without however any evidence of an enclosing membrane. Bundles of filaments in the lumen of adjacent
sieve elements do not appear to be continuous through the
sieve plate pores. Similarly oriented P-protein also existed

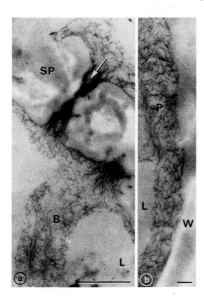

Fig.1. - Freeze-substituted sieve element from the surface 12 μm layer of celery phloem, frozen onto a copper surface at liquid nitrogen temperature, while still attached to the plant at both ends (Dempsey et 1975). (a) The sieve plate pore (arrow) shows an unobstructed central channel, with P-protein, occurring as a peripheral lining. The P-protein within the lumen (L) of the sieve element is confined to a discrete bundle (B), with no indication of a limiting membrane. (Bar represents 0.5 μm). (b) A parietal layer (P) of P-protein lines the lateral walls (W) of the sieve elment. L - lumen. (Bar represents 0.1 μm).

as a parietal layer in the lumen of the sieve element, varying in thickness from 0.1 to 0.5 μm (Fig. 1b). In general, the sieve plate pores appear to have a central unobstructed channel, but are lined with a peripheral layer of P-protein filaments, which reduces the effective diameter of the pore by about one-third.

THE SIEVE ELEMENT MODEL

By using various tissue preparation procedures to study sieve element structure, we have found a variety of arrangements of P-protein in a single type of phloem tissue. The results fit a coherent pattern in which the effects of chemical fixatives and turgor release have dominated the P-protein structures observed. By avoiding these two evils, through the rapid freezing of attached phloem strands, we have produced electron micrographs of sieve elements which we believe are the most unmodified from the *in vivo* condition yet achieved. They reveal an internal organization of the sieve element as represented diagrammatically in Fig. 2.

The bundles of P-protein filaments within the lumen must be very susceptible to disturbance. At depths slightly greater than 12 μm from the well-frozen surface, where small (25 nm) ice crystals occur in the sieve element walls and in the adjacent cells, bundles of P-protein filaments are not found near the sieve plate. The extreme susceptibility of the bundles of P-protein to physical and chemical

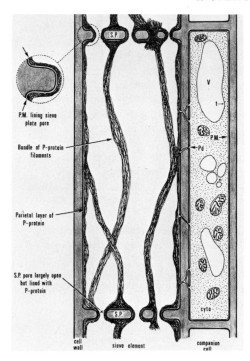

P.M. lining sieve plate pore

Bundle of P-protein filaments

Parietal layer of P-protein

S.P. pore largely open but lined with P-protein

cell wall   sieve element   companion cell

Fig. 2 - Model for the organization of structures within celery sieve element. SP - sieve plate; PM - plasma membrane; V - vacuole; t - tonoplast; Pd - plasmodesma; cyto-cytoplasm (from Dempsey et al. 1975).

manipulation would also explain their absence in glutaraldehyde-fixed thin-sections.

In the past it has been difficult to reconcile the established physiological characteristics of sieve element translocation with structures observed by electron microscopy. However, calculations of pressure gradients along the lines of those discussed by Weatherley (1972), indicate that the disposition of P-protein that we have proposed for the lumen and pores of mature sieve elements is compatible with the Münch mass flow theory of translocation, for known rates of translocation over reasonable distances. On the other hand, the highly organized arrangement of the P-protein suggests that it may have some important role to play in phloem translocation.

REFERENCES

Dempsey, G.P. and Bullivant, S. (1976a). A copper block method for freezing non-cryoprotected tissue to produce ice-crystal-free regions for electron microscopy. I. Evaluation using freeze-substitution. *J. Microscopy* (In press).

Dempsey, G.P. and Bullivant, S. (1976b). A copper block method for freezing non-cryoprotected tissue to produce ice-crystal-free regions for electron microscopy. II.

Evaluation using freeze-fracturing with a cryo-ultra-microtome. *J. Microscopy* (In press).

Dempsey, G.P., Bullivant, S. and Bieleski, R.L. (1975). The distribution of P-protein in mature sieve elements of celery. *Planta (Berl.) 126*, 45-59.

Weatherley, P. (1972). Translocation in sieve tubes. Some thoughts on structure and mechanism. *Physiol. Veg. 10*, 731-742.

CHAPTER 23

Nutrients and Metabolites of Fluids Recovered from Xylem and Phloem: Significance in Relation to Long-distance Transport in Plants

J.S. Pate

Botany Department, University of Western Australia, Nedlands, W.A., Australia

INTRODUCTION

A logical and direct way of examining long distance transport of solutes in plants is to recover the liquid contents of conducting elements and identify and measure the amounts of the compounds present. This approach has been applied to a number of plants, using several techniques, and, when coupled to isotope studies, it can provide useful information on how various metabolities and nutrients enter specific transport pathways and the manner in which they subsequently move within the plant body.

COLLECTION OF XYLEM FLUIDS

Two basically different techniques exist for recovering liquid from xylem. In the first 'tracheal' sap from xylem elements is extracted from stem segments by centrifugation (Gottleib 1943), liquid displacement (Gregory 1966), or application of a mild vacuum to one end of the segment whilst cutting a succession of short segments off the other end (Bennett et al. 1927; Bollard 1953; Kessler 1966). Extraction proves easiest with woody species especially with tropical lianes possessing large vessels (Gessner 1965). In herbaceous plants the vacuum extraction technique is restricted to species with solid stems (Bollard 1960).

The second technique exploits release from xylem under root pressure. Plants decapitated at ground level are favoured, but bleeding sap can be obtained higher up a shoot from cut petioles (Pate et al. 1964). Xylem of deciduous trees bleeds most prolifically after leaf fall in autumn or in spring just before bud burst: cutting a branch or boring a hole in the trunk is used to collect the sap (see Zimmermann 1961; Die and Willemse 1975).

Several problems exist with the techniques. Damaged cells may yield adventitious solutes to bleeding sap, and to counteract this the first drops to be exuded should be discarded. Contamination is less if sap is collected from a central cylinder of xylem enclosed by a sleeve fitted inside the bark or cortex. The period of collection should be short to minimize starvation reactions in root cells loading the xylem (Koster 1963, Pate and Greig 1964). Microbial activity in the collecting sap should be avoided (e.g. see Schnathorst 1970). The principal danger with the vacuum displacement technique is release of solutes from compartments other than xylem elements. This is thought to be serious by some investigators (Morrison 1965; Hardy 1969; Hardy and Possingham 1969), but less important by others (Bollard 1953; Cooper et al. 1972; Die and Willemse 1975). The magnitude of the vacuum applied and the anatomical nature of the tissues are probably critical in this respect.

Root bleeding sap has much higher solute levels than tracheal sap of the same plant, as is to be expected from the differing rates of water flux in detopped and whole plants.

Gradients in concentrations of different xylem solutes have been demonstrated from base to top of woody and herbaceous stems, (Klepper and Kaufmann 1966; Jones 1971; Cooper et al. 1972), so composition and solute concentrations relate strictly to the point of collection. Solute gradients are attributed to selective exchange between the ascending xylem stream and the bordering walls and protoplasts, perfusion studies indicating that both adsorption on cell walls and irreversible uptake into living cells are involved (Die and Vonk 1967; Hill-Cottingham and Lloyd-Jones 1973). Autoradiography shows that vascular parenchyma can absorb xylem-delivered solutes (Wooding and Northcote 1965; Pate and O'Brien 1968). Transfer cells located at nodes are particularly active in this respect (Pate and Gunning 1972). A specific region of a stem can be shown to absorb certain solutes from xylem whilst releasing others (Pate et al. 1964; Hardy and Possingham 1969; Pate et al. 1970). Upper parts of roots can also withdraw solutes from the xylem stream (Shone et al. 1969) so bleeding sap collected at ground level does not necessarily portray exactly what was discharged to the xylem by the distal feeding parts of the root.

Xylem fluid composition varies with time of day (Die 1959; Pate and Greig 1964), plant age and seasonal cycle (Bollard 1953; Hofstra 1964; Pate 1971, 1973), and the plants nutritional state (Dubinia 1965; Ozerol and Titus 1968; Selvendran 1970; Ivankov and Ingversen 1971; Ezeta and Jackson 1975). It can also be altered radically if the plant is diseased

(Singh and Smalley 1969; Schnathorst 1970; Beever 1970). A single sap analysis is therefore of little significance unless matched against the range of variation typical of the species.

COLLECTION OF PHLOEM EXUDATES

The delicate nature of phloem tissue and the readiness with which sieve plates become blocked with slime or callose make it very difficult to collect fluids from sieve elements. The techniques employed rely on the sieve tube contents being under turgor pressure and therefore likely to exude when cut or punctured.

The first technique, originating from Hartig (1860), employs incisions in the outer bark of a tree, the exuded sap being collected into capillaries. Too deep a cut results in loss of exudate to the xylem (Zimmermann 1960, 1969; Ziegler 1974). The technique has been applied to a few herbaceous Dicotyledons. An alternative technique is to cut transversely through an organ (e.g. inflorescence stalk, fruit tip, or petiole) and thus initiate bleeding. Examples of species of plants and regions of plants known to produce phloem exudates are listed in Table 1. Sealing of sieve pores is sluggish

TABLE 1. Examples of Plants known to Bleed from Cut Phloem and Organs used to Obtain the Exudate

Woody Species
 incisions in bark   —over 400 species known to bleed, mainly deciduous trees (Crafts and Crisp 1971; Ziegler 1974).

 cut distal tips of fruits   —*Spartium, Jacksonia, Genista* (Pate et al. 1974).

Arborescent Monocotyledons
 cut inflorescence stalks
 or floral apices   —*Yucca, Cocos, Arenga, Corypha* (Die 1974, Die and Tammes 1975).

Herbaceous Dicotyledons
 incisions in stem   —*Cucurbita* (Crafts and Crisp 1971); *Ricinus* (Milburn 1971, 1974).

 cut distal tips of fruit   —*Lupinus* (Pate et al. 1974).
 cut root phloem   —*Beta* (Fife et al 1962).
 cut inflorescence stalks   —*Pisum* (Lewis and Pate 1973); *Brassica* (Pate 1973).

 cut petioles   —*Phaseolus* (King and Zeevaart 1974); *Cucurbita* (Bieleski 1969).

255

in all these phloem bleeders leaving a period of minutes or even hours during which exudate flows, albeit at an ever diminishing rate. Bleeding can usually be reinitiated by recutting the phloem close to the original cut - a series of sap collections can then be taken from a specific region at several times in the life of the plant (Canny 1973; Die 1974; Sharkey and Pate 1975).

A second technique utilizes feeding aphids. They are anaesthetized in a gentle stream of carbon dioxide, their bodies severed, and phloem exudate collected from cut mouthparts using fine capillaries (Mittler 1953, 1958; Weatherley and Johnson 1968; Zimmermann 1969). Exudation occurs over a long period, but volumes obtained from a single aphid are obviously likely to be much lower than from a cut or incision. The technique is limited by the dearth of suitable large aphids, and by the restricted feeding locations of aphids on a plant. *Salix, Tilia, Juniperus, Picea* and *Heracleum* are genera from which stylet exudates have been obtained (see Canny 1973). Sectioning of tissues containing embedded aphid mouthparts shows that the stylet tips penetrate individual sieve elements (see Zimmermann 1961; Evert et al. 1968; Evert et al. 1973; Bornman and Botha 1973). The exudate obtained is therefore regarded as giving a most reliable sample of sieve elements contents.

Aphids are also used for collection of 'honey dew' (Mittler 1958; Hill 1963; Eschrich 1967; Lepp and Peel 1970; Cleland 1974) the technique having the advantages that the amounts of solute obtained are relatively great and that aphids of any size can be used. However, 'honey dew' is an excretory product and its solute composition is likely to be different from sieve elements on which the aphid is feeding.

There are several reasons why phloem exudates from incision or cutting are unlikely to give an accurate picture of the plants' translocatory fluid. Artefacts might arise from the following:-

a) *Discharge of non-mobile and structural constituents due to pressure release on cutting:-* The high protein level (10 mg ml^{-1}) in phloem exudate of Cucurbitaceae (Crafts and Crisp 1971) is an extreme example of this (Thaine 1964; Kollmann et al. 1970). Enzymatically-active protein, and other types of protein, (e.g. lectins) (Crafts and Crisp 1971; Lehman 1973; Kauss and Ziegler 1974; Ziegler 1974) may originate in this manner.

b) *Osmotic attraction of water to cut sieve elements from neighbouring tissues:-* This will cause a progressive dilution of phloem sap as is commonly observed after excision (Tingley 1944; Zimmermann 1961). Not all species respond in this manner; studies on *Ricinus* (Milburn 1971, 1974) and *Lupinus* (Pate et al. 1974), report an exponential decline with time in bleeding rate but no noticeable fall off in concentration of solutes. The water potential of adjacent tissues at the time of cutting has a major influence on solute concentrations (Hall and Milburn 1973). Temperature has a pronounced influence on bleeding performance of phloem (see Canny 1973).

c) *Loss of water from damaged tissue surfaces to the more concentrated drop of exudate:*- This, together with (b) may explain why sugar concentrations in phloem exudates of a plant may vary widely with the type of tissue and organ which is cut. Sucrose concentrations in exudates of *Lupinus*, for instance, vary from 60 mg ml^{-1} in young stems to up to 100 – 140 mg ml^{-1} in fruits, only that collected from the phloem-rich vascular tissue of the beak of the fruit having a concentration approaching that (150 – 170 mg ml^{-1}) of the exudate collected from stylets of aphids feeding on fruits of the species (Pate et al. 1974). Concentration values for the stylet exudates are regarded to be closest to that of the intact sieve tube.

d) *Exchange of solutes with damaged tissue surfaces irrigated with the phloem exudate:*- Direct loss of cell contents is a general source of artefact; a less obvious one is the possible elution of ions (e.g. Ca^{++}) from cut wall and damaged cytoplasm by the potassium-rich exudate. Discharge from xylem or laticiferous tissue are other likely sources of contamination.

e) *Loss of phloem functioning following cutting:*- Pressure release leads to solutes escaping from sieve elements at rates initially many times faster than the rate of translocation *in vivo* (e.g. see results on *Lupinus*, Sharkey and Pate 1975). The pressure surge following cutting is likely to result in a considerable length of phloem being suddenly emptied and, since non-aqueous contents may then block the sieve plates, further translocation may become greatly impeded (Canny 1973). Nevertheless in certain plants phloem bleeding resumes at almost its original rate when phloem is recut, and ^{14}C labelled assimilates from a distant source continue to arrive in quantity in the exudate (Die et al. 1973; Sharkey and Pate 1975).

f) *Changes in source of loading of assimilates following cutting:*- This effect is well displayed in *Yucca*, in which cut ends of isolated inflorescences bleed sugar profusely, despite their acting before detachment as sinks for assimilates (Die and Tammes 1966).

The above classes of artefact are less likely to apply to sieve tube sap collected from aphid stylets. The localized

257

nature of the puncture and the impervious nature of the cuticle
of the mouthparts will minimize contamination from surrounding
tissues, and since exudation proceeds for several days with
little change in solute flux, the translocation process is unlikely to have been radically altered (Canny 1973). Nevertheless the stylet, in acting as an 'artifical' sink may
cause mobilization of solutes and water from sources not previously supplying the system (Weatherly et al. 1959; Peel
1965).

Assuming that phloem exudates do indeed provide useful
information on translocated solutes, it is still important
to relate each single analysis to the broader picture of
variations within and between plant species. Although solute
composition of phloem appears to be more effectively buffered
against changes in the plant's nutrition than is composition
of xylem (Pate et al. 1975), seasonal and developmental
variations are sufficiently large to suggest that the translocation stream dispenses solutes at varying rates and proportions at different times and locations (Hill 1962; Zimmermann
1960, 1969; Leckstein and Llewellyn 1975). It is only when
these parameters have been explored that generalizations can
be established about the translocation behaviour of a species.

SOLUTES OF XYLEM FLUIDS

Xylem sap is acidic (pH 5.2 - 6.5) and usually has a
total solids content of $1 - 20$ mg ml^{-1} (Bollard 1960; Fife
et al 1962; Pate 1975), a third or so of this inorganic
(Bollard 1960). Essential mineral elements taken up from
soil are all classed as xylem mobile (Table 2), K^+, Ca^{++},
Mg^{++}, Na^+ being major cations, PO_4^{---} Cl^-, SO_4^{--} major anions.
The ionic composition of the xylem varies with the composition
of the rooting medium and with the current ionic status of
the root (see Collins and Reilly 1968; Weatherley 1969).

The phosphorus and sulphur of xylem sap may be partly
organic, phosphorus as phosphorycholine and other organic
phosphates (Tolbert and Wiebe 1955; Bieleski 1973), sulphur
as methionine, cysteine and glutathionine (Bollard 1960;
Pate 1965).

Nitrogenous constituents are major solutes of xylem, and
usually one nitrogen-rich organic molecule dominates the
spectrum of compounds present (Pate 1971). The principal
solute depends on the species (Pate 1973) and may be an amide
(glutamine or asparagine), a ureide, an alkaloid, or a non-
protein amino acid (Table 2). A wide range of amino acids
may also be present. The nitrogenous constituents are
present at $0.1 - 5$ mg ml^{-1}. Somewhat higher levels are recorded for the xylem sap of root nodules of legumes (see Pate, this

volume). Species capable of assimilating nitrate in roots
have high levels of organic nitrogen relative to free nitrate.
Species not able so to do (e.g. *Xanthium, Stellaria, Trifolium*)
have most of their xylem nitrogen as free nitrate, (Pate 1973),
but, if ammonium is supplied, amounts of amide are formed
in the roots and leave in the xylem (Wallace and Pate 1967;
Pate, unpublished data). The xylem of herbaceous plants is
most rich in nitrogen at flowering, (Hofstra 1964; Pate et al.
1965; Avundzhyan 1965), in fruit trees at full blossom
(Bollard 1953; Beever 1970).

Sugars occur in tracheal or bleeding sap of trees and
woody climbers (Olofinboba 1969; Hardy and Possingham
1969; Beever 1970), especially in sugar maple (*Acer saccharum*)
in winter (Sauter et al. 1973). The relatively high levels (6
mg ml^{-1}) of sorbitol in xylem of apple trees during winter
dormancy is said to be related to frost hardiness (Williams and
Raese 1974). Xylem sap of herbaceous plants tends to be much
less rich in carbohydrate than sap of woody species.

Organic acids of xylem are regarded as exported products
of root metabolism and evidence of cycling of carbon through
roots (Kursanov 1961, 1963). Citrate is implicated as a
chelating agent for transport of iron (Schmid and Gerloff
1961; Clark et al. 1973).

Substances exhibiting auxin-, gibberellin-, abscisin-,
or cytokinin-like activity have been found in xylem sap. Few
identifications of individual compounds have been made, but
ʸⁿ (ᴬ? ᵢₙₚₙₐₗₙₐₗₐₚ1) ᵢₙₗₐₘₙₐₜᵢₙₑ ₗₙₙₐ ₗₙₐ ₙ ᵢₙₐₜᵢₙₐₙₐₗ ᵢₙ ₐₚₗₐₘ ᵢₙᶠ
Acer saccharum (Hall 1973) and indole-3, acetic acid (IAA) in
xylem of *Ricinus* (Hall and Medlow 1974). The role of xylem-
borne growth substances has been widely discussed (Kende 1965;
Bowen and Hoad 1968; Burrows and Carr 1969; Selvendran and
Sabaratnam 1971).

SOLUTES OF PHLOEM EXUDATES

Phloem exudates contrast with exudates of xylem in
being highly alkaline (pH 8.0 - 8.4) and possessing a high
solids content (50 - 300 mg ml^{-1}). Soluble carbohydrates
account for 80 - 90% of the exudate's dry matter (see Table
3). Sucrose is usually the main component (see Table 3)
especially in herbaceous species (Crafts and Crisp 1971) but
the oligosaccharides raffinose, stachyose and verbascose
predominate in certain genera and families, e.g. Bignoniaceae,
Celastraceae, Combretaceae, Oleaceae, Verbenaceae, Myrtaceae,
Onagraceae, Tiliaceae, Ulmaceae, Anacardiaceae, Buxaceae,
Clethraceae, Corylaceae and Rutaceae (Zimmermann 1969; Ziegler
1974). Sugar alcohols occur in certain species, sorbitol
in some Rosaceae (Webb and Burley 1962; Kluge 1967), mannitol
in Oleaceae and some Combretaceae (Kluge 1967), dulcitol in

TABLE 2.   Solutes Recovered from Fluids of Xylem and Phloem

Carbohydrates
    sucrose, monosaccharides, raffinose, stachyose, verbascose,
    succinate, mannitol, dulcitol, sorbitol, myoinositol

Organic acids
    citrate, malate, tartrate, oxalate, keto glutarate,
    diketogulonate, maleate, succinate, phloretic acid* (1)

Nitrogenous solutes
    amides - asparagine, glutamine, γ-N ethyl-glutamine* (2)
            γ-methylene glutamine*
    ureides - allantoic acid, allantoin
    protein amino acids - alanine, aspartate, glutamate,
            histidine, leucine, isoleucine, glycine,
            methionine, cystine, phenylalanine, serine,
            proline (trace), threonine, tyrosine, valine,
            arginine (trace only in phloem)
    non-protein amino acids - homoserine, O-acetyl homoserine(3)
            citrulline, γ-amino butyric acid
    peptides - glutathione,* alanyl aminbutyric acid,
            glycylketoglutaric acid, other unidentified
            peptides
    alkaloids - nicotine, * and other unidentified alkaloids*
    inorganic nitrogen - ammonium and nitrate (rarely in
            phloem)

Organic Phosphates
    *variety of sugar phosphates, UDP, UMP, UTP, UDPG, ADP,
            AMP, ATP, GTP,* phosphoglycerate, phosphoryl-
            choline, phosphorylethanolamine

Inorganic Ions
    SO_4^{-2}, HCO_3^{-1}(4), PO_4^{-3}, HPO_4^{-2}, Cl^{-1}, K^{+1}, Ca^{+2}, Mg^{+2},
    Na^{+1}

Trace Elements
    B, Mn, Fe, Zn, Cu, Mo

Vitamins
    ascorbate, biotin, folic acid, nicotinic acid,
    pantothenic acid, pyridoxine, riboflavine, thiamine

Growth Regulating Substances
    abscisins, auxins, cytokinins, gibberellins, flower
    inducing and inhibiting substances, (5, 6) sterols

Cucumis (Kluge 1967). Seasonal and diurnal variations in the proportions of phloem sugars have been studied in incision exudates of *Frazinus* (Zimmermann 1971).

Amides, amino acids and ureides represent principal organic solutes of nitrogen in phloem sap (Table 2). There is species specificity in relation to the major solute, glutamine predominating in *Yucca* phloem exudates (Tammes and Die 1964), asparagine in *Lupinus, Pisum* and *Spartium* (Lewis and Pate 1973; Pate et al. 1974), allantoin and allantoic acid in *Acer* (Ziegler and Schnabel 1961), citrulline in *Cucurbita* (Eschrich 1963), *Alnus, Betula, Carpinus, Ostrya* and *Juglans* (Ziegler and Schnabel 1961). The major nitrogenous solute in xylem is often the same as in phloem (see Pate 1971), suggesting that free transfer between pathways might occur (see labelling studies in a later section). As well as one or more of these nitrogen-rich compounds phloem carries a variety of amino acids. Glutamic and aspartic acids, valine, serine, threonine, glycine and alanine probably rank higher in general abundance than others; proline occurs in unusually higher amount in *Tilia* and *Robinia* (Ziegler 1974). Amino compounds total 20 - 80 mg ml^{-1}, high values being recorded for deciduous tree species at leaf senescence (Weatherley et al. 1959), or in legumes at fruit filling (Pate et al. 1975; Pate (this volume)). Seasonal changes in amino acid composition of phloem exudates are described for *Salix* spp. (Mittler 1953; Peel and Weatherley 1959; Lockstein and Llewellyn 1975) and *Robinia* (Ziegler 1956). Total amino compounds in fruit tip phloem sap of *Lupinus* vary widely with time of day, age of fruit and type and concentration of nitrogen in the rooting medium. But the relative balance between individual constituents is remarkably constant (Pate et al. 1974, 1975; Atkins et al. 1975; Sharkey and Pate 1975). Individual amino acids can be from 3 - 47 times more concentrated in phloem than in xylem (Pate et al. 1975).

Note: Substances underlined recorded only for phloem; substances asterisked* recorded only for xylem. All other compounds constituents of both xylem and phloem.

Main Source References: Bollard 1960; Zimmermann 1960, 1969; Crafts and Crisp 1971; Pate 1971, 1973 1975; Canny 1973; Bieleski 1973; Ziegler 1974.

Other Specific References: (1) Grochowska 1967; (2) Selvendran 1970; (3) Lewis and Pate 1973; (4) Hall and Baker 1972; (5) Cleland 1974; (6) Cleland and Ajami 1974.

Protein occurs in phloem sap in amounts ranging from 0.5 - 10 mg ml^{-1} but it is unlikely to be a mobile constituent. Discussion of the enzymatic activity of phloem exudates is considered outside the scope of this paper. A good reference to the subject is Ziegler (1974).

Nitrate is usually recorded as absent from phloem sap (MacRobbie 1971; Crafts and Crips 1971). Ammonium is unlikely to be present in the highly alkaline contents of sieve tubes, and if recorded it is likely to have arisen by breakdown of amides during or after collection of the sap.

Organic acids recorded for phloem exudates are listed in Table 2. Malate, present in *Ricinus* exudate at 2 - 3.2 mg ml^{-1} (Hall and Baker 1972) is obviously an important anion.

Mineral elements of phloem exudates are shown in Tables 2 and 3. Potassium (1.2 - 4.4 mg ml^{-1}) is invariably the major cation. Sodium content is much less (0.002 - 0.300 mg ml^{-1}). The other common cation is Mg^{++} (0.06 - 0.20 mg ml^{-1}). Ca^{++} is usually at very low level (0.002 - 0.014 mg ml^{-1}), but phloem exudates of *Ricinus* (Hall and Baker 1972) and *Lupinus* and *Spartium* (Pate et al. 1974), may show higher levels (0.02 - 0.08 mg ml^{-1}).

Phloem exudates possess high K : Na and Mg : Ca ratios (Table 3) features shared with fruits and young leaves - organs fed principally by phloem. Conversely, the high calcium levels of mature leaves reflect continued arrival in the transpiration stream but little export in the phloem (Pate 1975). Indeed calcium may be present in phloem at a lower level than in xylem (e.g. data for *Lupinus*, Table 3).

Sulphate, chloride phosphate and bicarbonate are major inorganic anions of phloem exudate of *Ricinus* (Hall and Baker 1972), phosphate is likely to be the agent buffering phloem contents to high pH. Significant fractions of phloem phosphorus are organic, a variety of constituents having been identified (Bieleski 1969, 1973; Table 2). The ATP of phloem exudates has been studied in *Salix* (Gardner and Peel 1969), *Cucurbita* (Bieleski 1969), *Yucca* (Kluge et al. 1970) and *Tilia* (Becker et al. 1971). Its high concentration and the ease with which it exchanges with ^{32}P-phosphate (Becker et al. 1971; Ziegler 1974) suggests importance as a translocated source of high energy phosphate.

Trace elements of phloem exudates are shown in Table 2. It is not certain whether they occur in ionic, bound or chelated form. Specific elements are from 5 to 20 times more concentrated in phloem sap than in xylem of *Lupinus* (Pate et al. 1975).

The manganese and calcium of phloem are stated by Goor and Wiersma (1974) to be present almost at the limit of their

TABLE 3. Comparisons of Solute Concentrations in Xylem and Phloem Fluids in Three Species of Plants

Constituents mg ml⁻¹	Quercus rubra (Die and Willemse 1975)		Lupinus albus (Pate, Sharkey and Lewis 1974, 1975; Hocking and Pate, unpublished)		Ricinus communis (Hall, Baker and Milburn 1971; Hall and Baker 1972)	
	xylem vessel bleeding sap	bark incision phloem exudate	stem xylem tracheal sap	Fruit tip phloem exudate	xylem root bleeding sap	stem incision phloem exudate
dry matter	1.4	220	n.a.*	n.a.	2	100–125
sucrose	0.128	140	absent	80–190	absent	80–106
potassium	0.177	2.1	0.090–0.179	1.54–2.26	0.40–0.56	2.3–4.4
magnesium	0.0115	0.077	0.027–0.039	0.085–0.124	n.a.	0.109–0.122
sodium	0.004	0.037	0.060–0.082	0.12–0.14	0.21–0.28	0.046–0.276
calcium	0.018	0.049	0.017–0.095	0.021–0.063	0.23–0.45	0.020–0.092
amino acids	n.a.	n.a.	0.7–2.5	15–30	n.a.	5.2
phosphorus	0.005	0.052	0.10–0.20	0.30–0.80	n.a.	0.35–0.55**
pH	4.9–5.0	7.5	6.3	7.9	6.0	8.0–8.2

* n.a. - analysis not available
** as phosphate

263

solubility. They suggest that high pH and high phosphate levels restrict solubility.

The presence of growth substances, sterols and vitamins in phloem (Table 2) suggests that the translocation stream may carry factors regulating growth and development. The presence of B class vitamins (Ziegler and Ziegler 1962) is of interest in view of their growth stimulating properties in isolated cultures of roots. The demonstration of flower inhibiting and flower promoting substances in honey dew of aphids feeding on *Xanthium* (Cleland 1974; Cleland and Ajami 1974) fits well with the concept that such substances are translocated in phloem. The presence of gibberellins, auxins, cytokinins and abscisins in xylem and phloem implies that these substances mught circulate in plants, but this cannot be tested properly until the specific endogenous compounds involved in transport have been identified.

LABELLING STUDIES AND THE ORIGIN OF SOLUTES OF XYLEM AND PHLOEM

Only a minute proportion of the vast number of isotopic studies on transport in plants has involved analysis of labelled products in exudates of xylem and phloem, yet from these few experiments much has been learned of the origin of transported solutes and the transformations and processes which lead to export from one region of the plant to another.

Several experimental strategies have been used, some applicable to only one species, others of much wider application. Figure 1 illustrates these and cites examples of how labelled substrates have been applied, and transport fluids collected. The table accompanying the figure (Table 4) provides information on the principal labelled products recovered in each study. The information presented is not exhaustive: many studies on the feeding of labelled ions to roots are excluded, and the author's personal bias to study of organic solutes is admitted. Several general points can be made:-

1) A distinction must exist between those studies in which a substrate is fed through a pathway and at a concentration likely to simulate its normal entry into the plant, and those studies in which neither the mode of presentation nor the class of substrate applied can be considered natural or to have relevance to normal plant functioning. Thus the $^{14}CO_2$ feeding of leaves, feeding of ^{32}phosphate, ^{35}sulphate, or ^{15}nitrate to roots and the introduction of naturally-occurring substrates to cut shoots through the transpiration stream are examples of physiologically real experiments, whereas the feeding of labelled mineral ions or exotic organic substances to surfaces of leaves, bark or stem, the introduction of unnatural substances to cut shoots through xylem, and

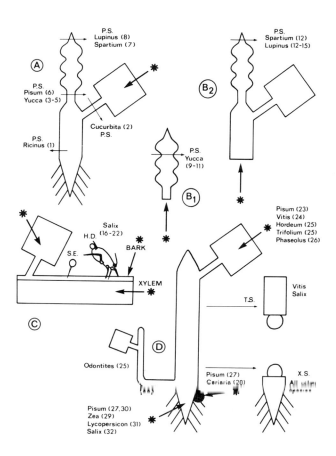

Fig. 1. Experimental strategies (A - D) used in labelling of transport fluids of xylem and phloem. Asterisks denote the region of the plant to which the isotope was applied. Regions from which transport fluids were collected and genera used in the studies are indicated. The numbers refer to references listed in Table 4, in which further details of the experiments are given.
T.S. - trachael sap; X.S. - xylem bleeding sap; P.S. - phloem incision sap; S.E. - aphid stylet exudate; H.D. - aphid honey dew.

the presentation to roots of substrates not normally present in the rooting medium are clearly 'unnatural' experiments in the sense implied above.

2) In many of the studies listed in Table 4 direct access of a solute to a conducting channel is demonstrated (e.g. the passage of sulphate and phosphate to xylem, and the

TABLE 4. Examples of Labelling Studies Involving the Collection and Analysis of Labelled Xylem or Phloem Exudates (see also Fig. 1)

(i) *STRATEGY A* (Fig.1)   *Labelling of phloem (incision) sap (P.S.) after feeding isotope to leaves*

Literature ref. and Number in Fig.1	Genus used	Labelled substrate	Principal compounds labelled in phloem exudate*
1. Hall, et al (1971)	*Ricinus*	$^{14}CO_2$	^{14}C sugars (NI)**, time course of labelling
2. Bieleski (1969)	*Cucurbita*	$^{32}PO_4$	variety of sugar phosphates, nucleotides and other compounds, inorganic phosphate
3. Tammes and Die (1964)	*Yucca*	$^{14}CO_2$	sucrose, organic and amino acids (NI)
4. Die, et al. (1973)	*Yucca*	$^{14}CO_2$	sucrose, time course of transfer to phloem
5. Kluge, et al. (1970)	*Yucca*	$^{32}PO_4$	inorganic phosphate and variety of organic compounds including ATP
6. Lewis and Pate (1973)	*Pisum*	$^{14}CO_2$	disaccharide, SER, ALA, GLY, ASP, GLU, γABU, HOM.***
7. Pate, et al. (1974)	*Spartium*	$^{14}CO_2$	sucrose, SER, ALA, GLU, ASN, GLN, organic acids (NI)
8. Pate, et al (1975)	*Lupinus*	$^{14}CO_2$	sucrose, organic acids (NI), GLU, SER, ALA, ASP, GLN, ASN, VAL

266

TABLE 4. Continued

(ii) *STRATEGY B* (Fig.1) *Labelling of phloem (incision) sap (P.S.) after feeding label via xylem to inflorescence stalk (B_1) or base of shoot (B_2)*

Literature ref. and Number in Fig.1	Genus used	Labelled substrates	Principal compounds labelled in phloem exudate*
9. Die (1968)	*Yucca*	$^{32}PO_4$	inorganic phosphate and variety organic compounds (NI)**
10. Tammes, et al. (1967)	*Yucca*	^{14}C fructose, glucose	sucrose
11. Tammes, et al. (1973)	*Yucca*	^{14}C-monosaccharides, sucrose	sucrose
		^{14}C-glycerate, glycollate	in same compound as supplied
		^{14}C-sorbitol, sorbose	mainly in same compound as applied
		^{14}C-glycerol	
		C-GLN	GLN
12. Pate, et al. (1974) (1975)	*Spartium*	^{14}C-ASN	ASN, (90% of ^{14}C), ASP, GLU, organic acids and sugars (NI)

TABLE 4. Continued

(ii) *STRATEGY B* (Fig.1)   *Labelling of phloem (incision) sap (P.S.) after feeding label via xylem to inflorescence stalk (B_1) or base of shoot (B_2)*

Literature ref. and Number in Fig.1	Genus used	Labelled substrates	Principal compounds labelled in phloem exudate*
		^{14}C-ASP	ASN, ASP, GLN, THR, SER, organic acids and sugars (NI).
		^{14}C-GLN	GLN, ASN, GLU, ASP, SER, VAL, organic acids and sugars (NI).
	Lupinus	$^{15}NO_3$	ASP, ASN, GLN, GLU, SER, ILE, LEU, THR, ALA, VAL
13. Sharkey and Pate (1975)	*Lupinus*	variety ^{14}C amino acids	complex cross transfer to phloem (see Pate, this volume)
14. Hocking and Pate (unpublished)	*Lupinus*	$^{35}SO_4$	mainly $^{35}SO_4$, some organic S compounds
15. Atkins, et al (1975).	*Lupinus*	$^{14}C, ^{15}N$ (amide)-ASN	over 90% of label still as ASN

268

TABLE 4. Continued

(iii) *STRATEGY C* (Fig. 1) *Labelling of aphid stylet exudate (S.E.) or aphid honey dew (H.D.) after bark, stem or leaf feeding of isotope to Salix spp.*

Literature ref. and Number in Fig.1	Labelled substrates	Material* collected	Principal compounds labelled in Transport Channel
16. Canny (1961)	$^{14}CO_2$	S.E.	sucrose
17. Kluge, (1970)	$^{32}PO_4$	S.E.	inorganic phosphate, variety of sugar phosphates and nucleotides
18. Peel (1970)	(leaf) $^{32}PO_4$, $^{35}SO_4$ (xylem)	H.D.	^{35}S and ^{32}P label transferred
19. Lepp and Peel (1970)	^{14}C sucrose $^{86}RbCl$	S.E.	^{14}C and ^{86}Rb label transferred
20. Gardner and Peel (1969)	^{14}C–GLU, ^{14}C–ASP	S.E.	GLU, (ASN + Sucrose?)
21. Bowen and Wareing (1969)	^{14}C kinetin, ^{14}C–GA$_3$(bark xylem)	H.S.	^{14}C transferred, metabolism suspected during or before transfer
22. Lepp and Peel (1971)	^{14}C–IAA bark	H.D.	IAA, and indolyl–aspartic acid

TABLE 4. Continued

(iv) *STRATEGY D* (Fig.1) *Labelling of root bleeding sap (X.S.) or tracheal (xylem) sap (T.S.) after application of labelled substrate to leaf or root*

Literature ref. and Number in Fig.1.	Genus used	Labelled substrates collected	Material*	Principal compounds labelled in transport channel
23. Pate (1962)	*Pisum*	$^{14}CO_2$	X.S.	ASN, ASP, GLN***
24. Hardy (1969)	*Vitis*	$^{14}CO_2$	T.S.	GLN, malate
25. Govier, et al (1967)	*Odontites* as root hemipara-site on *Hordeum* or *Trifolium*	$^{14}CO_2$	X.S.	*Odontites* on *Trifolium* ASN, ASP, GLU, GN, ALA, γABU *Odontites* on *Hordeum* ARG, GLN, ASN, LYS, HIS, ALA
26. Pate (unpublished)	*Phaseolus*	$^{14}CO_2$	X.S.	allantoin, allantoic acid, ASN, ASP
27. Oghoghorie and Pate (1972)	*Pisum*	$^{15}N_2$, $^{15}NO_3$	X.S.	organic N labelled (NI)** 60% xylem export in 3 days
28. Silvester (1968)	*Coriaria*	$^{15}N_2$	X.S.	GLN
29. Ivankov and Ingversen (1971)	*Zea*	$^{15}NO_3$	X.S.	GLU, GLN, ARG
30. Pate (1965)	*Pisum*	$^{35}SO_4$	X.S.	sulphate, MET, CYS, gluta-thione

TABLE 4. Continued

(iv) *STRATEGY D* (Fig.1) *Labelling of root bleeding sap (X.S.) or tracheal (xylem) sap (T.S.)*
after application of labelled substrate to leaf or root

Literature ref. and Number in Fig.1.	Genus used	Labelled substrates	Material* collected	Principal compounds labell-ed in transport channel
31. Die (1963)	*Lycopersicon*	^{14}C-GLU	X.S.	GLN, γABU, GLU, ASP
32. Morrison (1965)	*Salix*	^{32}PO$_4$	X.S.,T.S.	inorganic phosphate, organic-P compounds (NI) (labelling of X.S. different from T.S.).

*   Site of collection on plant is shown in Fig.1.

**   NI, chemical species not identified

*** Abbreviations for amino compounds are as suggested in Biochem. J. 126:773-780 (1972)

direct transfer of xylem-borne asparagine to phloem). In
other instances access is much more restricted and products
derived from the radiosubstrate are the main compounds available for export (e.g. the loading of products of nitrate
reduction, but not nitrate, into phloem; the conversions of
hexoses to sucrose before transport in phloem, and the inter-
conversions and metabolism of amino acids in transfer from
xylem to phloem). In these latter cases the element represented in isotopic form must be classed as mobile, but mobility
involves the generation of a new set of solutes to the one
supplied. Complexities of this sort are not always obvious
from a simple examination of the qualitative composition of
conducting channels.

3) It is usual for an author to list only those compounds in xylem or phloem fluid which achieve high specific
activity or acquire significant proportions of the fed isotope.
(Table 1). Equally important is the negative evidence that
the study presents. Substances present in large amount and
containing the same element as the isotope applied, yet not
substantially labelled in the experiment, must clearly have
arisen from physiological activities different from those
traced by the isotope. For instance ^{14}C is distributed
asymmetrically amongst amino acids of phloem after $^{14}CO_2$ feeding of leaves, and the compounds with no label or of low
specific activity might arise, for example, by transfer
from incoming xylem, by release of previously stored nitrogenous solutes, or by breakdown of protein in tissues adjacent
to the translocation pathway. Other experimental strategies
(e.g. Fig. 1B) are needed to test these possibilities.
Similarly, in the case of xylem loading in roots, labelling
studies should be designed to distinguish solutes arising by
root uptake, from those cycling through roots from shoots
or originating from pools already established within the
root.

CONCLUSIONS

Analyses of fluids recovered deliberately from xylem
and phloem give a fairly reliable picture of the kinds and
amounts of solutes engaged in long-distance transport in
plants. Artefacts relating to the collection techniques
limit detailed interpretation of composition and concentrations.

Despite their anatomical proximity the transport fluids
of phloem and xylem differ widely in pH, and in levels and
relative balance of solutes. These differences engender
nutritional repercussions in the organs served by these
transport systems.

Much has still to be learned regarding the origins and mechanisms of loading of xylem- and phloem-borne solutes. Labelling studies have shown that a transfer channel may carry materials mobilized locally at the site of loading, substances transferred from an adjacent transport pathway, or products of recent assimilation or uptake or raw materials from the environment. If experiments of this kind were applied widely among plants and to many classes of solutes a much clearer picture would emerge of the co-ordination and regulation of the plant's transport activity.

REFERENCES

Atkins, C.A., Pate, J.S. and Sharkey, P.J. (1975). Asparagine in legume seed nutrition. *Plant Physiol. 56*, 807-812.

Avundzhyan, E.S. (1965). Variation of the amino acid composition of the bleeding sap of plants during ontogenesis. *Fiziol. Rast. 12*, 930-932.

Becker, D., Kluge, M. and Ziegler, H. (1971). Der Einbau von $^{32}PO_4^{\equiv}$ in organische Verbindungen durch Siebröhensaft. *Planta 99*, 154-162.

Beever, D.J. (1970). The relationship between nutrients in extracted xylem sap and the susceptibility of fruit trees to silverleaf disease caused by *Stereum purpureum* (Pers.) Fr., *Ann. appl. Biol. 65*, 85-92.

Bennett, J.P., Anderssen, F.G. and Milad, Y. (1927). Methods of obtaining trachael sap from woody plants. *New Phytol. 26*, 316-323.

Bieleski, R.L. (1969). Phosphorus compounds in translocating phloem. *Plant Physiol. 44*, 497-502.

Bieleski, R.L. (1973). Phosphate pools, phosphate transport, and phosphate availability. *Ann. Rev. Plant Physiol. 24*, 225-252.

Bollard, E.G. (1953). The use of trachael sap in the study of apple-tree nutrition. *J. Exp. Bot. 4*, 363-368.

Bollard, E.G. (1960). Transport in the xylem. *Ann. Rev. Plant Physiol. 11*, 141-166.

Bornman, C.H. and Botha, C.E.G. (1973). The role of aphids in phloem research. *Endeavour 32* (117), 129-133.

Bowen, M.R. and Hoad, G.V. (1968). Inhibitor content of phloem and xylem sap obtained from willow *(Salix viminalis)* entering dormancy. *Planta (Berl.) 81*, 64-70.

Bowen, M.R. and Wareing, P.F. (1969). The interchange of ^{14}C-kinetin and ^{14}C-gibberellic acid between the bark and xylem of willow. *Planta (Berl.) 89*, 108-125.

Burrows, W.J. and Carr, D.J. (1969). Effects of flooding the root system of sunflower plants on the cytokinin content in the xylem sap. *Physiol. Plant. 22*, 1105-1112.

Canny, M.J. (1961). Measurements of the velocity of translocation. *Ann. Bot. 26*, 181-196.

Canny, M.J. (1973). Phloem translocation. Cambridge University Press, Cambridge.

Clark, R.B., Tiffin, L.O. and Brown, J.C. (1973). Organic
acids and iron translocation in maize genotypes. *Plant
Physiol. 52*, 147-150.

Cleland, C.F. (1974). Isolation of flower-inducing and flower-
inhibiting factors from aphid honeydew. *Plant Physiol.
54*, 894-903.

Cleland, C.F. and Ajami, A. (1974). Identification of the
flower-inducing factor isolated from aphid honeydew as
being Salicylic acid. *Plant Physiol. 54*, 904-906.

Collins, J.C. and Reilly, E.J. (1968). Chemical composition
of the exudate from excised maize roots. *Planta (Berl.)
83*, 218-222.

Cooper, D.R., Hill-Cottingham, D.G. and Shorthill, M.J. (1972).
Gradients in the nitrogenous constituents of the sap
extracted from apple shoots of different ages. *J. Exp.
Bot. 23*, 247-254.

Crafts, A.S. and Crisp, C.E. (1971). Phloem transport in
plants. Freeman, San Francisco, U.S.A.

Die, J. van. (1959). Diurnal rhythm in the amino acid content
of xylem exudate from tomato plants bleeding under
constant environmental conditions. *Koninkl. Ned. Akad.
Wetersch. C 62*, 50-58.

Die, J. van (1963). Pathways of translocation and metabolic
conversions of root absorbed $^{14}C(U)$ L-glutamic acid in
tomato plants. *Acta. Bot. Neerl. 12*, 269-280.

Die, J. van (1968). The use of phloem exudates from *Yucca
flaccida* Haw. in the study of the translocation of
assimilates. Vorträge aus dem Gesamtgebiet der Botanik
N.f. (2) 27-30.

Die, J. van (1974). The developing fruits of *Cocos nucifera*
and *Phoenix dactylifera* as physiological sinks importing
and assimilating the mobile aqueous phase of the sieve
tube system. *Acta. Bot. Neerl. 23*, 521-540.

Die, J. van. and Tammes, P.M.L. (1966). Studies on phloem
exudation from *Yucca flaccida* Haw. III. Prolonged
bleeding from isolated parts of the young inflorescence.
Koninkl. Ned. Akad. Weterschap. Proc. (C) 69, 648-654.

Die, J. van and Tammes, P.M.L. (1975). Phloem exudation
from Monocotyledonous axes. In: Encyclopedia Plant
Physiol. (N.S.) 1 Transport in Plant I Phloem transport.
(Eds. A.Pierson and M.H. Zimmermann. Springer. Berlin -
Heidelberg - New York, Chapter 8).

Die, J. van and Vonk, R.C. (1967). Selective and stereo-
specific absorption of various amino acids during xylem

translocation in tomato stems. *Acta. Bot. Neerl. 16,* 147-152.

Die, J. van., Vonk, C.R. and Tammes, P.M.L. (1973). Studies on phloem exudation from *Yucca flaccida* Haw. XII. Rate of flow of ^{14}C-sucrose from a leaf to the wounded inflorescence top. Evidence for a primary origin of the major part of the exudate sucrose. *Acta. Bot. Neerl. 22,* 446-451.

Die, J. van and Willemse, P.C. M. (1975). Mineral and organic nutrients in sieve tube exudate and xylem vessel sap of *Quercus rubra* L. *Acta. Bot. Neerl. 24,* 237-239.

Dubinia, I.M. (1965). Pathways of the primary incorporation of inorganic forms of nitrogen into root metabolism. *Fiziol. Rast. 12,* 577-583.

Eschrich, W. (1963). Der Phloemsaft von *Cucurbita ficifolia*. *Planta (Berl.) 60,* 216-224.

Eschrich, W. (1967). Bidirektionelle translokation in Siebröhren. *Planta (Berl.) 73,* 37-49.

Evert, R.F., Eschrich, W., Eichhorn, S.E. and Limbach, S.T. (1973). Observations on penetration of barley leaves by the aphid *Rhopalosiphum maidis* (Fitch). *Protoplasma 77,* 95-110.

Evert, R.F., Eschrich, W., Medler, J.T. and Alfieri, F.J. (1968). Observations on penetration of Linden branches by stylets of the apid *Longistigma caryae*. *Amer. J. Bot. 55,* 860-874.

Fife, i, J.M. and Jackson W.A. (1975). Nitrate translocation by detopped corn seedlings. *Plant Phyoiol. 5C,* 148-156.

Fife, J.M., Price, F. and Fife, D.C. (1962). Some properties of phloem exudate collected from root of sugar beet. *Plant Physiol. 37,* 791-792.

Gardner, D.C.J. and Peel, A.J. (1969). ATP in sieve tube sap from willow. *Nature 222,* 774.

Gessner, F. (1965). Untersuchungen über den Gefäss-Saft tropischer Lianen. *Planta (Berl.) 64,* 186-190.

Goor, B.J. and Wiersma, D. (1974). Redistribution of potassium, calcium, magnesium and manganese in the plant. *Physiol. Plant. 31,* 163-168.

Gottleib, D. (1943). The presence of a toxin in tomato wilt. *Phytopath 36,* 126-135.

Govier, R.N. Nelson, M.D. and Pate, J.S. (1967). Hemiparastic nutrition in Angiosperms. 1. The transfer of organic compounds from host to *Odontites verna* (Bell.) Dum. (Scrophulariaceae). *New Phytol. 66,* 285-297.

Gregory, C.F. (1966). An apparatus for obtaining fluid from xylem vessels. *Phytopath. 56,* 463.

Grochowska, M.J. (1967). Occurrence of free phloretic acid
    (p-Hydroxy-dihydro-cinnamic acid) in xylem sap of the
    apple tree. *Bull. Acad. Polon. Sci. Sér. Sci. biol. 15,*
    455-459.
Hall, R.H. (1973). Cytokinins as a probe of developmental
    processes. *Ann. Rev. Plant Physiol. 24,* 415-444.
Hall, S.M. and Baker, D.A. (1972). The chemical composition
    of *Ricinus* phloem exudate. *Planta (Berl.) 106,* 131-140.
Hall, S.M., Baker, D.A. and Milburn, J.A. (1971). Phloem
    transport of ^{14}C labelled assimilates in *Ricinus*. *Planta
    (Berl.) 100,* 200-207.
Hall, S.M. and Medlow, G.C. (1974). Identification of IAA in
    phloem and root pressure saps of *Ricinus communis* L.
    by mass spectrometry. *Planta (Berl.) 119,* 257-261.
Hall, S.M. and Milburn, J.A. (1973). Phloem transport in
    Ricinus. Its dependence on the water balance of the
    tissues. *Planta (Berl.) 109,* 1-10.
Hardy, P.J. (1969). Selective diffusion of basic and acidic
    products of CO_2 fixation into the transpiration stream
    in grapevine. *J. Exp. Bot. 20,* 856-862.
Hardy, P.J. and Possingham, J.V. (1969). Studies on trans-
    location of metabolites in the xylem of grapevine
    shoots. *J. Exp. Bot. 20,* 325-335.
Hartig, T. (1860). Beiträge zur physiologischen Forstbotanik.
    Allg Forst Jagdztg. 36, 257-300.
Hill, G.P. (1962). Exudation from aphid stylets during the
    period from dormancy to bud break in *Tilia americana* (L)
    J. Exp. Bot. 13, 144-151.
Hill, G.P. (1963). The source of sugars in sieve tube sap.
    Ann. Bot. 27, 79-87.
Hill-Cottingham, D.G. and Lloyd-Jones, C.P. (1973). A tech-
    nique for studying the adsorption, absorption and
    metabolism of amino acids in intact apple stem tissue.
    Physiol. Plant. 28, 443-446.
Hofstra, J.J. (1964). Amino-acids in the bleeding sap of
    fruiting tomato plants. *Acta. Bot. Neerl. 13,* 148-158.
Ivankov, S. and Ingversen, J. (1971). Investigations on the
    assimilation of nitrogen by maize roots and the transport
    of some major nitrogen compounds by xylem sap. III.
    Transport of nitrogen compounds by xylem sap. *Physiol.
    Plant. 24,* 355-362.
Jones, O.P. (1971). Effect of rootstocks and interstocks on
    the xylem sap composition in apple trees. Effects on
    nitrogen, phosphorus and potassium content. *Ann. Bot.
    35,* 825-836.
Kauss, H. and Ziegler, H. (1974). Carbohydrate-binding
    proteins from the sieve tube sap of *Robinia pseudoacacia*
    L. *Planta (Berl.) 121,* 197-200.

Kende, H. (1965). Kinetin-like factors in the root exudate of sunflowers. *Proc. natn. Acad. Sci. U.S.A. 53*, 1302-1307.

Kessler, K.J. (1966). Xylem sap as a growth medium for four tree wilt fungi. *Phytopathology 56*, 1165-1169.

King, R.W. and Zeevaart, J.A.D. (1974). Enhancement of phloem exudation from cut petioles by chelating agents. *Plant Physiol. 53*, 96-103.

Klepper, B. and Kaufmann, M.R. (1966). Removal of salt from xylem sap by leaves and stems of guttating plants. *Plant Physiol. 41*, 1743-1747.

Kluge, M. Untersuchungen über Kohlenhydrate und Myo-inosit in Siebröhrensaften von Holz. Doctoral Dissertation, Darmstadt Technical University. (1967).

Kluge, M., Becker, D. and Ziegler, H (1970). Untersuchungen über ATP und andere organische Phosphorverbindungen im Siebröhrensaft von *Yucca flaccida* und *Salix triandra* *Planta (Berl.) 91*, 68-79.

Kollmann, R., Dörr. I.and Kleinig, H. (1970). Protein filaments – structural components of the phloem exudate. *Planta (Berl.) 95*, 86-94.

Koster, A.L. (1963). Changes in metabolism of isolated root systems of soybean. *198*, 709.

Kursanov, A.L. (1961). The transport of organic substances in plants. *Endeavour 20*, 19-25.

Kursanov, A.L. (1963). Metabolism and the transport of organic substances in the phloem. *Adv. Bot. Res. 1*, 209-278. Academic Press, London.

Leckstein, P.M. and Llewellyn, M. (1975). Quantitative analysis of seasonal variation in the amino acids in phloem sap of *Salix alba* L. *Planta (Berl.) 124*, 89-91.

Lehman, J. (1973). Untersuchungen am Phloemexsudat von *Cucurbita pepo* L. 1. Enzymaktivitäten von Glykolyse, Gärung and Citrat-Cyclus. *Planta (Berl.) 114*, 41-50.

Lepp, N.W. and Peel, A.J. (1970). Some effects of IAA and kinetin on the movement of sugars in the phloem of willow. *Planta (Berl.) 90*, 230-235.

Lepp, N.W. and Peel, A.J. (1971). Patterns of translocation and metabolism of ^{14}C-labelled IAA in the phloem of willow. *Planta (Berl.) 96*, 62-73.

Lewis, O.A.M. and Pate, J.S. (1973). The significance of transpirationally derived nitrogen in protein synthesis in fruiting plants of pea (*Pisum sativum* L.) *J. Exp. Bot. 24*, 596-606.

MacRobbie, E.A.C. (1971). Phloem translocation. Facts and Mechanisms. A comparative survey. *Biol. Rev. 46*, 429-481.

Milburn, J.A. (1971). An analysis of the response in phloem exudation on application of massage to *Ricinus*. *Planta (Berl.) 100*, 143-154.

Milburn, J.A. (1974). Phloem transport in *Ricinus*: Concentration gradients between source and sink. *Planta (Berl.) 117*, 303-319.

Mittler, T.E. (1953). Amino acids in the phloem sap and their excretion by aphids. *Nature 172*, 207.

Mittler, T.E. (1958). Sieve-tube sap *via* aphid stylets. In: The Physiology of Forest Trees. (ed. K.V. Thimann) New York, Ronald Press Company.

Morrison, T.M. (1965). Xylem sap composition in woody plants. *Nature 205*, 1027.

Oghoghorie, C.G.O. and Pate, J.S. (1972). Exploration of the nitrogen transport system of a nodulated legume using ^{15}N. *Planta (Berl.) 104*, 35-49.

Olofinboba, M.O. (1969). Seasonal variations in the carbohydrates in the xylem of *Antiaris africana*. *Ann. Bot. 33*, 339-349.

Ozerol, N.H. and Titus, J.S. (1968). Translocation of nitrogenous compounds in one year old apple trees. *Proc. Amer. Soc. Hort. Sci. 93*, 7-15.

Pate, J.S. (1962). Root-exudation studies on the exhange of ^{14}C-labelled organic substances between the root and shoot of the nodulated legume. *Plant and Soil 17*, 333-356.

Pate, J.S. (1965). Roots as organs of assimilation of sulfate. *Science 149*, 547-548.

Pate, J.S. (1971). Movement of nitrogenous solutes in plants. In. Nitrogen-15 in soil-plant studies. International Atomic Energy Agency, Vienna. (IAEA-P1-341-13).

Pate, J.S. (1973). Uptake, assimilation and transport of nitrogen compounds by plants. *Soil Biol. Biochem. 5*, 109-119.

Pate, J.S. (1975). Exchange of solutes between phloem and xylem and circulation in the whole plant. Encyclopedia of Plant Physiol. (N.S.). 1. Transport in Plants I Phloem Transport., eds. M.H. Zimmermann and J.A. Milburn, Springer-Verlag, Berlin, Heidelberg, New York, Chapter 19.

Pate, J.S. and Greig, J.M. (1964). Rhythmic fluctuations in the synthetic activities of the nodulated root of the legume. *Plant and Soil 21*, 163-184.

Pate, J.S. and Gunning, B.E.S. (1972). Transfer Cells. *Ann. Rev. Plant Physiol. 23*, 173-196.

Pate, J.S., Gunning, B.E.S. and Millikan, F.F. (1970). Function of transfer cells in the nodal regions of stems, particularly in relation to the nutrition of young seedlings. *Protoplasma 71*, 313-334.

Pate, J.S. and O'Brien, T.P. (1968). Microautoradiographic study of the incorporation of labelled amino acids into insoluble compounds of the shoot of a higher plant. *Planta (Berl.) 78*, 60-71.

Pate, J.S., Sharkey, P.J. and Lewis, O.A.M. (1974). Phloem bleeding from legume fruits - a technique for study of fruit nutrition. *Planta (Berl.) 120*, 229-243.

Pate, J.S., Sharkey, P.J. and Lewis O.A.M. (1975). Xylem to phloem transfer of solutes in fruiting shoots of legumes, studied by a phloem bleeding technique. *Planta (Berl.)*. *122*, 11-26.

Pate, J.S., Walker, J., and Wallace, W. (1965). Nitrogen-containing compounds in the shoot system of *Pisum arvense* L. II. The significance of amino-acids and amides released from nodulated roots. *Ann. Bot. 29*, 475-493.

Pate, J.S. Wallace, W. and Die, J. van. (1964). Petiole bleeding sap in the examination of the circulation of nitrogenous substances in plants. *Nature 204*, 1073-1074.

Peel, A.J. (1965). The effect of changes in the diffusion potential of the xylem water on sieve tube exudation from isolated stem segments. *J. Exp. Bot. 16*, 249-260.

Peel, A.J. (1970). Further evidence for the relative immobility of water in sieve tubes of willow. *Physiol. Plantarum 23*, 667-672.

Peel, A.J. and Weatherley, P.E. (1959). Composition of sieve tube sap. *Nature 184*, 1955 1956.

Sauter, J.J. Iten, W. and Zimmermann, M.H. (1973). Studies on the release of sugar into the vessels of sugar maple (*Acer saccharum*). *Can. J. Bot. 51*, 1-8.

Schmid, W.E. and Gerloff, G.C. (1961). A naturally occurring chelate of iron in xylem exudate. *Plant Physiol. 36*, 226-231.

Schnathorst, W.C. (1970). Obtaining xylem fluid from *Gossypium hirsutum* and its uses in studies on vascular pathogens. *Phytopath. 60*, 175-176.

Selvendran, R.R. (1970). Changes in the composition of the xylem exudate of tea plants (*Camellia sinensis* L.) during recovery from pruning. *Ann. Bot. 34*, 825-833.

Selvendran, R.R. and Sabaratnam, S. (1971). Composition of the xylem sap of tea plants (*Camellia sinensis* L.). *Ann. Bot. 35*, 679-682.

Sharkey, P.J. and Pate, J.S. (1975). Selectivity in xylem to phloem transfer of amino acids in fruiting shoots of white Lupin (*Lupinus albus* L.). *Planta (Berl.) 127*, 251-262.

Shone, M.G.T., Clarkson, D.T. and Sanderson, J. (1969). The

absorption and translocation of sodium by maize seedlings. *Planta (Berl.) 86*, 301-314.

Silvester, W.B. (1968). Nitrogen fixation by *Coriaria*.
Ph.D. Thesis, University of Canterbury, New Zealand.

Singh, D. and Smalley, E.B. (1969). Nitrogenous compounds in
the xylem sap of American elms with Dutch elm disease.
Can. J. Bot. 47, 1061-1065.

Tammes, P.M.L. and Die, J. van (1964). Studies on phloem
exudation from *Yucca flaccida* Haw. I. Some observations
on the phenomenon of bleeding and the composition of the
exudate. *Acta. Bot. Neerl. 13*, 76-83.

Tammes, P.M.L. Vonk, C.R. and Die, J. van (1967). Studies on
phloem exudation from *Yucca flaccida* Haw. VI. The
formation of exudate-sucrose from supplied hexoses in
excised inflorescence parts. *Act. Bot. Neerl. 16*,
244-246.

Tammes, P.M.L., Vonk, C.R. and Die, J. van (1973). Studies
on phloem exudation from *Yucca flaccida* Haw. XI. Xylem
feeding of 14C-sugars and some other compounds, their
conversion and recovery from the phloem exudate. *Acta.
Bot. Neerl. 22*, 233-237.

Thaine, R. (1964). Protoplast structure in sieve tube
elements. *New Phytol. 63*, 236-243.

Tingley, M.A. (1944). Concentration gradients in plant
exudates with reference to the mechanism of translocation.
Amer. J. Bot. 31, 30-38.

Tolbert, N.E. and Wiebe, H. (1955). Phosphorus and sulfur
compounds in plant xylem sap. *Plant Physiol. 30*,
499-504.

Wallace, W. and Pate, J.S. (1967). Nitrate assimilation in
higher plants with special reference to the cocklebur
(*Xanthium pennsylvanicum* Wallr.). *Ann. Bot. 31*, 213-
228.

Weatherley, P.E. (1969). Ion movement within the plant and
its integration with other physiological processes. In:
Ecological aspects of the mineral nutrition of plants.
(ed. Rorison, I.H.) Blackwells, Oxford.

Weatherley, P.E. and Johnson, R.P.C. (1968). The form and
function of the sieve tube: A problem of reconciliation.
Int. Rev. Cytol. 24, 149-192.

Weatherley, P.E. Peel, A.J. and Hill, C.P. (1959). The
physiology of sieve tube sap. *J. Exp. Bot. 10*, 1-16.

Webb, K.L. and Burley, J.W.A. (1962). Sorbitol translocation
in apple. *Science 137*, 766.

Williams, M.W. and Raese, J.T. (1974). Sorbitol in tracheal
sap of apple as related to temperature. *Physiol Plant.
30*, 49-52.

Wooding, F.B.P. and Northcote, D.H. (1965). An anomalous
wall thickening and its possible role in the uptake of
stem-fed tritiated glucose by *Pinus pinea*.
J. Ultrastructure Res. 12, 463-472.

Ziegler, H. (1956). Untersuchungen über die Leitung und
Sekretion der Assimilate. *Planta (Berl.) 47*, 447-500.

Ziegler, H. (1974). Biochemical aspects of phloem transport.
Symp. Soc. exp. Biol. XXVIII. "Transport at the Cellular
Level" Cambridge University Press.

Ziegler, H. and Schnabel, M. (1961). Über Harnstoffderivate in
Siebröhrensaft. *Flora 150*, 306-317.

Ziegler, H. and Ziegler, I. (1962). The water soluble vitamins
in the sieve tube sap of some trees. *Flora 152*, 257-278.

Zimmermann, M.H. (1960). Transport in the phloem. *Ann. Rev.
Plant Physiol. 11*, 167-189.

Zimmermann, M.H. (1961). Movement of organic substances in
trees. *Science 133*, 73-79.

Zimmermann, M.H. (1969). Translocation of nutrients. In :
The physiology of plant growth and development (ed.
Wilkins, M.B.) McGraw Hill, Maidenhead, England.

Zimmermann, M.H. (1971). Transport in the phloem. In: Trees,
structure and function. (eds. M.H. Zimmermann and
C.L. Brown) Springer : Berlin-Heidelberg-New York.

CHAPTER 24

Movement of Viruses in Plants: Long Distance Movement of Tobacco Mosaic Virus in *Nicotiana glutinosa*

Katie Helms and I.F. Wardlaw
CSIRO Division of Plant Industry, Canberra, A.C.T., Australia

INTRODUCTION

The vascular system of plants can transport particles which vary in size from small ions and sugars to more complex particles such as proteins and lipids. In addition to transporting particles which occur in normal plants, the vascular system commonly transports disease producing agents such as viruses. These vary in shape and size. They may be isometric, bullet shaped, or either flexible or rigid rods. Their size varies from about 200 Å in diameter for an isometric virus to about 12,500 x 133 Å for a long flexible rod, viz., beet yellows virus (Matthews 1970)

Events which occur when substances such as ions and sugars are applied to leaves, differ from those which occur when viruses are inoculated. When ions and sugars are applied, some of the particles are absorbed into cells and move symplastically and apoplastically into the phloem (Geiger - this volume) where they are transported to various parts of the plant. The course of the movement can be followed by radioactive tracers and the translocated material can be recovered soon afterwards at a distance from the point of application. Some plant viruses also can be recovered soon after they are inoculated at a distance from the point of inoculation. These include viruses which are thought to multiply mainly in the phloem and to move long distances in the sieve tubes e.g. curly top of sugar beet. Such viruses can be inoculated directly into the phloem by a phloem feeding vector and can be recovered by a virus-free vector feeding in the phloem.

More typical plant viruses appear to be unrestricted in their host tissue relations. When tobacco mosaic virus (TMV) is inoculated to leaves by the standard method of mechanical inoculation, some of the inoculated particles multiply in the

epidermal cells and some of the newly formed particles move
to adjacent mesophyll and epidermal cells. This process of
multiplication and invasion is repeated until eventually the
virus reaches the vascular bundles, where it is transported
to various regions of the plant. Like most viruses, TMV is
thought to be transported in the phloem, but some viruses are
known to be transported in the xylem. Usually, a virus is
located in a new region of the plant only after it has multiplied sufficiently to be detected in a biological assay, or
after it has produced characteristic symptoms.

There are only a few viruses e.g. beet yellows and beet
western yellows for which a full study has been made of the
relationship of particles to channels of communication
through cell walls. The movement of viruses between adjacent
parenchyma cells and between parenchyma cells and sieve
tubes, is thought to occur via plasmodesmata. This possibility is supported by numerous electron micrographs of isometric virus particles within these structures and also by
electron micrographs of two elongate viruses viz. beet
yellows (Esau and Hoefert 1971) and potato virus Y (Weintraub
et al. 1974). However, available information on the structure of plasmodesmata in normal plants (Robards 1971) suggests
that without modification they may not be large enough to
allow passage of some viruses. Moreover, detailed pictures
of beet yellows virus within plasmodesmata provide no evidence of two features of Robards' model, viz., the endoplasmic reticulum at the ends of the plasmodesmatal openings and
the central core (Esau and Hoefert 1972). Weintraub et al.
(1974) reported that part of the core may be present in
transverse sections of plasmodesmata which contain potato
virus Y. Esau (1968) considered that if the central core is
present, it could provide a substantial impediment to the
passage of virus. Therefore, it is of interest that Gunning
and Robards (this volume) have suggested that the core may be
an artifact produced during preparation.

Most viruses are thought to move long distances in the
phloem and there is no doubt that sieve pores, which are
formed as a modification of plasmodesmata, are sufficiently
large to provide a satisfactory channel for this movement.

The question arises as to whether viruses move short
distances and long distances as complete particles or as
naked nucleic acid. The following remarks refer only to
long distance movement in the phloem. Evidence from electron micrographs indicates that at least some viruses are
present in sieve tubes as complete particles, but their
presence in these cells does not constitute evidence that
they are translocated as complete particles. Since it is

unlikely that viruses can multiply in mature sieve tubes, virus particles observed in these structures could be either the progeny of virus particles which invaded and multiplied in young sieve elements, or virus particles which entered the sieve tubes from adjoining companion cells or parenchyma cells. For similar reasons, the presence of ribonuclease resistant particles in phloem exudate (Kluge 1967) does not prove that viruses move in the phloem as complete particles. Nevertheless, until naked virus nucleic acid is demonstrated within sieve tubes, it is reasonable to conclude that at least some viruses move long distances in the phloem as complete particles.

The purpose of the experimental work to be described was to obtain detailed information on the long distance movement of TMV in *Nicotiana glutinosa* and to compare the speed of movement of TMV and assimilate. This host-virus interaction was used in previous work in which we examined the effect of temperature on the production of systemic symptoms (Helms and Wardlaw, unpublished). Only systemic symptoms are formed at 36°C, whereas both systemic symptoms and local lesions develop at lower temperatures.

MATERIALS AND METHODS

Plant Growth, Virus Inoculation and Assay. - Plants of *Nicotiana glutinosa* L. were grown in pots of 10 cm diam. in a mixture of perlite and vermiculite in the Canberra phyto-tron, under natural light and with 8 h day/16 h night temperatures of 27/22°C. Pots were watered with Hoaglands' nutrient in the morning and water in the afternoon. To simplify the experimental system we used plants which were partially defoliated just before floral buds became apparent, i.e. about 8 weeks after the seeds were sown. Plants used for mechanical inoculation were defoliated as follows: (a) the apical shoot and all leaves were removed except the upper almost expanded leaf immediately below the excised shoot; in these plants there was one major sink viz. the roots; (b) the apical shoot was retained and all leaves were removed except one vigorous lower leaf about 4 cm from the base of the stem; in these plants there were two major sinks, viz. the apical shoot and the roots. In both (a) and (b) the stem and developing axillary buds provided additional minor sinks. The single large leaf in each plant was inoculated with a high concentration of the common strain of TMV (about 30 µg/ml) by the standard method of mechanical inoculation. Plants used for jet-injection were defoliated except for the apical leaves. The central part of the stem was twice jet-injected (5 mg/ml) using a Pan-jet gun which is normally used in veterinary and dental work (F.H. Wright

Dental M.F.G. Co. Ltd., Kingsway, West Dundee, Scotland).
After inoculation the plants were kept at 36°C and under continuous light of 64 Wm^{-2} (400-700 nm).

The method used for identifying long distance movement
of virus in petioles and stems was based on that of Samuel
(1934). At various intervals after inoculation, the petioles
and stems were cut into consecutive segments (1 cm in length)
using a clean razor for each segment. The segments were
incubated in tubes containing one tenth concentration of the
major salts of Hoaglands' nutrient solution. The tubes were
kept at 25°C under continuous light of 32 Wm^{-2} for 4-5 days
when each section was ground with celite and phosphate buffer
of pH 7 and mechanically inoculated to a single leaf of plants
of *N. glutinosa*. After 3-5 days numbers of lesions were
counted.

Heat-ringing of Petioles and Stems. - Plants were partially
defoliated as described above and petioles or stems of half
of the plants in each test were ringed by passing steam
through a curved piece of capillary tubing which was closely
applied to the tissues. Plants partially defoliated as in
(a) were ringed on the petiole of the upper leaf for study
of downward movement, and plants defoliated as in (b) were
ringed on the young part of the stem for study of upward
movement. The heat treatment killed all living cells in a
ring of tissue (including the internal and external phloem),
leaving only a core of xylem. The heated region was supported
by a wrapping of aluminium foil. The ringed plants were kept
for 24 h at 27/22°C. Then the leaf on each plant was mechanically inoculated in the standard manner, or the abaxial,
proximal end of the mid vein of the lamina was jet-injected.
The mechanically inoculated plants were kept under the standard post-inoculation conditions of 36°C with continuous light
and the jet-injected plants were kept at 27/22°C with natural
light. The apical leaves and/or the axillary leaves were
subsequently examined for development of symptoms and samples
of the tissue were tested for infectivity on *N. glutinosa*.
In one test with mechanically inoculated plants, infectivity
tests were made 48 h after inoculation. At this time the
small axillary bud in the axis of the ringed petiole and a
proximal segment of the ringed petiole were incubated for
4-5 days as described earlier and then tested for infectivity.

*Time-course Multiplication of Virus in Mechanically Inoc-
ulated Leaves.* - Five plants of *N. glutinosa* were decapitated
and defoliated except for one upper leaf and one lower leaf.
These were mechanically inoculated with TMV as described
earlier. At successive intervals, one disc (18 mm diam.) was
cut from both the upper and lower leaf of each plant and the

discs were frozen. Subsequently they were ground and assayed for infectivity on leaves of *N. glutinosa*, using a balanced half leaf design. The experiment was repeated on three occasions.

Movement of ^{14}C-photosynthate. - $^{14}CO_2$ generated from 100 mg of $Ba^{14}CO_3$, containing 100 μCi of ^{14}C, was circulated for 5 min over the upper leaves of six partially defoliated plants placed together in a Perspex (plexiglas) assimilation chamber. Air temperatures were held at 36°C and light at 64 Wm^{-2}. A second group of six plants was treated identically, 30 min after the first. Subsequently, individual plants were harvested at various intervals of time after the start of $^{14}CO_2$ uptake; they were then analysed for the distribution of ^{14}C. At harvest, each plant was divided into treated leaf, petiole, stem and roots; then the samples were oven dried at 80°C for 48 h, weighed, and ground in a Wiley Mill to pass a 40 mesh sieve. Thirty milligram (30 mg) aliquots were counted with a thin window gas flow counter (Wardlaw 1965) to determine radioactivity in each sample (cpm/30 mg). For each sample, this value was multiplied by the dry weight of the part being measured to determine the relative total activity.

RESULTS

Long Distance Movement of TMV Through Living Tissue. - There is little critical experimental evidence to support the generally accepted conclusion that TMV moves long distances in the phloem. Evidence often quoted is that of Caldwell (1931), who showed that TMV did not pass through steam-killed internodes of *Nicotiana tabacum* (var. White Burley) and that of Bennett (1940a), who showed that excised rings which broke phloem continuity delayed but did not prevent spread of TMV in Turkish tobacco; excised rings prevented spread of TMV in certain plants of *Nicotania glauca*.

The aim of the following experiments was to find out if ringing by steam prevents movement of TMV in *N. glutinosa*. In one test with mechanically inoculated plants, stems of 10 plants were ringed, while 10 comparable plants were kept as controls. Within 2 weeks of inoculation, virus had moved upwards to the apical leaves of all control plants, but not into the apical leaves of any ringed plants. In another test, the petioles of the upper leaves of 10 plants were ringed, while 10 comparable plants were kept as controls. Within 48 h virus had moved down to the proximal end of the petioles and into the axillary buds of all control plants, but not into those of any ringed plants.

In one test with jet-injected plants the petioles of the upper leaves of 12 plants were ringed, while 12 comparable plants were kept as controls. Fourteen days after inoculation, virus had moved through the petioles and stems of 8 of the 12 control plants but not through the petioles of any of the ringed plants.

Since TMV failed to move either upwards or downwards through the xylem, one can conclude that it normally moves long distances through living cells. These are almost certainly phloem sieve tubes because the estimated speed of movement (data presented elsewhere in the text) is greater than could be expected in cell to cell movement of tissues such as the cortex.

Time-course Multiplication of TMV in Upper and Lower Leaves. - When plants were inoculated mechanically on an upper leaf or a lower leaf, virus moved downwards from the upper leaf earlier than it moved upwards from the lower leaf (Helms and Wardlaw, unpublished). Moreover, movement of the virus was detected earlier when leaves of relatively young plants rather than relatively old plants were inoculated. This suggested that the earlier movement from upper leaves could be due to virus being available for translocation earlier in upper, younger leaves than in older, lower leaves. This possibility was examined by comparing the time-course multiplication of virus in upper and lower leaves. Data in Fig. 1 show that multiplication in an upper leaf occurred between 6 and 12 h after inoculation, whereas that in a lower leaf was detected about 18 h after inoculation. One can conclude that virus can enter sieve tubes and hence be available for translocation earlier from an upper leaf than from a lower leaf.

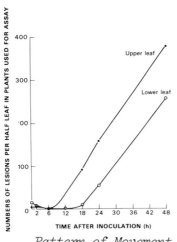

Fig. 1. - Comparison of time-course multiplication of TMV in upper and lower leaves of plants of Nicotiana glutinosa *inoculated at a concentration of TMV of 30 µg/ml and kept after inoculation at 36°C and under continuous light of 64 Wm^{-2}.*

Pattern of Movement of TMV in Petioles and Stem Following Mechanical Inoculation of Leaves. - The distribution of virus in plants inoculated on either an upper or a lower leaf was

examined at 18, 24, 42 or 48 h after inoculation. Data for
6 representative plants are shown in Fig. 2 and data for additional plants will be presented elsewhere. Following early
movement from the inoculated leaves, virus was located in some

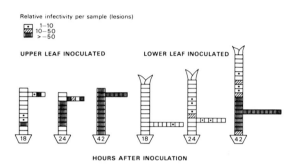

Fig. 2. - *Diagramatic representation of the location of
TMV in petioles and stems of plants of* Nicotiana glutinosa
*inoculated on an upper or a lower leaf by the standard
method of mechanical inoculation and examined at intervals
after inoculation. Plants were kept after inoculation
under the conditions described for Fig. 1. Each section
of the diagrams represents 1 cm.*

segments of the petiole and stems but not in others, i.e.
virus moved through some regions of the petioles and stems
without causing infection. There was no evidence of a grada-
tion in concentration as virus moved from the inoculated
leaf towards the apical shoot or the roots. Virus was detected in more segments of plants inoculated on an upper leaf
than on a lower leaf. Movement of virus from an upper leaf
or from a lower leaf was detected within 18 h. Within 24 h,
virus had moved either downwards towards the roots from an
upper leaf, or upwards towards the apical shoot from a lower
leaf. Within 42 h it was detected throughout the petiole and
throughout all or most of the stem.
  *Pattern of Movement of TMV Following Jet-Injection into
Stems*. - In studies of long distance movement of viruses such
as TMV, it would be advantageous if short distance movement between the inoculated cells of the epidermis and the phloem could
be separated experimentally from long distance movement within the phloem. In previous work (Helms and Wardlaw, unpublished), we showed that the Pan-jet gun, which injects a
stream of virus under high pressure, could be used to inocu-
late *N. glutinosa* with TMV. In the present experiments we
injected the stems of plants by this method. We hoped to in-

sert virus directly in the phloem and so avoid short distance, cell to cell movement which normally occurs following inoculation. If some of the sieve tubes in which virus gained entry were suitably injured, the virus might move long distances within a short period. Data in Fig. 3 provide no evidence that this occurred. Long distance movement of virus was detectable at 18 or 24 h after inoculation. There was no evidence that virus had been translocated immediately following jet-injection. If this had occurred, one could have expected it to be present in relatively distant regions of the stem. Instead, there was evidence of a high concentration of virus at the point of injection and a marked decrease in concentration as virus moved away. The data suggest that the observed long distance movement of virus was preceded by multiplication – presumably in cells of the cortex and parenchyma.

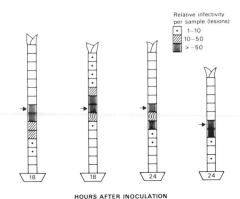

Relative infectivity per sample (lesions)
- 1–10
- 10–50
- > –50

HOURS AFTER INOCULATION

Fig. 3. - Diagramatic representation of the location of TMV in plants of Nicotiana glutinosa *inoculated in the centre of the stem (position indicated by arrow) by jet-injection and examined at 18 or 24 h after inoculation. Plants were kept after inoculation under the conditions described in Fig. 1.*

Estimate of Speed of Long Distance Movement of Virus and Assimilate. - Estimates were made of the mean speed of long distance movement of TMV in a number of plants of *N. glutinosa* (Helms and Wardlaw, unpublished). The highest speed was derived from a plant inoculated on a lower leaf and examined after 24 h. The distance moved towards the apex through petiole and stem was 21 cm. Since data from Fig. 1 show that virus multiplication in a lower leaf was not detected until about 18 h after inoculation, the mean speed of movement of virus is $21/(24-18) = 3.5$ cm/h.

The speed of movement of assimilate in plants of *N. glutinosa* was calculated by comparing the change with time of radioactivity (% of plant total ^{14}C) in the upper leaf with that in the roots (Fig. 4). The delay in arrival of ^{14}C in the roots, in relation to the loss from the leaf was estimated to be 14 min and since the mean distance was 14 cm the mean speed of translocation of assimilate was 1 cm/min, or 60 cm/hr. A similar estimate of speed was obtained for a number of plants using ^{32}P (Helms and Wardlaw, unpublished).

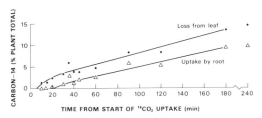

TIME FROM START OF $^{14}CO_2$ UPTAKE (min)

Fig. 4. - Movement of ^{14}C-assimilate from an upper leaf to the roots in plants of Nicotania glutinosa kept at 36°C and under continuous light of 64 Wm^{-2}. Root and leaf data are shown for individual plants harvested at successive times from the start of $^{14}CO_2$ uptake.

DISCUSSION

Because of the number of variables involved, previous estimates of speed of long distance movement of TMV have varied considerably e.g. 1.3 cm/h for tobacco (Bennett 1940b) and 17.8 cm/h for tomato (Kunkel 1939). To increase the reproducibility of our data, plants were grown under controlled conditions. They were partially defoliated before inoculation and after inoculation they were kept under constant light and under high temperature, which promotes development of symptoms. Under these conditions, movement of virus was detected in some plants 18 h after inoculation. For reasons to be discussed later, the speed of long distance movement is likely to be highest when virus first enters the sieve tubes. Our highest estimate was 3.5 cm/h. This is faster than that of cell to cell movement of virus through parenchyma of the stem and leaf, which has been estimated as between 0.01 and 1 mm/day (Schneider 1965), and approximates that of cell to cell movement of some plant hormones.

Virus movement is generally thought to be correlated with movement of assimilate (Bennett 1956), and therefore it was of interest to compare the speed of movement of TMV and of assimilate in plants kept under the same environmental conditions. The estimated speed of movement of assimilate in *N. glutinosa* (60 cm/h), is comparable with that for a number of different species (Canny 1973) and much higher than that for TMV. We do not think that inaccuracies associated with the timing of entry of virus into the sieve tubes could account for the magnitude of the difference in speed between TMV and assimilate. Assuming that TMV is translocated as complete virus particles, we suggest that once it has entered the sieve tubes, it can move passively at the same rate as the assimilate.

However, because of its size and shape, its early movement is
likely to be impeded by obstacles such as protein filaments
and sieve plates. As more virus particles move into the sieve
tubes, speed is likely to be reduced still further.

The observed non-uniform distribution of virus within the
petioles and stems soon after movement is initiated has been
observed previously (e.g. Samuel 1934; Kunkel 1939) and is in
accord with the possibility that virus particles move inter-
mittently into the phloem and pass through regions of the
petioles and stems without multiplying.

Robb (1964) considered that the rate of movement of virus
from an inoculated leaf into the stem was in some part associated with the maturity of the inoculated leaf. Our data
accord with this idea since they show that virus multiplied
earlier and moved earlier from relatively young leaves than
from relatively old leaves.

Early stages in the movement of virus was detectable
about 18 h after inoculation, irrespective of whether the
plants were inoculated by jet-injection, or by the standard
method of mechanical inoculation. Thus there was no evidence
that some virus particles moved long distances immediately
following jet-injection. Although the injury associated with
jet-injection allowed entry and multiplication of virus in
some cells, it apparently was unsuitable for promoting entry
and immediate movement of virus in sieve tubes.

REFERENCES

Bennett, C.W. (1940a). Relation of food translocation to
    movement of virus of tobacco mosaic. *Jour. Agric. Res.*
    60, 361-390.
Bennett, C.W. (1940b). The relation of viruses to plant
    tissues. *Bot. Rev. 9*, 427-473.
Bennett, C.W. (1956). Biological relations of plant viruses.
    Ann. Rev. Plant Physiol. 7, 143-170.
Caldwell, J. (1931). The physiology of virus diseases in
    plants. II. Further studies on the movement of mosaic in
    the tomato plant. *Ann. Appl. Biol. 18*, 279-298.
Canny, M.J. (1973). Phloem translocation. Cambridge University
    Press.
Esau, K. (1968). Viruses in plant hosts. The 1968 John Charles
    Walker lectures. The University of Wisconsin Press,
    Madison, Milwaukee and London.
Esau, K. and Hoefert, L.L. (1971). Cytology of beet yellows
    virus infection in *Tetragonia*. II. Vascular elements in
    infected leaf. *Protoplasma 72*, 459-476.
Esau, K. and Hoefert, L.L. (1972). Ultrastructure of sugar
    beet leaves infected with beet western yellows virus.

J. Ultrastructure Res. 40, 556–571.

Kluge, M. (1967). Viruspartikel im Siebröhrensaft von *Cucumis sativus* L. nach Infektion durch das Cucumisvirus2A. *Planta (Berl.) 73*, 50–61.

Kunkel, L.O. (1939). Movement of tobacco-mosaic virus in tomato plants. *Phytopathology 29*, 684–700.

Matthews, R.E.F. (1970). Plant Virology, Academic Press. New York and London.

Robards, A.W. (1971). The ultrastructure of plasmodesmata. *Protoplasma 72*, 315–323.

Robb, S.M. (1964). The movement of tomato aucuba mosaic virus in tomato. *New Phytol. 63*, 267–273.

Samuel, G. (1934). The movement of tobacco mosaic virus within the plant. *Ann. Appl. Biol. 21*, 90–111.

Schneider, I.R. (1965). Introduction, translocation and distribution of viruses in plants. *Advances in Virus Research 11*, 163–221.

Wardlaw, I.F. (1965). The velocity and pattern of assimilate translocation in wheat plants during grain development. *Aust. J. Biol. Sci. 18*, 269–281.

Weintraub, M., Ragetli, H.W.J. and Lo, E. (1974). Potato virus Y particles in plasmodesmata of tobacco leaf cells. *J. Ultrastructure Res. 46*, 131–148.

CHAPTER 25

Influence of Stomatal Behaviour on Long Distance Transport

W.G. Allaway
*School of Biological Sciences, The University of Sydney,
N.S.W., Australia*

STOMATA AS SOURCES AND SINKS WITHIN THE LEAF

Are stomata in symplasmic contact with the cells of the epidermis and the rest of the leaf? Many authors have failed to find plasmodesmata between mature guard cells and epidermal cells, although they looked for them with the electron microscope (e.g. Allaway and Setterfield 1972; Singh and Srivastava 1973; Humbert et al. 1975). Plasmodesmata have been reported from guard cells during differentiation (e.g. Landré 1969; Kaufman et al. 1970).

It is interesting to imagine stomata becoming isolated ~~lalunalu uu thuy uumulutu uhuiu davilupmunt, nn tho phnonon of~~ plasmodesmata would make it easier to believe in the large pressure differences that must exist across the guard cell wall when the stoma is open. Some estimates of this pressure differential are tabulated in Table 1: in this table I have assumed the water potential of the whole tissue to be zero. This is an unrealistic assumption except in the cases where measurements were made on isolated epidermal strips; so the estimates of pressure difference should perhaps be regarded as maximum values. It is tempting to imagine that under such pressures the membranous structures inside plasmodesmata would not be able to stop the guard cell contents squirting out into the neighbouring cells. Perhaps this is too naive a view, however, since Pallas and Mollenhauer (1972) have reported plasmodesmatal connections between guard cells and epidermal cells. Guard cells are not specialized with "transfer-cell" wall-labyrinths for transfer from the apoplast (e.g. Allaway and Setterfield 1972).

Whether or not mature guard cells are part of the symplasm, we know that there are flows of material between them and the rest of the leaf: water and ions are well known examples (see below). Simple calculations based only on

TABLE 1. Notional Pressure Differences Across Guard Cell Walls

Species	Pressure differences in kPa		
	stoma closed	stoma open	
(A) Reliable values			
Pelargonium hortorum	70	2000	Bearce & Kohl (1970)[1]
Chrysanthemum morifolium	800	2400	" " " " [1]
Vicia faba		4500	Raschke (1975)[2]
(B) Values less reliable			
Vicia faba	1750	3300	Humble & Raschke (1971)[3,4]
	800	1800	Allaway & Hsiao (1972)[3,4]
	220	380	Stålfelt (1967)[3,5]
	180	920	in Meidner & Mansfield (1968)[3,5]
Ranunculus bulbosus	-400	500	" " " [3,5]
	0 to 1050	500 to 2850	" " " [3,5]

[1] These data are perhaps the most reliable, since they were obtained by the freezing-point depression method. No corrections for volume were needed. Mean values given.

[2] Data obtained by rapid plasmolysis and so probably reliable (Fischer 1973). Epidermal cells presumed broken, and therefore not allowed for: pressure difference in the intact leaf therefore probably not so large. Allowance has been made for a twofold decrease in guard cell volume between full opening and plasmolysis.

[3] Data all probably underestimates because they were obtained by slow plasmolysis (Fischer 1973). No corrections have been made for volume changes, so values are likely to be overestimates of the situation in the leaf, especially for open stomata.

[4] Epidermal cells broken, not allowed for. In the intact leaf the pressure difference would therefore be smaller. Mean values given.

[5] Epidermal cells not broken: their measured values subtracted.

measurement of chlorophyll content of epidermis led Pallas and Dilley (1972) to the conclusion that ATP produced photosynthetically in the guard cells could be enough for the energy requirements for potassium pumping in stomatal opening. It is surprising that in this study the guard cell chloroplasts were found to contain as much chlorophyll per chloroplast as those from other plant cells, since guard cell chloroplasts are small (Allaway and Milthorpe 1976) and generally rather pale green. However the authors minimised contamination from other cells by sonicating the preparation; two unsonicated preparations of my own contained about 80 times their quantity of chlorophyll. Photophosphorylation as the major direct source of ATP for stomatal opening does not allow a function to the very many mitochondria of guard cells (e.g. Miroslavov 1966); these mitochondria must be the energy source in the dark stomatal opening of Crassulacean Acid Metabolism (CAM) plants. In their paper Pallas and Dilley used the ratio 2 ATP per 1 K^+ transported as a basis of their justification for the sufficiency of guard cell photophosphory-lation. Theoretically, perhaps, 0.5 ATP per K^+ might be sufficient just for the potassium uptake of the cells. However, there will be other demands on the ATP and reducing power of the cells: particularly for the manufacture of organic anions for charge balancing. Perhaps 2 mol ATP per mole of organic acid formed would be a minimum estimate, and as Table 2 shows perhaps 1.1 to 1.4 x 10^{-12} mol of such anions must be made in stomatal opening over 3 h. The extra amount of ATP needed then is at least 0.7 to 1 x 10^{-12} mol per guard cell per hour - approximately the same as the rate of production as calculated by Pallas and Dilley. This does not account for the ordinary metabolism of the cells, which also must consume some ATP. It would not be wise to place too much emphasis on calculations derived merely from chlorophyll measurements: but they are the only data we have on this aspect at present, and they do not support the idea of the photophosphorylative self-suffi-ciency of guard cells during the period of the actual stomatal opening movement. I think it likely that guard cells run at a photophosphorylative energy deficit during the 3 h of the opening movement. This energy deficit could be made up in the short term by the mitochondria consuming some of the starch reserves of the guard cells. It is well known that guard cells often show large reserves of starch which often (but not always) disappear during stomatal opening (cf. Meidner and Mansfield 1968). Guard cell plastids often have some develop-ment of grana (e.g. Allaway and Setterfield 1972), which suggests that ATP and NADPH could be made in them as a result of light-driven electron flow.

Continuing this approach, perhaps it is worth speculat-ing as to whether stomata could make enough organic anions for ionic balance from their own carbon dioxide fixation. Table 2 shows phosphoenolpyruvate (PEP) carboxylase rates and CO_2 fixation rates, which suggest that sufficient carbon could be fixed in the first 3 h of light to account for the forma-tion of enough organic anions, without an import requirement. The PEP required could come from the starch reserves via glycolysis. Starch as a source for organic acid provides me with the only explanation for the accumulation of malate in

TABLE 2. Organic Anions in Guard Cells During the First
3 Hours of Stomatal Opening

Concentration change of K^+ in guard cells with opening	about 400 mol m^{-3}	(e.g. Allaway & Hsiao 1972; *V. faba*)
Concentration change of Cl^-	20 to 80 mole m^{-3}(?)	(Humble & Raschke 1971, *V. faba*; Raschke & Fellows 1971, *Z. mays*)
Concentration increase in malate	about 100 mol m^{-3}	(Allaway 1973; *V. faba*)
Concentration of negative charges required in guard cells additionally to balance charge, therefore	about 120 to 180 mol m^{-3}	(assumes both negative charges of malate to be effective in ion balance)
Volume of guard cell (open)	about 5×10^{-15} m^3	(Humble & Raschke 1971; Allaway 1973; Fischer & Hsiao 1968; *V. faba*)
Quantity of malate required per guard cell	about 0.5×10^{-12} mol	
Quantity of other additional negative charge needed (presumably organic) per guard cell	about 0.6 to 0.9×10^{-12} mol	
PEP carboxylase in guard cells	about 3×10^{-12} mol.h^{-1} per guard cell	(*C. communis*) Calculated from data of Willmer et al. (1973).
	about 1×10^{-12} mol. h^{-1} per guard cell	(Allaway unpublished; *V. faba*)
Light CO_2 fixation, rate, mol CO_2 per hour per guard cell	1.5, 5.6×10^{-12}	(*V. faba, C. cyanea*, Pearson & Milthorpe 1974)

guard cells on leaves treated with CO_2-free air (Figure 1).
Guard cells on isolated epidermal strips often have a
rather unusual pattern of $^{14}CO_2$ fixation products (Fig. 2;
cf. Pearson and Milthorpe 1974; Willmer and Dittrich 1974).

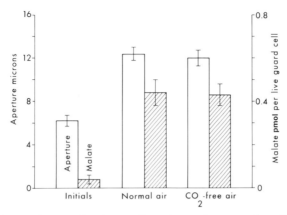

Fig. 1. - Accumulation of malate in guard cells of
Vicia faba *in CO_2- free air. Leaves were placed for*
three hours in the experimental treatments, and then
epidermal strips were made, rolled, rinsed, extracted
and analysed for malate by the same procedures as
Allaway (1973). Histogram shows means of 13 samples
with error bars equal to twice the standard error of
the mean ("Initials" - 6 samples).

This experiment, in which strips were rolled so that all mesophyll chloroplasts should have been broken (Allaway and Hsiao
1972) suggests that at least a little C_3-type CO_2 fixation
can occur in guard cells, in contrast to Mooshke's data for
another species (this volume). If $^{14}CO_2$ fixation is allowed
with the intact leaf and then epidermis subsequently stripped
and analysed, it has been suggested (Raschke, this volume)
that a more normal labelling pattern is observed. It is not
possible to distinguish between the two interpretations of
this - namely, either that in making epidermal strips we
somehow impair the guard cell chloroplasts' ability to do
their normal CO_2 fixation, or that products of mesophyll CO_2
fixation are capable of being translocated into guard cells
from elsewhere in the leaf. There is a need for much more
thorough work on the interrelations between guard cells and
the other cells of the plant as far as metabolites are concerned.

Guard cells behave very unusually in alternately taking
up and excreting large quantities of ions. I now examine
whether this ion exchange is significant in terms of the
total potassium balance of the plant. In the guard cell of
V. faba, which is about 5×10^{-15} m^3 in volume, the change in
concentration of potassium is from about 100 to 500 mol m^{-3}
with stomatal opening (e.g. Allaway and Hsiao 1972). Each

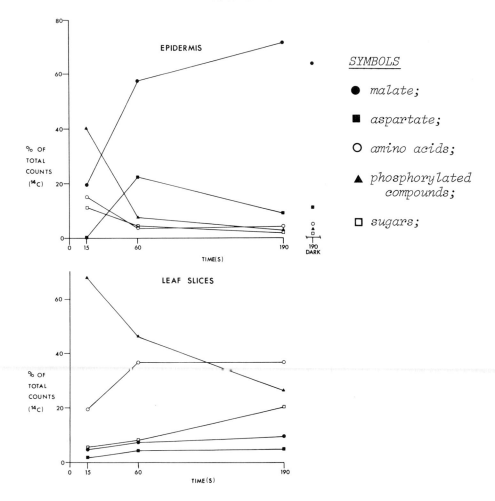

Fig. 2. - *Patterns of ^{14}C distribution during carbon fixation in rolled epidermal strips and leaf slices of* V. faba.
$^{14}CO_2$ *was supplied to the samples by immersing them in radioactive sodium bicarbonate solution for the times shown, in
darkness (shown at right side of epidermis graph) or light of
saturating intensity from a HPLR high-pressure mercury fluorescent lamp. After the times shown, samples were killed and
extracted in hot 80% ethanol and the extracts chromatographed.
The accumulation of label in malate in the epidermis preparation is striking, and the similiarity of the final pattern
after 190 s to dark fixation (although there was a twofold
stimulation of overall rate of fixation by light). In spite
of the accumulation of label in malate, the rapid decline in
label in phosphorylated compounds suggests a C_3-type method of
carbon fixation in epidermal strips.*

TABLE 3.   Changes in Potassium in Epidermis of
Commelina communis

	Guard cell	Inner	Outer	Terminal	Epidermal cell
		Subsidiary cells			
STOMATA CLOSED					
K activity[1] mol m^{-3}	95	156	199	289	448
Estimated[2] cell volume m^3 x 10^{-15}	3.2	6	22	8	59
Estimated amount of K per cell, pmol	0.3	0.9	4.4	2.3	26.4
STOMATA OPEN					
K activity[1] mol m^{-3}	448	293	98	169	73
Estimated[2] cell volume m^3 x 10^{-15}	7.6	6	22	8	59
Estimated amount of K per cell, pmol	3.4	1.8	2.2	1.4	4.3
Change in amount of K per cell with opening, pmol	3.1	0.9	−2.2	−0.9	−22.1
Net K change for stomatal complex, pmol per stomatal complex[3]			1.8		
Net K change for epidermis, pmol per stomatal complex[4]			−42.4		

[1] Mean values from Penny and Bowling (1974).
[2] For guard cells, from Pearson and Milthorpe (1974 –
C. cyanea). For epidermal and subsidiary cells, estimated
on the following basis:
thickness of all epidermal cells 1.89 times that of guard
cells (from Allaway and Milthorpe 1976 – *C. cyanea*);
superficial area of cells in the following proportions to
that of guard cells: inner subsidiary 0.96, outer subsidiary
3.66, terminal subsidiary 1.38; epidermal 9.68 (from 4
stomatal complexes measured on photomicrographs of Fujino
(1967)). Cell volumes estimated by proportion, taking cell

volume to be 3.2×10^{-15} m^3.

(3) Guard cells and all types of subsidiary cells only, two of each type per stomatal complex.

(4) Two epidermal cells included per stomatal complex (probably an underestimate of the real number per complex).

---

guard cell therefore acts, daily, as a source and sink for 2×10^{-12} mol of K$^+$. I selected a 4-week-old *V. faba* plant typical of those used in my experiments: it had 20 leaves each of roughly 6000 mm^2 and with about 150 guard cells per mm^2 (counting both sides of the leaf). Thus for the whole plant, during the first 3 h of light, roughly 36 µmol of K$^+$ is removed from the rest of the plant by the stomata. Cl$^-$ may be taken up to the extent of between about 2 and 7 µmol in the stomata of the whole plant (Humble and Raschke 1971; Raschke and Fellows 1971). A 4-week-old *V. faba* plant (total fresh weight 60.0 g; dry weight 6.98 g) on analysis showed a total K content of 4.2 mmol and Cl of 1.4 mmol, (well within the normal range for herbaceous plants – Baumeister 1958). The values indicate that the quantities of these ions involved in stomatal opening are of the order of one per cent of the total content of the plant – I suppose quite trivial in relation to the whole plant's ion content.

It is appropriate to note the work of Penny and Bowling (1974) who measured the concentrations of K$^+$ in various cells of *Commelina communis* epidermis. They showed large increases in K$^+$ activity in guard cells on stomatal opening, as expected, but also large losses of K$^+$ from other epidermal cells. These changes are summarized in Table 3. As the table shows, much more K$^+$ is lost from the subsidiary and epidermal cells than can be taken up by the guard cells: most of the K$^+$ for stomatal opening can come from within the stomatal complex. We are forced to suggest that K$^+$ must leave the epidermis and go to the rest of the leaf in the light. However this suggestion is in conflict with the measurements of Pearson (1975) who showed in *C. cyanea* that the amount of K$^+$ in the whole epidermis did not change between light and darkness. In *Zea mays* L. (Raschke and Fellows 1971) it was concluded that K$^+$ and Cl$^-$ were "shuttled" between the guard cells and subsidiary cells, and that no exogenous K$^+$ was required other than that contained within the stomatal complex.

INFLUENCE OF STOMATA ON WATER AND NUTRIENT FLOW THROUGH THE PLANT

I am not going into detail about the well documented effect of stomatal opening in allowing access of CO_2 to the leaf and thereby creating the source of photosynthate. In-

stead, I deal with the effects of the water flow through the
plant, that results from stomatal opening. I am ignoring
guttation and root pressure. Many substances are transported
through the plant in the transpiration stream, and they are
dealt with elsewhere in this volume. It is often stated (e.g.
Crafts 1968) that the xylem sap flow is well over the amount
needed to carry enough mineral nutrients up the plant. As an
illustration of this, Table 4 shows transpiration rates and
expressed sap concentrations of K and Cl for *V. faba* plants
in light and darkness. We are not at all surprised in the
cases of K and Cl to find that the transpiration stream is
easily sufficient to transport the whole plant's content of
these ions, and to support a relative growth rate far in excess of any likely to be achieved (cf. Evans 1972). Potassium
and chloride, the ions that were convenient for me to measure,
are widely mobile within the plant in phloem and other
tissues, as well as in xylem; but there are other elements
which are more or less only mobile in xylem. These elements,
such as calcium (Ziegler 1976) depend for their translocation
to any organ on the existence of a transpiration stream to
that organ, and therefore on the existence and behaviour of
stomata on that organ. A number of physiological disorders
of fruits have been correlated with calcium deficiency or im-
balance in the whole plant, which is more strongly expressed
in fruit since transpiration is low in maturing fruit (e.g.
blossom-end rot in tomatoes, Wiersum 1966). In spite of some
recently published data showing that Ca can in some cases be
transported in phloem (e.g. Tromp 1975) there is considerable
evidence in many cases for a close correlation of Ca transport with transpiration rate (Fischer 1958). As a result of
this dependence on water flow, together with poor ability to
be re-translocated in phloem, such elements as calcium and
boron often accumulate in transpiring leaves, while K, which
can be re-translocated, does not (Baumeister 1958). Nutrition with calcium and other xylem-transported minerals is
particularly a problem for subterranean fruits and tubers
growing in low-nutrient soils. For example, because of the
absence of a transpiration stream to them lack of Ca adversely affects fruit development in peanut (*Arachis hypogaea* -
e.g. Campbell et al. 1975); in this particular work there is
evidence that boron can be transported to developing peanuts
without the assistance of xylem flow. A number of other
cases of this have been reviewed by Ziegler (1976), an interesting one of which is the case of plants parasitic on the
phloem of their host (Ansiaux 1958): in three species of
phloem parasites the mean ratio of Ca to K was 0.082, while
in their three herbaceous hosts it was 1.6. The parasites

---

did not have access to the transpiration flow of the hosts and so could not accumulate transpiration-dependent Ca.

TABLE 4.   Transpiration Rates and Sap Concentrations of *V. faba* in Light and Dark

		Dark (stomata closed)	Light (stomata open)
Rate of transpiration in the conditions of the experiment mg m^{-2} s^{-1}		6.9	42.6
K concentration in sap expressed by Scholander bomb, mol m^{-3}		9.9	5.1
Cl concentration, mol m^{-3}		2.5	1.7
Rate of translocation for a standard 4-week old plant of leaf area 0.12 m^2, fresh weight 60.0 g. (nmol s^{-1})	K Cl	8.2 2.1	26.1 8.7
Total content of the plant millimoles	K Cl	4.2 1.4	
Approximate time to transport the whole ionic content of the plant at the above rates if constant (days)	K Cl	6 8	2 2
Relative growth rate (week^{-1}) that this transport could sustain	K Cl	1.2 0.9	3.8 3.8

It seemed interesting to me to think about CAM plants in this context: these plants are characterized by low transpiration rates all the time, and especially low during water stress (e.g. Ting et al. 1967). It would be interesting if CAM plants were found to be particularly sensitive to lack of calcium, or perhaps especially resistant to boron or lead toxicity as a result. Unfortunately for this hypothesis, the only data I have so far found on the nutrient requirements of CAM crops (in this case Pineapple) have indicated that the plants respond adversely to high soil calcium levels, and prefer a soil rather on the acid side (Collins 1960). I think this suggestion might repay further investigation, however. The ratio of CO_2 uptake to transpiration is very high in CAM plants (CO_2 uptake/transpiration = about 0.02 in pineapple and *Agave americana*) compared with mesophytes (e.g. 0.006 in tobacco and sunflower) (Neales et al. 1968). It is

easy to calculate - in a simple-minded way - from these values
and the usual ranges of concentrations of ions in xylem sap
(the only ones I found were those of Bollard 1953, from apple
trees), the proportionality between ion content and dry weight
accumulation that might be expected in these plants. To be
more realistic, perhaps, we should use the water use efficiency of the plants, but I have not found any values of this
for CAM plants. In C3 plants it may be about 0.002 at ordinary temperatures (cf. Björkman 1971), and so I am taking
0.007 for a likely value for CAM plants' water use efficiency,
by the same proportion. If it is assumed that xylem sap concentrations are like those of apple, then on average (over
the whole year) there might be about 1.6 mol m^{-3} K, and 0.35
mol m^{-3} P (Bollard 1953), with perhaps Ca about the same as
P (from values for bleeding of several species, Curtis 1944).
Doing these calculations, which extend the data well beyond
their area of validity, we nevertheless find that the content
of minerals per g dry weight of C3 plant tissue (assuming 10%
dry weight to be ash - Baumeister 1958) works out to values
within the measured range: 3.1% K, 0.55% P, and 0.7% Ca (cf.
Strigel 1912). However for CAM plants the values are 0.79%
K, 0.14% P, and 0.18% Ca - all smaller than the range of
values from Strigel's work. A single measurement I made gave
a leaf K content of 2.7% of dry weight for *Kalanchoë
diagremontiana*. This is within the range of contents for non-
CAM species: if CAM plant mineral contents are similar, then,
to those of other plants, it may be necessary to think that
the ions that make up the mineral content cannot all be transported in the xylem to the top of the plant. This is very
speculative, but at least it indicates an area that might repay study.

INFLUENCE OF STOMATAL BEHAVIOUR ON PHLOEM TRANSPORT

As well as the direct effect of stomatal opening in
allowing access of CO_2 to the leaf, and therefore in making
available a source of photosynthate for translocation, there
are more indirect effects that I wish to consider. Firstly
I shall remark on the effects of water stress on phloem
transport, and then I shall try to put this together with
information on the daily changes in plant water potential to
make some speculations on the course of translocation throughout the day as affected by stomata.

Reviews of translocation list many papers stating that
translocation is reduced under conditions of water stress
(e.g. Crafts 1968). Wiebe and Wihrheim (1962) showed that
the removal of ^{14}C-labelled assimilate from leaves of sunflower within 2 h of first feeding was reduced about five-

fold by a fall in water potential of about 1000 J kg^{-1}. There
has, however, been argument as to whether this decrease in
translocation results from decline of photosynthesis (leading
to a reduced source), decline in phloem loading, or reduction
in rate of translocation itself. There are data supporting
all three points of view. Munns and Pearson (1974), showed
that, while translocation was reduced by the imposition of a
drought treatment on potato plants, the rate of photosynthesis
was also reduced and to a similar degree: they suggested
that the cause of the translocation decline was the reduction
in available photosynthate, and no need was seen for an effect
of stress on the mechanisms of loading or translocation. Peel
and Weatherley (1963) found that directly altering the water
potential of xylem (by applying pressure or suction) changed
the pressure in phloem in the stem, and this can be interpreted as suggesting that Münch pressure-flow system of translocation would be slowed down by water stress. The data of
Hartt (1967) in sugar cane showed a greater reduction in
translocation than in photosynthesis, resulting from the imposition of water stress. And an elegant experiment by
Wardlaw (1969) showed in *Lolium temulentum* that water stress
led to a decrease in ability to load photosynthate into the
phloem, but that the rate of transport once in the phloem was
likely to be increased: the overall result is seen in a reduction in rate of transport away from leaves, as usual. It
is difficult for me to draw these different points of view
together, except into the all-embracing hypothesis that water
stress affects different parts of the transport system in
different species - and perhaps this hypothesis is not a great
help in increasing our understanding.

However its effect comes about, there is an effect of
water stress in reducing the transport of photosynthate out
of the leaf. We can therefore readily predict that in the
long term under water stress plant growth and translocation
will be decreased - and this of course is a familiar observation. But I want to explore the likelihood of effects of
water stress becoming apparent on transport of photosynthate
in a plant during the course of the normal changes that occur
throughout the day. Cowan (1965) has presented a theoretical
treatment of the changes in water potential of a plant as it
undergoes droughting through a number of days, indicating
that the water potential of a leaf can be expected to fall
considerably during the course of a day, even if stress is
not present. This results simply from the opening of the
stomata and loss of water from the leaf, together with the
resistances to water flow in plant and soil. Can we expect
real values for this fall in water potential to be extreme

enough to cause inhibition of transport at some time of day in a normal, otherwise unstressed, plant? Turner and Begg (1973) found in maize, sorghum and tobacco that leaf water potentials fell to -1200 to -1500 J kg^{-1} near mid-day in some leaves of well-watered plants: in stressed plants the leaf water potential fell to -1500 to -2200 J kg^{-1} (Turner 1974). According to the data of Wiebe and Wihrheim (1962) these water potentials are sufficient to inhibit photosynthate movement drastically. I have not been able to find a conversion between relative turgidity and water potential for *Lolium temulentum*, but if we assume that Wardlaw's (1969) plants had roughly a similar relationship as sunflowers (rather unlikely), then inhibition of vein loading is expected at noon even in well-watered plants, as a result of the normal level of water potential in the plants. I wish to suggest that photosynthates accumulate in leaves during the day and are translocated away at night not just simply because of an excess of maximal photosynthesis rate over maximal translocation rate, but additionally because in the middle of the day, when photosynthesis is at its maximum, a state of water potential exists in the plant which may slow down the processes of vein loading and translocation. Pearson (1974), using *V. faba* in growth chambers, has suggested that only about half the photosynthate formed in the light is translocated away in the same light period, with about 20% translocated in subsequent dark and light periods; I think it likely that in plants growing in the field, where water potentials in leaves are likely to be more extreme, perhaps even more of the material may be kept for translocation in the dark. Of course, if water stress is so great as to cause stomatal closure during the day, then source strength will be decreased as well. Further examination of the relation between leaf water potentials and photosynthate removal throughout the day is required.

I have tried to suggest in the last two sections that the behaviour of stomata should not be ignored in studies of translocation. The influence of stomata on translocation is a secondary one resulting from the stomata letting water out of the leaf. In its flow it draws along the ions that have been selected for it by the root tissues, and they will accumulate at the point of evaporation, unless they can be retranslocated, or else are absorbed from the xylem on the way by other tissues. Because of the resistances to water flow, stomatal opening sets up reductions in leaf water potential, whose effects merit further consideration in studies of translocation throughout the day. We should take care to be sure that these effects are taken into account in studies we make on long-distance transport.

ACKNOWLEDGEMENTS

I thank the Australian Research Grants Committee for support for some of my own research presented in this paper.

REFERENCES

Allaway, W.G. (1973). Accumulation of malate in guard cells of *Vicia faba* during stomatal opening. *Planta (Berl.)* 110, 63-70.

Allaway, W,G. and Hsiao, T.C. (1972). Preparation of rolled epidermis of *Vicia faba* L. so that stomata are the only viable cells: analysis of guard cell potassium by flame photometry. *Aust. J. Biol. Sci.* 26, 309-318.

Allaway, W.G. and Milthorpe, F.L. (1976). Structure and functioning of stomata. In "Water deficits and plant growth" Vol. IV, T.T. Kozlowski, ed., Academic Press, New York (in press).

Allaway, W.G. and Setterfield, G. (1972). Ultrastructural observations on guard cells of *Vicia faba* and *Allium porrum*. *Can. J. Bot.* 50, 1405-1413.

Ansiaux, J.R. (1958). Sur l'alimentation minérale des phanérogames parasites. *Bull. Acad. Roy. Belg., Cl. Sci.* 44, 787-793.

Baumeister, W. (1958). Die Aschenstoffe. In "Encyclopaedia of Plant Physiology" Vol. IV, W. Ruhland, ed., Springer, Berlin, pp. 5-36.

Bearce, B.C. and Kohl, H.C. (1970). Measuring osmotic pressure of sap within live cells by means of a visual melting point apparatus. *Plant Physiol.* 46, 515-519.

Björkman, O. (1971). In "Photosynthesis and photorespiration", M.D. Hatch, C.B. Osmond and R.O. Slatyer, eds. Wiley Interscience, New York.

Bollard, E.G. (1953). The use of tracheal sap in the study of apple-tree nutrition. *J. Exp. Bot.* 4, 363-368.

Campbell, L.C., Miller, M.H. and Loneragan, J.F. (1975). Translocation of boron to plant fruits. *Aust. J. Plant Physiol.* 2, 481-487.

Collins, J.L. (1960). The pineapple. Leonard Hill, London.

Cowan, I.R. (1965). Transport of water in the soil-plant-atmosphere system. *J. Appl. Ecol.* 2, 221-239.

Crafts, A.S. (1968). Water deficits and physiological processes. In "Water deficits and plant growth" Vol. II, T.T. Kozlowski, ed., Academic Press, New York, pp. 85-133.

Curtis, L.C. (1944). The influence of guttation fluids on pesticides. *Phytopathology* 34, 196-205.

Evans, G.C. (1972). The quantitative analysis of plant growth. Blackwell, Oxford.

Fischer, H. (1958). Der Transport der Mineralstoffe. In "Encyclopaedia of plant physiology" Vol. IV, W. Ruhland, ed., Springer, Berlin, pp. 289-306.

Fischer, R.A. (1973). The relationship of stomatal aperture and guard-cell turgor pressure in *Vicia faba*. *J. Exp. Bot. 24*, 387-399.

Fischer, R.A. and Hsiao, T.C. (1968). Stomatal opening in isolated epidermal strips of *Vicia faba*. II. Response to KCl concentration and the role of potassium absorption. *Plant Physiol. 43*, 1953-1958.

Fujino, M. (1967). Role of adenosinetriphosphate and adenosinetriphosphatase in stomatal movement. *Sci. Bull. Fac. Educ. Nagasaki Univ., 18*, 1-47.

Hartt, C.E. (1967). Effect of moisture supply upon translocation and storage of ^{14}C in sugarcane. *Plant Physiol. 42*, 338-346.

Humbert, C., Louguet, P. and Guyot, M. (1975). Étude ultrastructurale comparée des cellules stomatiques de *Pelargonium* x *hortorum* en relation avec un état d'ouverture ou de fermeture des stomates physiologiquement défini. *C.R. Acad. Sci. Paris 280D*, 1373-1376.

Humble, G.D. and Raschke, K. (1971). Stomatal opening quantitatively related to potassium transport. Evidence from electron probe analysis. *Plant Physiol. 48*, 447-453.

Kaufman, P.B., Petering, L.B., Yocum, C.S. and Baic, D. (1970). Ultrastructural studies on stomata development in internodes of *Avena sativa*. *Amer. J. Bot. 57*, 33-49.

Landré, P. (1969). Premières observations sur l'évolution infrastructurale des cellules stomatiques de la moutarde (*Sinapis alba* L.) depuis leur mise en place jusqu'à l'ouverture de l'ostiole. *C.R. Acad. Sci., Paris 269D*, 943-946.

Meidner, H. and Mansfield, T.A. (1968). The physiology of stomata. McGraw-Hill, London.

Miroslavov, E.A. (1966). Electron microscopic studies of the stomata of rye leaves *Secale cereale* L. *Bot. Zh. 51*, 446-449.

Munns, R. and Pearson, C.J. (1974). Effect of water deficit on translocation of carbohydrate in *Solanum tuberosum*. *Aust. J. Plant Physiol. 1*, 529-537.

Neales, T.F., Patterson, A.A. and Hartney, V.J. (1968). Physiological adaptation to drought in the carbon assimilation and water loss of xerophytes. *Nature 219*, 469-472.

Pallas, J.E. and Dilley, R.A. (1972). Photophosphorylation can provide sufficient adenosine 5'-triphosphate to drive K^+ movements during stomatal opening. *Plant Physiol. 49*, 649–650.

Pallas, J.E. and Mollenhauer, H.H. (1972). Physiological implications of *Vicia faba* and *Nicotiana tabacum* guard cell ultrastructure. *Amer. J. Bot. 59*, 504–514.

Pearson, C.J. (1974). Daily changes in carbon-dioxide and photosynthate translocation in leaves of *Vicia faba*. *Planta (Berl.) 119*, 59–70.

Pearson, C.J. (1975). Fluxes of potassium and changes in malate within epidermis of *Commelina cyanea* and their relationship with stomatal aperture. *Aust. J. Plant Physiol. 2*, 85–89.

Pearson, C.J. and Milthorpe, F.L. (1974). Structure, carbon dioxide fixation and metabolism of stomata. *Aust. J. Plant Physiol. 1*, 221–236.

Peel, A.J. and Weatherley, P.E. (1963). Studies in sieve-tube exudation through aphid mouthparts. II. The effects of pressure gradients in the wood and metabolic inhibitors. *Ann. Bot. 27*, 197–211.

Penny, M.G. and Bowling, D.J.F. (1974). A study of potassium gradients in the epidermis of intact leaves of *Commelina communis* L. in relation to stomatal opening. *Planta (Berl.) 119*, 17–25.

Raschke, K. (1975). Stomatal action. *Ann. Rev. Plant Physiol. 26*, 309–340.

Raschke, K. and Fellows, M.P. (1971). Stomatal movement in *Zea mays*: shuttle of potassium and chloride between guard cells and subsidiary cells. *Planta (Berl.) 101*, 296–316.

Singh, A.P. and Srivastava, L.M. (1973). The fine structure of pea stomata. *Protoplasma 76*, 61–82.

Stålfelt, M.G. (1967). The components of the CO_2-induced stomatal movements. *Physiol. Plant. 20*, 634–642.

Strigel, A. (1912). Mineralstoffaufnahme verschiedener Pflanzenarten aus ungedüngtem Boden. Über den Einfluss der botanischen Natur, der Herkunft und Erntezeit auf die chemische Zusammensetzung von Wiesenheu. *Landwirtsch. Jb. 43*, 349–371.

Ting, I.P., Thompson, M-L.D. and Dugger, W.M. (1967). Leaf resistance to water vapor transfer in succulent plants: effect of thermoperiod. *Amer. J. Bot. 54*, 245–251.

Tromp, J. (1975). The effect of temperature on growth and mineral nutrition of fruits of apple, with special reference to calcium. *Physiol. Plant. 33*, 87–93.

Turner, N.C. (1974). Stomatal behavior and water status of maize, sorghum and tobacco under field conditions. II. At low soil water potential. *Plant Physiol. 53*, 360–365.

Turner, N.C. and Begg, J.E. (1973). Stomatal behavior and water status of maize, sorghum and tobacco under field conditions. I. At high soil water potential. *Plant Physiol.* *51*, 31-36.

Wardlaw, I.F. (1969). The effect of water stress on translocation in relation to photosynthesis and growth. II. Effect during leaf development in *Lolium temulentum*. *Aust. J. Biol. Sci.* *22*, 1-16.

Wiebe, H.H. and Wihrheim, S.E. (1962). The influence of internal moisture stress on translocation. In "Radioisotopes in soil-plant nutrition studies", International Atomic Energy Agency, Vienna, pp. 279-287.

Wiersum, L.K. (1966). Calcium content of fruits and storage tissues in relation to the mode of water supply. *Acta Bot. Neerl.* *15*, 406-418.

Willmer, C.M. and Dittrich, P. (1974). Carbon dioxide fixation by epidermal and mesophyll tissues of *Tulipa* and *Commelina*. *Planta (Berl.)* *117*, 123-132.

Willmer, C.M., Pallas, J.E. and Black, C.C. (1973). Carbon dioxide metabolism in leaf epidermal tissue. *Plant Physiol.* *52*, 448-452.

Ziegler, H. (1976). Nature of transported substances. In "Encyclopaedia of plant physiology". N.S. Vol. I, U. Lüttge and M.G. Pitman, eds. (in press).

CHAPTER 26

Water Transport through Plants:
Current Perspectives

M. R. Kaufmann

*Department of Plant Sciences, University of California,
Riverside, Calif. 92502 U.S.A.*

Considerable attention has been given to physical and
thermodynamic characteristics of short and long distance water
flow in plants. For thoughtful and thought-provoking articles
the reader is referred to Cowan (1972), Cowan and Milthorpe
(1968), Dainty (1969), Newman (1974), Richter (1973), Slatyer
(1967), and Tyree (1971). Without wishing to detract in any
way from the value of these contributions, there is need for
a somewhat simplified overview of the process of water transport, particularly for understanding in the sense of plant-
environment interaction how the development of water deficits
is related to edaphic, atmospheric, and physiological factors
in the natural environment (Kaufmann and Hall 1974). This
paper attempts to indicate some considerations important in
understanding water flow and the development of water stress
in the natural or field environment as well as in carefully
controlled artificial environments. Emphasis is placed upon
interpreting experimental data on the basis of current
theories of water flow.

Van den Honert (1948) proposed a model for water transport through the soil-plant-atmosphere continuum (Eq. 1):

$$\frac{dm}{dt} = \frac{\psi_1 - \psi_0}{r_R} = \frac{\psi_2 - \psi_1}{r_X} = \frac{\psi_3 - \psi_2}{r_L} = \frac{\psi_4 - \psi_3}{r_G} \quad (1)$$

where dm/dt is rate of water transport, ψ is water potential
at different positions in the continuum, and r is the flow
resistance in the root (R), xylem (X), leaves (L), and air
(G). Rawlins (1963), Dainty (1969), and others have demonstrated that using a water potential (ψ) difference in the vapor
phase results in a resistance which varies with ψ, and that it
is more convenient to use vapor pressure differences in this
segment of the continuum.

WATER TRANSPORT AND THE DEVELOPMENT OF PLANT WATER STRESS

LEAF WATER POTENTIAL MODEL

An analysis of factors influencing ψ_{leaf} has been undertaken in a number of laboratories, since ψ_{leaf} is measureable
and serves as a sensitive integrator of simultaneous edaphic,
atmospheric, and physiological effects on water flow. According to conventional transport equations, flux is dependent
upon the driving force and resistances to flow. Elfving et
al. (1972) and Kaufmann and Hall (1974) pointed out that
ψ_{leaf} should be considered as a variable which is dependent
on water supply to the leaf and transpirational flux, not as
an independent variable controlling water flow through the
plant. The relationship between ψ_{leaf} and transpirational
flux of water through plants is shown in Fig. 1. Data are

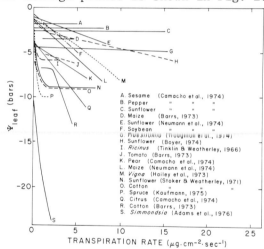

*Fig.1. Relationship between ψ_{leaf} and transpiration rate
for well-watered plants having healthy root systems at
normal growing temperatures. Curves D, I, J, M, N, O,
and R were determined with plants growing in solution
culture. The rest were determined with plants growing
in well-watered soil, perlite, or sand.*

presented for studies in which the water supply was adequate
and root systems were well-aerated, healthy, and not at ab-
normally low or high temperatures. Thus soil water potential
(ψ_{soil}) and soil and plant resistances to flow were presumed
to be non-limiting. Several observations may be made about
these data. First, in most instances ψ_{leaf} decreased as
transpirational flux increased, but there were notable exceptions where ψ_{leaf} remained nearly constant even at very high
fluxes. Secondly, several distinctly different curves have

been observed for the same species studied in different laboratories (maize and sunflower in particular, also cotton). Third,
when flux was zero considerable variation occurred in ψ_{leaf}
among the species examined.

Huber (1924), working without van den Honert's model, and
later Elfving et al. (1972), Kaufmann and Hall (1974), Richter
(1973), and a number of others working with the model have
written equations for ψ_{leaf} having the general form:

$$\psi_{leaf} = \psi_{soil} + \psi_g - (flux) \, (r_{soil \, to \, leaf}) \qquad (2)$$

where ψ_{soil} is the total soil water potential, ψ_g is a gravita-
tional component (frequently omitted for smaller plants), flux
is the transpiration rate (generally assumed to equal flow in
each segment of the continuum if steady state conditions prevail), and $r_{soil \, to \, leaf}$ is the resistance to flow in the
liquid phase from soil to leaves. To the physiologist and
ecologist this type of equation has significant appeal because
it incorporates most factors known to influence ψ_{leaf}.

It generally has been assumed that the slope of the
curve relating ψ_{leaf} to transpiration rate represents the
resistance to flow ($r_{soil \, to \, leaf}$ in Eq. 2). These considerations, if correct, lead to the conclusion from Fig. 1 that
Simmondsia, Engelmann spruce, and citrus are species having
rather high resistances to flow whereas sesame and pepper have
very low resistances, and sunflower and maize are variable.
It also follows that at 0 flux ψ_{leaf} should equal ψ_{soil} (plus
ψ_g). However, several factors suggest the need for caution
in applying this model.

Richter (1973) showed that branched pathway models for
water flow through the plant may lead to different conclusions
than those from single-pathway models. He notes that different resistances, potential drops, and flux densities may
exist for each level of the pathway. Hence resistances
obtained from the slopes of curves such as those in Fig. 1
may only be gross estimates of resistances for the soil-plant
system, and the numerical values observed may not exist in
any portion of the pathway.

Perhaps more important is the question of the driving
force for water flow. It is widely assumed that in rapidly
transpiring plants water moves passively along water potential
gradients established by transpirational loss of water from
leaves. Water absorption coupled with active solute accumulation in the root, first described by Renner (see Kramer
1969) and investigated extensively by Brouwer (1965), Kramer
(1969), and others, has been assumed to be of little importance except in slowly transpiring plants. Recent theoretical

papers published simultaneously by Dalton et al. (1975) and
Fiscus (1975) re-examined the interaction between water and
solute movement into the root xylem. In both papers the
authors concluded that hydrostatic and osmotic gradients for
water flux into the root interacted strongly over a range of
conditions. The small (or non-existent) decreases in ψ_{leaf}
observed as flux increased in some species (Fig. 1) may be
related to a coupled solute and water transport. Hence
resistances calculated from the relationship of ψ_{leaf} and
water flux would include parameters that affect fluxes of
both water and solutes.

From these considerations, it seems appropriate to
revise Eq. 2 to include a solute effect. Thus we may write:

$$\psi_{leaf} = f\ (\psi_{soil},\ \psi_g) - f(Ts,\ r_{soil},\ r_{plant}, \text{coupled solute and water flow)}$$

(3)

In this equation, ψ_{leaf} is expressed as a function of two sets
of factors, the first, an equilibrium component including the
static effects of soil water supply (ψ_{soil}) and hydrostatic
head (ψ_g), the second a flux component in which mass flow
from soil to leaves (Ts), flow resistances (or hydraulic
conductivities), and the coupling of solute and water flow are
treated. It is not important here to develop Eq. 3 explicitly
with an incorporation of hydraulic conductivities, reflection
coefficients, active ion uptake components, etc. This can be
done by extending the work of Dalton et al. (1975) and Fiscus
(1975) (A.E. Hall and G. Hoffman, personal communication).

Thus the differences in the relationship between ψ_{leaf}
and transpirational flux observed in Fig. 1 may have several
causes. Plant resistances to water flow may vary, but also
there may be significant differences in osmotic effects,
including coupled solute and water flow into roots and
also solute removal along the xylem. Osmotically-induced uptake of water is known to vary widely among contrasting species
(Kramer 1969). The differences in the relationship between
ψ_{leaf} and flux within a species (maize, sunflower, and cotton)
may be related to differences in growth and treatment conditions among the studies. Such variations might lead to
dissimilar contributions of the osmotic and hydrostatic
components for water movement (Brouwer 1965). To illustrate,
A.E. Hall (personal communication) observed exudation rates
from detopped sunflower root systems at rates ranging from
10 to 20 percent of the transpiration rate observed for intact plants (e.g. 10 $\mu g\ cm^{-2}\ sec^{-1}$) prior to detopping.
In contrast, Kramer (1939) observed in a different sunflower
study that the root exudation rate was only 0.4 percent of the

transpiration rate. The percentages are naturally sensitive to the levels of transpiration, but it is clear that in one case osmotically-induced water absorption may have been a significant factor, and in the other case it was of little importance.

Presumably the variation in ψ_{leaf} among species at 0 flux (Fig. 1) represents some plant factor influencing the equilibrium portion of Eq. 3, but further research is needed before this factor can be described adequately. Differences in ψ_{leaf} at the intercept are predicted by Dalton et al.'s (1975) model, but it is hard to imagine how root ion relations or growth (Boyer 1974) could explain ψ_{leaf} values of no higher than -8 to -10 bars in *Simmondsia* plants enclosed in a dark, humid chamber (Adams, Bingham, Kaufmann and Yermanos (unpublished); Fig. 1).

Factors controlling the relative importance of osmotic and hydrostatic components of water absorption have not yet been resolved, although the papers by Dalton et al. (1975) and Fiscus (1975) suggest that ion absorption rates, membrane leakage, etc. may be important factors. Those studies indicating minimal changes in ψ_{leaf} with changing tanspiration (Fig. 1) involved herbaceous plants examined under controlled environmental conditions. Studies on woody perennials have indicated that ψ_{leaf} decreased at increased transpiration rates under both laboratory and field conditions. Even within a woody species, however, different relationships between ψ_{leaf} and transpiration may be observed. Camacho et al. (1974b) grew citrus seedlings in a greenhouse, maintaining an adequate supply of water at all times. For these unstressed plants little change in ψ_{leaf} was observed from low to intermediate transpiration rates. At higher fluxes, ψ_{leaf} decreased, but ψ_{leaf} was less negative than observed in field trees at equivalent transpiration rates. The seedlings were then subjected to a series of 3 drying cycles, simulating field irrigation and drought periods. The relationship of ψ_{leaf} and flux after stressing the seedlings was shifted and corresponded to the field observations. This curve shift may have several causes. The contribution of an osmotic component to water transport may have changed, the root system may have been reduced by drought damage, or the geometry of the soil-root interface may have been altered by an irreversible decrease in root diameter caused by cortex collapse. The latter possibility is discussed later. It is puzzling that drying cycles produced varying results with sunflower. Kramer (1950) and Neumann et al. (1974) reported increases in root resistance to water absorption after wilting. However, Levy and Kaufmann (unpublished data) wilted sunflower plants

similar to those used by Camacho et al. (1974a) and found no
shift in the relationship of ψ_{leaf} and flux.

WATER MOVEMENT INTO ROOTS

Taking the plant as a whole, what factors affect water
absorption by roots sufficiently to influence water flow in
plants and the development of plant water stress? Equations 2
and 3 predict that soil drying will affect ψ_{leaf} at all fluxes,
whereas a change in root resistance will have an effect only
at fluxes above 0. A test of the effects of ψ_{soil} on ψ_{leaf}
predicted by Eq. 2 requires data collected over a range of
transpiration rates. Considerable evidence has been published
indicating that as ψ_{soil} decreases, ψ_{leaf} and transpiration
decrease simultaneously (e.g. Doley and Trivett 1974; Hansen
1974). This relationship is seemingly opposite to those shown
in Fig. 1 and results from a continued decrease in ψ_{soil}.
These and other studies generally involved uniform shoot
environmental conditions and do not provide an adequate range
of transpiration rates for a given ψ_{soil}.

Comparisons of ψ_{leaf} and transpiration rates for different
levels of ψ_{soil} for white oak and *Simmondsia* are given in Fig.
2. Similar data for citrus were presented by Elfving et al.

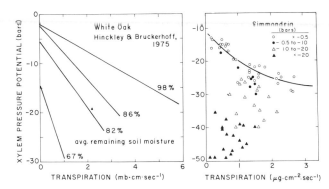

*Fig.2. - (Left): Relationship of xylem pressure potential
and transpiration rate for white oak having different
levels of soil water supply. Transpiration rate was
estimated as the product of the vapor pressure gradient
from leaf to air (mbar) and leaf conductance. Taken
from Hinckley and Bruckerhoff (1975). Reproduced by permission of the National Research Council of Canada.
(Right): Relationship of xylem pressure potential and
transpiration rate for* Simmondsia chinensis *having different levels of* ψ_{soil}. *Taken from Adams, Bingham,
Kaufmann and Yermanos - unpublished).*

(1972). Leaf water potential for well-watered plants

(estimated in these experiments by xylem pressure potential)
decreased as transpiration increased. As the supply of soil
water decreased ψ_{leaf} decreased at all fluxes from near 0 to
high rates. The white oak data exhibited both a decrease in
ψ_{leaf} at the intercept (0 flux) and a more negative slope as
soil water content decreased, the latter an effect of reduced
soil hydraulic conductivity. The data in Fig. 2 and those of
Elfving et al. (1972) for citrus support the hypothesis that
changes in ψ_{soil} have a predictable effect on the relationship between ψ_{leaf} and transpiration.

The relationships between ψ_{leaf} and transpiration for cold
soil temperatures are different from the relationships for
dry soil conditions. Reductions in soil temperature to below
$10^{\circ}C$ resulted in progressively lower values of ψ_{leaf} in
Engelmann spruce (again estimated as xylem pressure potential)
at equivalent fluxes (Fig. 3). Similarly, Elfving et al. (1972)

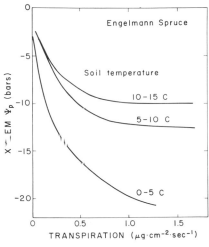

*Fig.3.-Effect of soil tempera-
ture on the relationship between
xylem pressure potential and
transpiration rate for Engelmann
spruce trees in the controlled
environment. Adapted from
Kaufmann (1975).*

observed strong soil temperature effects in citrus as soil
temperature decreased from 15° to $5^{\circ}C$. In both spruce and
citrus, however, ψ_{leaf} at 0 flux appeared to be uninfluenced
by soil temperature. Assuming changes in root resistance to
be the cause, Kaufmann (1975) attempted to separate water
viscosity effects from apparent changes in liquid flow
resistance by comparing the relative increase in viscosity and
resistance as temperature decreased. In spruce, flow resistances and viscosity increased by the same proportion as
temperature decreased from 12.5 to $7.5^{\circ}C$, and then resistance
increased more sharply (Fig. 4). In contrast, citrus flow
resistances appeared to increase more than viscosity at
temperatures at least as high as $10^{\circ}C$. From these data it
appears that resistance to liquid flow from soil to leaves is

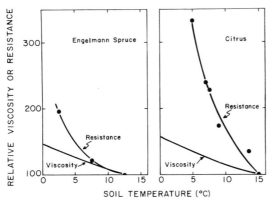

Fig.4.- Effects of reduced temperature on relative viscosity of water and on relative soil-plant resistances for water flow to leaves of Englemann spruce and citrus. Taken from Kaufmann (1975).

influenced in citrus, a subtropical species, at a higher temperature than in spruce, a subalpine species.

Perhaps the importance of root contact with the soil has been under-estimated with regard to water absorption by roots. While an expanding root system is generally regarded to be in intimate contact with the soil, under certain conditions root shrinkage may drastically affect the root-soil interface geometery. Huck et al. (1970) followed root diameter changes in cotton and observed that on a dry, sunny day root diameter decreased by 40 percent from the maximum. Kaufmann made observations (unpublished) of root diameter during his study of water stress effects on growth of pine seedlings (Kaufmann 1968). When plants were subjected to drying cycles, roots became suberized, the cortex collapsed, and root diameter decreased to 50 or 60 percent of the diameter before the stress. Calculations based upon his data led Elfving et al. (1972) to conclude that significantly higher root resistances to water absorption existed during a second drying cycle than during the first. Perhaps part of the shift in the relation-ship between ψ_{leaf} and transpiration in citrus (Camacho et al. 1974b) results from changes in root diameter resulting from periods of water stress. Clearly more research is needed to determine the importance of changes in root diameter in the the process of water absorption.

WATER MOVEMENT TO LEAVES

Considering water flow through the soil-plant-atmosphere continuum, water flux in the stem is probably best understood, largely because most stem transport occurs in dead xylem tissue and is affected only slightly if at all by physiological factors. Thus xylem flow has been described as a physical process consisting primarily of laminar flow driven by gradi-ents in hydrostatic pressure, as described by Poiseuille's law (see Zimmermann and Brown, 1971).

Considerable research has been directed toward the question of the relative magnitude of flow resistances in different
portions of the soil–plant–atmosphere continuum. Studies
generally have indicated that root resistance for liquid flow
is higher than stem or leaf resistances (Kramer 1969; Slatyer
1967). Several pieces of evidence suggest that significant
stem and leaf resistances also occur. Measurements have been
made of ψ_{leaf} for sunlit and shaded portions of the crown of
spruce and citrus (Kaufmann 1975) and cotton (Klepper et al.
1971). In the shade ψ_{leaf} was less negative than in the sun.
In spruce and citrus these differences are related to unequal
transpiration rates caused by different vapor pressure gradients and stomatal conductances. Presumably the ψ_{leaf} differences in cotton are caused by similar variations in transpiration. Hinckley and Scott (1971) failed to observe differences
in ψ_{leaf} between the north and south sides of Douglas-fir, but
their trees may have been shaded somewhat on the south side.
Consistent ψ_{leaf} differences (typically 1 to 4 bars in spruce,
3 to 5 bars in citrus, 1 to 3 bars in cotton) indicate that
significant flow resistances exist within the shoot after the
water flow pathway divides. Tyree et al. (1974) concluded
that about two-thirds of the shoot resistance in hemlock
(*Tsuga canadensis*) was in the stem xylem, and that resistances
in the minor branches were much higher than in the trunk
xylem. Hellkvist et al. (1974) concluded that large reductions in ψ_{leaf} for Sitka spruce were caused largely by high
xylem resistances in the shoot. Begg and Turner (1970)
observed that root resistance in tobacco was variable at
different times of the day but was always considerably lower
than the resistance between the stem and leaf. They concluded
that the major resistance in tobacco occurred in the petiole
or at the nodal connection between the stem and leaf. Un-
doubtedly anatomical differences among species influence the
relative resistances in each part of the transport pathway.

WATER LOSS FROM LEAVES

Water movement through plants is controlled primarily by
transpiration. The effect of any atmospheric, edaphic, or
physiological factor on water flux occurs because of an effect
on either the vapor gradient or resistances for water vapor
flux from within the leaf to bulk air outside the leaf.
For recent reviews of the factors controlling stomatal behavior, the reader is directed to papers by Hall et al. (1976)
and Raschke (this volume).

It has long been recognized that stomata respond to
changes in light, temperature, carbon dioxide, and leaf water
stress, and models of photosynthesis and transpiration
generally take these factors into account. Until recently

the possible role of humidity in regulating stomatal aperture
remained controversial (see Lange et al. 1971). During the
last several years, however, data have been collected, using
intact plants under both field and laboratory conditions,
which clearly indicate that leaf-to-air humidity gradients
have a strong effect on stomatal aperture (Camacho et al.1974a;
Hall et al. 1975; Hall and Kaufmann 1975; Schulze et al.
1972). In Fig. 5, data are summarized showing that as the
absolute humidity difference from leaf to air increased,
stomatal conductance decreased. In sesame and sunflower,
stomatal closure occurred with no measureable change in bulk

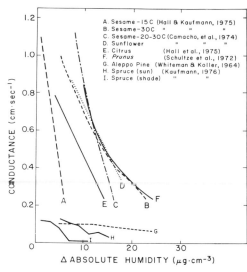

Fig.5.- Effect of leaf-to-air absolute humidity difference on stomatal conductance. Conifers had much lower conductances than the other species.

leaf water potential (Camacho et al. 1974a). This suggests
that stomatal closure in response to humidity gradients may
be a sensitive mechanism for the control of water loss, capable of operating independently from the classical water
potential feedback mechanism.

Stomatal response to humidity is influenced by other
factors. Hall and Kaufmann (1975) provided evidence
showing that at lower temperatures stomata of sesame were
closed with small increases in the leaf-to-air absolute
humidity difference, whereas at higher temperatures the closing
response was more gradual (Fig. 5). In Engelmann spruce,
Kaufmann (1976) observed that in full sunlight stomatal
conductance decreased as the absolute humidity difference
increased, but in shade the closing response was much more
pronounced (Fig. 5). It also was observed that stomatal
response to humidity in spruce was influenced by the
level of ψ_{leaf}. When ψ_{leaf} (estimated by xylem pressure

potential measurements) fell below –15 bars, stomatal conductances were decreased at all levels of absolute humidity
difference and at 2 levels of photosynthetically active radiation. Levy and Kaufmann (unpublished data) compared the
response of citrus stomata to high vapor pressure gradients
induced by high temperatures for citrus seedlings kept well-
watered and for seedlings which had been subjected to 3 drying
cycles (Fig. 6). Previously unstressed seedlings had minimal
stomatal closure when exposed to high vapor gradients and
increased leaf temperatures whereas stomata of pre-stressed

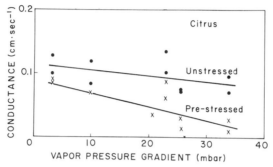

*Fig.6.- Relationship of stomatal conductance and vapor
pressure gradient for previously unstressed citrus
seedlings and for seedlings subjected to a series of
drying cycles. The high vapor pressure gradients were
achieved largely by an increase in temperature.
Unpublished data of Levy and Kaufmann.*

seedlings closed substantially. Their data suggest that
plants subjected to drying cycles might have a more sensitive
stomatal mechanism for controlling transpiration.

A large number of reports indicate that stomatal cycling
occurs in plants. Most studies have been conducted under
controlled environmental conditions with artificial light,
and cycling often was induced by rapid changes in humidity,
temperature, or light. This had led to the concern that
cycling is in some way an artifact of unnatural environments
and that it may have little physiological or ecological
importance in the natural environment. Several pieces of
evidence suggest, however, that in citrus stomatal cycling
does occur naturally, although its causes and function remain
unclear. Levy and Kaufmann (unpublished data) studied
stomatal behavior of citrus under orchard conditions and on
several occasions observed apparent oscillations of stomatal
conductance (Fig. 7). In their field observations ψ_{leaf} was
not well-correlated with changes in conductance, although with

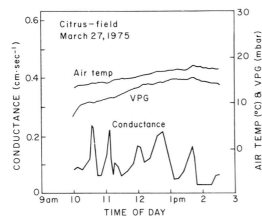

Fig. 7. - Apparent cycling of stomatal aperture in 12-yr old field-grown citrus trees. The period of oscillation was slightly less than one hour. Unpublished data of Levy and Kaufmann.

4-year-old potted trees in large containers simultaneous changes in stomatal conductance, leaf-air temperature difference, ψ_{leaf}, and trunk diameter occurred with a similar period. Trunk diameter and ψ_{leaf} changes often lagged behind stomatal conductance changes by a few minutes. An examination of older dendrograph records of citrus trunk diameter supports Levy and Kaufmann's field observations of stomatal cycling. Records indicated that distinct cycling in trunk diameter occurred with a 1-hour period in 30-year-old trees at several locations in southern California at different times of the year. Perhaps additional observations of stomatal cycling in citrus and other species exposed to natural conditions will reveal the physiological or ecological significance of transient stomatal behaviour.

ACKNOWLEDGEMENTS

Much of the author's research cited here was supported by National Science Foundation Grant No. GB-39856. The helpful comments of Drs. A.E. Hall and R.T. Leonard are greatly appreciated.

REFERENCES

Barrs, H.D. (1973). Controlled environment studies of the effects of variable atmospheric water on photosynthesis, transpiration and water status of *Zea mays* L. and other species. In: "Plant Response to Climatic Factors". *Proc. UNESCO Uppsala Symp.* pp 249-258.

Begg, J.E. and Turner, N.C. (1970). Water potential gradients in field tobacco. *Plant Physiol.* 46, 343-346.

Boyer, J.S. (1974). Water transport in plants: mechanism of apparent changes in resistance during absorption. *Planta (Berl.)* 117, 187-207.

Brouwer, R. (1965). Water movement across the root. In: "The State and Movement of Water in Living Organisms". *Proc. 19th Symp. Soc. Exp. Biol.* pp 131-149.

Camacho-B., S.E. Hall, A.E. and Kaufmann, M.R. (1974a). Efficiency and regulation of water transport in some woody and herbaceous species. *Plant Physiol. 54*, 169-172.

Camacho-B., S.E. Kaufmann, M.R. and Hall, A.E. (1974b). Leaf water potential response to transpiration by citrus. *Physiol. Plant. 31*, 101-105.

Cowan, I.R. (1972). An electrical analogue of evaporation from, and flow of water in plants. *Planta (Berl.) 106*, 221-226.

Cowan, I.R. and Milthorpe, F.L. (1968). Plant factors influencing the water status of plant tissues. In: "Water Deficits and Plant Growth" ed. by T.T. Kozlowski, Academic Press, New York 1, 137-193.

Dainty, J. (1969). The water relations of plants. In: "The Physiology of Growth and Development" ed. by M.B. Wilkins, McGraw-Hill, New York, pp 421-452.

Dalton, F.N., Raats, P.A.C., and Gardner, W.R. (1975). Simultaneous uptake of water and solutes by plant roots. *Agron. J. 67*, 334-339.

Doley, D. and Trivett, N.B.A. (1974). Effects of low water potentials on transpiration and photosynthesis in Mitchell grass (*Astrebla lappacea*). *Aust. J. Plant Physiol. 1*, 539-550.

Ellving, D.C., Kaufmann, M.R. and Hall, A.E. (1972). Interpreting leaf water potential measurements with a model of the soil-plant-atmosphere continuum. *Physiol. Plant. 27*, 161-168.

Fiscus, E.L. (1975). The interaction between osmotic- and pressure-induced water flow in plant roots. *Plant Physiol. 55*, 917-922.

Hailey, J.L., Hiler, E.A., Jordan, W.R., van Bavel, C.H.M. (1973). Resistance to water flow in *Vigna sinensis* L. (Endl.) at high rates of transpiration. *Crop Sci. 13*, 264-267.

Hall, A.E., Camacho-B, S.E. and Kaufmann, M.R. (1975). Regulation of water loss by citrus leaves. *Physiol. Plant 33*, 62-65.

Hall, A.E. and Kaufmann, M.R. (1975). Regulation of water transport in the soil-plant-atmosphere continuum. In: "Perspectives of Biophysical Ecology" ed. by D.M. Gates and R.B. Schmerl, Springer-Verlag, Berlin, pp 187-202.

Hall, A.E., Schulze, E.-D. and Lange, O.L. (1976). Current perspectives of steady-state stomatal responses to environment. In: "Water and Plant Life--Problems and

Modern Approaches" ed. by O.L Lange, L. Kappen, and E.-D. Schulze, Springer-Verlag, New York, Ecological Studies Vol. (in press).

Hansen, G.K. (1974). Resistance to water flow in soil and plants, plant water status, stomatal resistance and transpiration of Italian ryegrass, as influenced by transpiration demand and soil water depletion. *Acta Agric. Scandinavica 24*, 83-92.

Hellkvist, J., Richards, G.P. and Jarvis, P.G. (1974). Vertical gradients of water potential and tissue water relations in Sitka spruce trees measured with the pressure chamber. *J. Appl. Ecol. 11*, 637-667.

Hinckley, T.M. and Bruckerhoff, D.N. (1975). The effect of drought on water relations and stem shrinkage of *Quercus alba*. *Can. J. Bot. 53*, 62-72.

Hinckley, T.M. and Scott, D.R. (1971). Estimates of water loss and its relation to environmental parameters in Douglas-fir saplings. *Ecology 52*, 520-524.

Honert, T.M. van den (1948). Water transport in plants as a catenary process. *Disc. Faraday Soc. 3*, 146-153.

Huber, B. (1924). Die Beurteilung des Wasserhaushaltes der Pflanze. Ein Beitrag zur vergleichenden Physiologie. *Jb. Wiss Bot. 64*, 1-120.

Huck, M.G., Klepper, B. and Taylor, H.M (1970). Diurnal variations in root diameter. *Plant Physiol. 45*, 529-530.

Kaufmann, M.R. (1968). Water relations of pine seedlings in relation to root and shoot growth. *Plant Physiol. 43*, 281-288.

Kaufmann, M.R. (1975). Leaf water stress in Engelmann spruce: influence of the root and shoot environments. *Plant Physiol. 56*, 841-844.

Kaufmann, M.R. (1976). Stomatal response of Engelmann spruce to humidity, light, and water stress. *Plant Physiol.* (in press).

Kaufmann, M.R. and Hall, A.E. (1974). Plant water balance - its relationship to atmospheric and edaphic conditions. *Agric. Meteorol. 14*, 85-98.

Klepper, B., Browning, V.D. and Taylor, H.M. (1971). Stem diameter in relation to plant water status. *Plant Physiol. 48*, 683-685.

Kramer, P.J. (1939). The forces concerned in the intake of water by transpiring plants. *Amer. J. Bot. 26*, 784-791.

Kramer, P.J. (1950). Effects of wilting on the subsequent intake of water by plants. *Amer. J. Bot. 37*, 280-284.

Kramer, P.J. (1969). Plant and Soil Water Relationships: A Modern Synthesis. McGraw-Hill, New York, 482 pp.

Lange, O.L. Lösch, R., Schulze, E.-D. and Kappen, L. (1971).
    Responses of stomata to changes in humidity. *Planta*
    (Berl) 100, 76-86.
Neumann, H.H., Thurtell, G.W. and Stevenson, K.R. (1974). In
    situ measurements of leaf water potential and resistance
    to water flow in corn, soybean, and sunflower at several
    transpiration rates. *Can. J. Plant Sci. 54*, 175-184.
Newman, E.I. (1974). Root and soil water relations. In"
    "The Plant and Its Environment" ed. by E.W. Carson, Univ.
    Press of Virginia, pp 363-440.
Rawlins, S.L. (1963). Resistance to water flow in the trans-
    piration stream. In: "Stomata and Water Relations in
    Plants" ed. by I. Zelitch, *Conn. Agric. Exp. Sta. Bull.*
    664, 69-85.
Richter, H. (1973). Frictional potential losses and total
    water potential in plants: a re-evaluation. *J. Exp.*
    Bot. 23, 983-994.
Schulze, E.-D., Lange, O.L., Buschbom, U., Kappen, L. and
    Evenari, M. (1972). Stomatal responses to changes in
    humidity in plants growing in the desert. *Planta (Berl.)*
    108, 259-270.
Slatyer, R.O. (1967). Plant-water Relationships. Academic
    Press, New York, 366 pp.
Stoker, R. and Weatherley, P.E. (1971). The influence of
    the root system on the relationship between the rate of
    transpiration and depression of leaf water potential,
    New Phytol 70, 317-334.
Tinklin, R. and Weatherley, P.E. (1966). On the relationship
    between transpiration rate and leaf water potential.
    New Phytol. 65, 509-517.
Troughton, J.H., Camacho-B., S.E. and Hall, A.E. (1974).
    Transpiration rate, plant water status, and resistance
    to water flow in *Tidestromia oblongifolia*. Carnegie
    Inst. Yearbook *73*, 830-835.
Tyree, M.T. (1971). The steady state thermodynamics of
    translocation in plants. In: "Trees: Structure and
    Function" by M.H. Zimmermann and C.L. Brown, Springer-
    Verlag, Berlin, pp 281-305.
Tyree, T., Caldwell, C. and Dainty, J. (1974). The location
    and measurement of the resistance to bulk water movement
    in the shoots of hemlock (*Tsuga canadensis*). In:
    "Membrane Transport in Plants" ed. by U. Zimmermann and
    J. Dainty, Springer-Verlag, Berlin, pp 84-89.
Whiteman, P.C. and Koller, D. (1964). Environemental control
    of photosynthesis and transpiration in *Pinus halepensis*.
    Israel J. Bot. 13, 166-176.
Zimmermann, M.H. and Brown, C.L. (1971) Trees: Structure and
    Function. Springer-Verlag, Berlin, 336 pp.

CHAPTER 27

Control of Translocation by Photosynthesis and Carbohydrate Concentrations of the Source Leaf

A.L. Christy* and C.A. Swanson**
*Agricultural Research Department, Monsanto Agricultural Products Company, St. Louis, Missouri 63166, U.S.A.
**Department of Botany, Ohio State University, Columbus, Ohio, 43210, U.S.A.

INTRODUCTION

Regulation of the rate of translocation from a source leaf appears to involve the integration of several source leaf processes and metabolite concentrations (Geiger and Batey 1967, Plaut and Reinhold 1969, Servaites and Geiger 1974). Important among these parameters is photosynthesis and the supply of sucrose, the principal molecular species translocated (Geiger et al. 1969, Geiger and Swanson 1965). In this paper we examine the relationship of translocation rate and velocity to the sucrose concentration and photosynthetic rate of the source leaf.

MATERIALS AND METHODS

Young sugar beet plants (*Beta vulgaris* L. cv. Klein Wanzleben) were cultured hydroponically as described previously (Geiger and Swanson 1965). The day before the experiment plants were selected for uniformity and trimmed to a simplified translocation system consisting of an 11 to 12 cm long source-leaf blade with an area of 0.7 to 1.0 dm^2, an 8 to 12 cm source leaf petiole, and a sink leaf 3.5 to 4.0 cm in length. Translocation to the beet was blocked by girdling the hypocotyl approximately 1 cm below the crown with an electrically heated nichrome wire.

Translocation rates were determined by steady state labeling (Geiger and Swanson 1965) with constant specific activity $^{14}CO_2$ metered into a gas flow system by an infusion pump. An infrared gas analyzer actuated the infusion pump and controlled the CO_2 concentration at the desired concentration ($\pm$ 3μl/l). Photosynthetic rates were determined by converting the running time of the infusion pump to mg CO_2 hr^{-1} and normalizing to unit area of the source leaf.

The apparent velocity of translocation was determined by pulse labeling (Geiger et al. 1969) with a minimum of 50 μc of $^{14}CO_2$ for 7 to 10 min. In these experiments, the photosynthetic rate was determined during the postlabeling period in an open flow system by measuring the decrease in CO_2 concentrations between influent and effluent streams using an infrared gas analyzer. Photosynthesis rates were determined at the prelabeling CO_2 concentrations.

In pulse labeling experiments the source leaf was sampled for carbohydrate analysis immediately after arrival of the ^{14}C-pulse in the sink leaf. Random samples of the source leaf were taken with an 0.34-cm^2 punch. In the pulse labeling experiments, each sample varied from 3 to 8 punches depending on the expected sucrose concentration. In steady state labeling experiments source leaf samples consisted of 4 punches.

The leaf samples were immediately placed in boiling 80% ethanol and extracted for 2 hr in a Soxhlet. The extract was passed through anion and cation exchange resins and the neutral fraction collected and freeze-dried. For analysis, the freeze-dried samples were dissolved in 1.5 ml H_2O and aliquots assayed for sucrose, glucose, and fructose using coupled enzyme systems (Bergmeyer and Klotzsch 1965). For determinations of the reserve polysaccharides, the ethanol extracted leaf material was extracted with 52% (v/v) perchloric acid (Geiger and Batey 1967) and aliquots assayed by the phenol-H_2SO_4 procedure of DuBois et al. (1956). In the present paper, only the sucrose and the polysaccharide concentrations are reported.

RESULTS AND DISCUSSION

Steady-state labeling experiments. - Steady-state labeling experiments were conducted to study the relationship of translocation rate to source leaf photosynthetic rate and carbohydrate concentrations. Nine experiments, each of 8-hr duration, were carried out; five involving a step-down in CO_2 concentration from 400 to 200 μl/l at the end of the first 4-hr period, and four involving a step-up from 400 to 600 μl/l. The first 4-hr period in each experiment is referred to as the control period; the second 4-hr period, as the experimental period. In two experiments of each series, leaf samples were harvested periodically for carbohydrate analysis. Approximately 2 to 3 min were required to increase or decrease the CO_2 concentration to the desired level. Since an additional 20 to 30 min were required to obtain reliable data from the running time of the infusion pump, the time course for the net photosynthetic rate during this transition period was estimated from the CO_2 concentration in the flow system.

The results of a typical step-up experiment are presented in Fig. 1. The net photosynthetic rate increased rapidly to a new steady-state rate when the CO_2 concentration was increased to 600 μl/l (Fig. 1B), and the translocation rate increased slowly stabilizing at a new steady state rate after approximately 200 min (Fig. 1A). The rate of accumulation of sucrose and polysaccharide increased when the CO_2 concentration was increased (Fig. 1C).

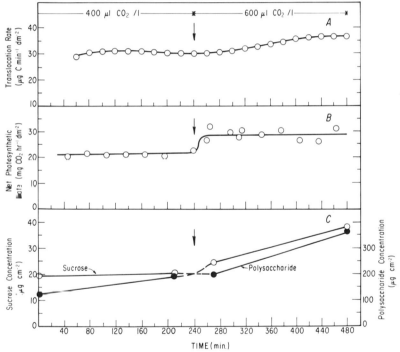

Fig. 1. - Time course of A: translocation rate, B: net photosynthetic rate and C: the changes in concentration of the carbohydrate pools following an increase in the CO_2 concentration from 400 μl/l to 600 μl/l. (Because of rapid changes occurring in pool sizes following the change in CO_2 concentration, the interpolated time concentration course during the interval between final harvest in the control period and the initial harvest in the experimental period is dashed to indicate its lesser certainty).

In the step-down experiments similar responses were noted in the time course of the photosynthetic rate and translocation rate (Fig. 2A and B). The translocation rate declined sharply (approximately 30%) during the first 15 to 20 min immediately following the decrease in the CO_2 concentration, plateaued for a period of 30 to 40 min, and then slowly declined to a new steady state rate which was maintained for the remainder of the experimental period. Similarly, the photosynthetic rate (Fig. 2B) and sucrose concentration (Fig. 2C) declined rapidly when the CO_2 concentration was lowered but quickly stabilized and were reasonably constant for the remainder of the experimental period. Servaites and Geiger (1974) reported a similar response in the translocation rate to a decrease in the photosynthetic rate.

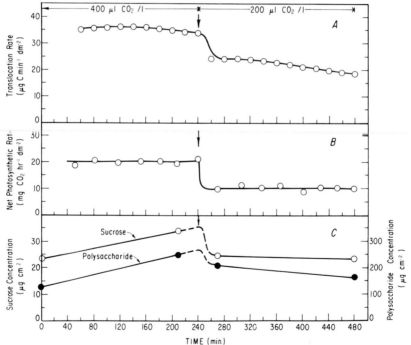

Fig. 2. - Time course of A: translocation rate, B: net photosynthetic rate and C: the changes in concentration of the carbohydrate pools following a decrease in the CO_2 concentration from 400 µl/l to 200 µl/l. (Because of rapid changes occurring in pool sizes following the change in CO_2 concentration, the interpolated time concentration course during the interval between final harvest in the control period and the initial harvest in the experimental period is dashed to indicate its lesser certainty).

Several differences are evident in the time courses of
the translocation rate, photosynthetic rate and sucrose concentration in response to step-up or step-down changes in the
CO_2 concentration. The photosynthetic rate in both types of
experiments rapidly adjust to a new steady state rate. While,
the time courses of the translocation rate and sucrose concentration are similar they differ markedly from the time
course of the photosynthetic rate. In the step-up experiments, the translocation rate and sucrose concentration increased slowly while in the step down experiments the translocation rate and sucrose concentration declined rapidly
during the first 20 to 30 min and then slowly over the remainder of the experimental period. These data suggest that
the translocation rate is controlled by the source leaf
sucrose concentration when the concentrations are in the
range of the values reported here. At times, the translocation rate appears to be coupled to the photosynthetic rate,
since the sucrose concentration is controlled in part by the
photosynthetic rate. The sucrose pool size would therefore
mediate the effects of changes in the photosynthetic rate on
translocation rate.

Pulse labeling experiments. - To further evaluate the
coupling of translocation kinetics to the source leaf photosynthetic rate and sucrose concentration the minimum apparent
velocity of the translocation stream was determined by pulse
labeling the source leaf with $^{14}CO_2$. In Figures 3 and 4 the
apparent velocity is plotted as a function of source leaf
photosynthetic rate and sucrose concentration, respectively.

Translocation velocity appears to be dependent on the
source leaf photosynthetic rate over the entire range of
photosynthetic rates measured (Fig. 3). However, velocity
appears dependent on sucrose concentration over only a
limited range of sucrose concentrations. Above a sucrose
concentration of 15 μg cm^{-2}, translocation appears to be uncoupled from the sucrose concentration.

The lack of correlation between velocity and source leaf
sucrose concentration above a concentration of 15 μg cm^{-2} may
be the result of the compartmentation of sucrose among transport and storage pools. Sucrose, accumulated in storage
pools when leaf sucrose concentrations were high, may be relatively unavailable for rapid mobilization and export from
the leaf. Table 1 summarizes the experimental parameters of
several of the experiments shown in Figs. 3 and 4. In experiments 6, 10, 20 and 26 the plants were under steady state
conditions immediately prior to labeling and show the coupling between sucrose concentration, photosynthesis and velocity. Plants with intermediate sucrose concentrations and

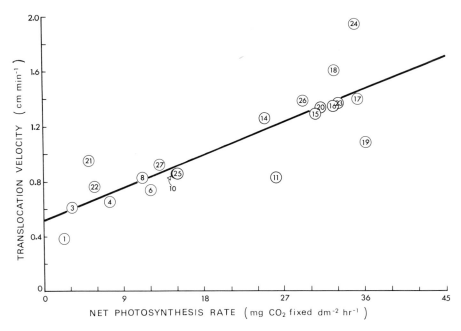

Fig. 3. - *Regression of translocation velocities on source leaf net photosynthetic rate. Numbers refer to experiment number in Table 1.*

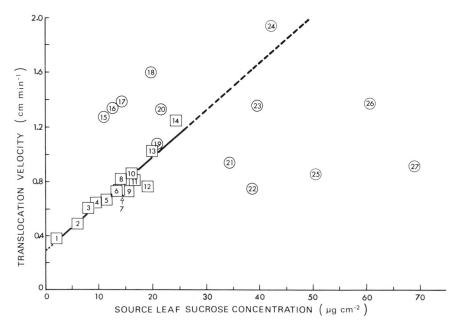

photosynthetic rates had intermediate velocities (Expt 6 and 10) and plants with high sucrose and high photosynthetic rates had high translocation velocities (Expt 20 and 26). Prolonged treatment with high CO_2 1100 µl/l followed by a short period at a lower CO_2 concentration (Expt 21, 22, 25 and 27) results in non-steady state conditions in the source leaf and apparent uncoupling of translocation velocity from the source leaf sucrose concentration and photosynthesis rate (Table 1). The depletion of the sucrose pool at a low CO_2 concentration (Expt 21 and 22) indicates that under conditions of low photosynthetic rates and relatively high sucrose concentrations, intermediate translocation velocities may be maintained by mobilization of sugars from storage pools.

Previous investigators have reported evidence of compartmentation of sucrose in leaves (Nelson 1963, Outlaw et al. 1975). One of these sucrose compartments may be present in the companion cells of minor veins as indicated by an accumulation of ^{14}C label (Geiger and Cataldo 1969) and high solute potential in these cells (Geiger et al. 1973). Outlaw et al. (1975) have reported data suggesting the presence of two sucrose pools in each of five tissue layers and one sucrose pool in the minor vein of *Vicia faba* leaves. With the present sampling technique we measured only the total source leaf sucrose concentration. A large percentage of sucrose in the leaf could have been in a storage pool and relatively unavailable for translocation.

The argument that the principal determinant of translocation velocity is the concentration of sucrose in a specific transport pool suggests that under steady state conditions, the transport pool will be in equilibrium with the total sucrose pool and photosynthesis and translocation velocity will be dependent on photosynthesis and sucrose concentration. Under non-steady state conditions the sucrose concentration in the transport pool would not be in equilibrium with the total sucrose pool and translocation velocity would appear to be uncoupled from the source leaf sucrose concentration.

Other investigators have reported evidence that indicates translocation is dependent on the source leaf sucrose concentration (Geiger and Batey 1967, Husain and Spanner 1966) and concurrent photosynthetic rate (Plaut and Reinhold 1969,

Fig. 4. - Regression of translocation velocities on sucrose concentration of the source leaf. Numbers refer to experiment number in Table 1. For experiments 1 to 14, normal or near normal levels of CO_2 were used; for experiments 15 to 27, high concentrations, or combinations of high and low concentrations of CO_2 were used.

TABLE 1. Comparison of data from several pulse labeling experiments showing the relationship of translocation velocity to source leaf photosynthetic rate and sucrose concentration.

Photosynthetic rates and sucrose concentrations were varied by changing the duration of light and dark exposures and the concentration of CO_2 during the pre-labeling periods. Illuminance was constant in all treatments at 2500 f.c. Experiment numbers correspond to numbers in Figure 3 and 4.

Expt.	Prelabel light period (hr)	CO_2 Conc. (µl/1) Prelabel	Prelabel[1]	Post label	Sucrose Conc. (µg/cm²)	Net Photosynthesis (mg CO_2/dm²/hr)	Translocation Velocity (cm/min)
6	1.0	405		405	13.1	12.1	0.74
10	1.5	405		405	15.9	14.9	0.86
20	3.0	956		956	21.5	31.1	1.33
26	6.0	956		956	60.4	29.0	1.38
25	24.5	1100	405 (30 min)	405	50.3	15.1	0.86
27	24.5	1100	405 (30 min)	405	68.8	13.1	0.92
21	12.5	1100	174 (30 min)	174	55.1[2] 34.1	5.2	0.95
22	12.5	1100	174 (30 min)	174	51.6[2] 38.5	5.8	0.76

[1] In certain experiments, the plants were exposed to two different CO_2 concentrations in the sequence shown during the pre-labeling light period.

[2] Sucrose concentration at the end of 12 hr. light period at CO_2 concentration of 1100 µl/1.

Servaites and Geiger 1974, Reinhold 1975). Several investigators have suggested that photosynthetically produced ATP may be involved in energizing translocation (Hartt and Kortschak 1967, Plaut and Reinhold 1969). The results presented in this paper indicate that the translocation rate may be coupled to the source leaf sucrose concentration. Since Husain and Spanner (1966) have shown that translocation from a darkened leaf can be maintained by supplying sucrose, it appears that photosynthesis serves only to supply translocate and does not directly energize translocation.

The results of both the steady state labeling and pulse labeling experiments indicate that the rate and velocity of translocation are dependent on source leaf sucrose concentration. We hypothesize that the instantaneous translocation rate is directly dependent on the instantaneous sucrose concentration of a transport pool. Although the location of this pool was not determined, it may be in the companion cells of the minor vein (Outlaw et al. 1975). The concentration of sucrose in the transport pool is considered to be controlled by the availability of sucrose for vein loading which, in turn, is regulated by the photosynthetic rate, rate of utilization of photosynthate, and rate of carbohydrate storage in the source leaf. Although in some cases the translocation rate may appear to be directly dependent on the photosynthetic rate, it is more likely coupled to photosynthesis through the transport pool. Thus, the dependence of the translocation rate on the photosynthetic rate is indirect and conditioned by the sucrose concentration.

REFERENCES

Bergmeyer, H.U. and Klotzsch, H. (1965). Sucrose. In: Methods of Enzymatic Analysis. Ed. by H.U. Bergmeyer, Academic Press, New York pp. 99-102.

Dubois, M., Gilles, K.A., Hamilton, J.K., Rabers, P.A. and Smith, F. (1956). Colorimetric method for determination of sugars and related substances. *Ann. Chem. 28*, 350-356.

Geiger, D.R. and Batey, J.W. (1967). Translocation of ^{14}C sucrose in sugar beet during darkness. *Plant Physiol. 42*, 1743-1749.

Geiger, D.R. and Cataldo, D.A. (1969). Leaf structure and translocation in sugar beet. *Plant Physiol. 44*, 45-54.

Geiger, D.R., Giaquanta, R., Sovonick, S.A. and Fellows, R.F. (1973). Solute distribution in sugar beet in relation to phloem loading and translocation. *Plant Physiol. 52*, 585-589.

Geiger, D.R., Saunders, M.A. and Cataldo, D.A. (1969). Translocation and accumulation of translocate in sugar beet petiole. *Plant Physiol.* 44, 1657–1665.

Geiger, D.R. and Swanson, C.A. (1965). Evaluation of selected parameters in a sugar beet translocation system. *Plant Physiol.* 40, 942–947.

Hartt, C.E. and Kortschak, H.P. (1967). Translocation in the sugarcane plant during the day and night. *Plant Physiol.* 42, 89–94.

Husain, A. and Spanner, D.C. (1966). The influence of varying concentrations of applied sugar on the transport of tracers in cereal leaves. *Ann. Bot.* 30, 549–561.

Nelson, C.D. (1963). Effect of climate on the distribution and translocation of assimilates. In: Environmental Control of Plant Growth. Ed. by L.T. Evans, Academic Press, New York, pp. 149–174.

Outlaw, W.H., Fisher, D.B. and Christy, A.L. (1975). Compartmentation in *Vicia faba* leaves. II. Kinetics of ^{14}C-sucrose redistribution among individual tissues following pulse-labeling. *Plant Physiol.* 55, 704–711.

Plaut, Z. and Reinhold, L. (1969). Concomitant photosynthesis implicated in the light effect on translocation in bean plants. *Aust. J. Biol. Sci.* 22, 1105–1111.

Reinhold, L. (1975). The effect of externally applied factors on translocation of sugars in the phloem. In: Phloem Transport ed. by S. Aronoff, J. Dainty, P.R. Gorham, L.M. Srivastava and C.A. Swanson, Plenum Press, New York, pp. 367–388.

Servaites, J.C. and Geiger, D.R. (1974). Effects of light intensity and oxygen on photosynthesis and translocation in sugar beet. *Plant Physiol.* 54, 575–578.

CHAPTER 28

Translocation in *Zea mays* Leaves

J.H. Troughton
Biophysics Division, Physics and Engineering Laboratory,
Department of Scientific and Industrial Research,
Lower Hutt, New Zealand.

INTRODUCTION

The mechanism of carbon translocation in the phloem has
been the subject of much debate and several opposing theories
are still under active consideration (Canny 1973, Fisher 1970).
To resolve this conflict several parameters associated with
translocation will have to be determined. One parameter is of
immediate interest, namely the influence of distance on the
shape of a carbon pulse moving in the sieve tubes. From
measurements of the half-width of the pulse it is possible to
derive information on the relative importance of the loading
zone, physiological conditions in the phloem or distance, in
determining pulse shape.

Experience with the short-lived, positron emitting carbon-
11 isotope has established that for some species and some
conditions, the resolution within individual pulses will allow
estimation of the half-width (More and Troughton 1973;
Troughton et al. 1974a and b). In addition, the speed of
translocation and changes in total counts within the pulse
with distance along the leaf can be obtained. The combination
of these techniques with C-14, sucrose concentration measurements, osmotic and turgor pressure measurements, will assist
in developing models to describe translocation under a
variety of environmental conditions and in numerous genotypes.

Previous results have indicated that the light level has
a significant effect on the speed of translocation in maize
leaves (Troughton 1975). It is also to be expected that
light may play an important role in determining pulse shape
through effects on events within the loading zone. For this
reason this paper concentrates on the effect of light on the
shape of the ^{11}C pulse in the phloem of *Zea mays* leaves.

MATERIALS AND METHODS

Zea mays (cv Morden 88) was used in all experiments. The plants were grown from seed for about four weeks in a controlled environment. The experiments were conducted in a controlled-environment cabinet with the same conditions as during growth; 12 hour photo period, light level of 250 Wm^{-2} (400-700 nm), 80 ± 10% RH, 28°C ± 1° and in normal air. The plants were in perlite and fed Hoaglands nutrient solution.

The fourth or fifth oldest mature leaf on a plant with seven leaves was used in the experiments. The leaf remained attached to the whole plant. Radioactive carbon dioxide was fed to a 2 cm wide strip of leaf, two thirds of the distance from the base to the tip of the leaf blade. Environmental conditions were monitored throughout all experiments.

Carbon-11 was produced by the $^{10}B(d,n)^{11}C$ reaction (More and Troughton 1973). This reaction produced ^{11}CO which was recovered from the target area by a differential pumping system and converted to $^{11}CO_2$ by passing the gas over CuO at 600°C. $^{11}CO_2$ gas was separated from other gases by a liquid oxygen trap. The isotope feeding procedure was that of pulse labelling, the gas being available to the leaf for about 4 minutes. The feeding chamber was flushed with $^{12}CO_2$ in normal air at the end of the period.

NaI scintillation counters were used to monitor the translocation of ^{11}C downstream and upstream from the feeding chamber. The counters were at 10 cm intervals and were collimated with lead to restrict the portion of the leaf in view of the counters to about 2 cm. The output from the detectors was amplified, passed to single channel analysers, counted on scalers and recorded on punch tape. All counters (usually five) were measured simultaneously, with counts being accumulated for 20 seconds every half minute. The results were corrected for half-life and background radiation and the corrected data displayed with a high speed plotter.

RESULTS

A most consistent and significant result was that the half-width of the pulse fed to the leaf was substantially changed during the loading process. The half-width during feeding was about 2 minutes whereas the half-width of the pulse at the first counter downstream from the fed region was always greater than 40 minutes. Although some changes in width would have occurred between the loading site and the first counter under most conditions, the bulk of this broadening effect would be associated with the uptake of carbon dioxide during photosynthesis, its subsequent metabolism and loading into the phloem. The shape of the loading pulse can be

340

determined from the efflux curves, and these support the general conclusion that the broadening is primarily due to events in the source region of the leaf.

Under steady state conditions it can be shown that the half-width of the pulse at the first counter, and therefore presumably during loading, will be relatively constant. This is indicated in Figure 1 for three positions along the leaf.

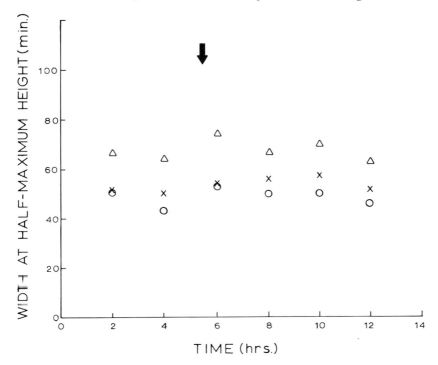

Fig. 1. - *The influence of shading a 10 cm length of leaf upstream from the feeding chamber on the half-width of the pulse at 3 counters, 10 (O), 20 (x) and 30 (△) cm downstream from the fed region. Arrow indicates time of shading.*

In this example the half-width of the pulse was between 40-50 minutes during 6 separate measurements over a 10 hour period. The same figure also indicates the same degree of consistency for two other counters on the same leaf.

The data in Figure 1 also shows an example of the effect of distance on the half-width of the pulse. In this example there was relatively no effect of distance between the first and second counters but there was a 30% increase in half-width between the second and third counters. This variation

in half-width with distance was consistent throughout the
length of the experimental period. This increase, however,
was small compared with the change in half-width of the pulse
during photosynthesis and subsequent loading events into the
phloem. These results illustrate the value of carbon-11 in
being able to distinguish between effects of the source region
and transport path on the pulse-shape. Where this distinction
between the two regions is not made, then it is difficult to
quantitatively assign significance to either of the two
regions as being determinants of pulse shape.

The influence of light level on the whole leaf on the
half-width of the pulse in the phloem is shown in Figure 2 for
the mean of three positions along the leaf. At low light
there was much variability in the half-width of the pulse,
whereas at high light the pulse widths were more consistent.
These data establish that the half-width decreases with increasing light level up to 240 Wm^{-2}. Extrapolation of this

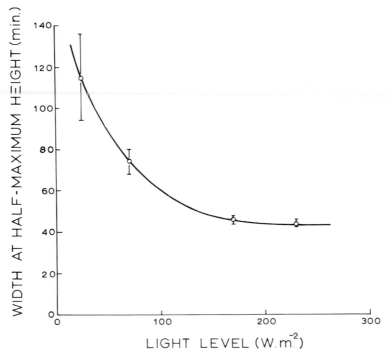

Fig. 2. - *The relationship between light level on the
whole plant and half-width of the pulse. The half-width
was measured at the counter 10 cm downstream from the
fed region.*

relationship to zero light yields a value of about 150 min. This is consistent with values obtained for leaves fed with radioactive CO_2 in the light but placed in darkness at the end of the feeding period.

The role of light level in determining half-width of the pulse was confirmed in other experiments. In particular it was shown that if the feeding chamber was placed in darkness immediately after feeding, then the half-width increased from about 50 min to > 120 min. This effect was within two hours, i.e. the interval between feedings. In that experiment the rest of the leaf was in high light and therefore the influence of light was localized in the source region of the leaf.

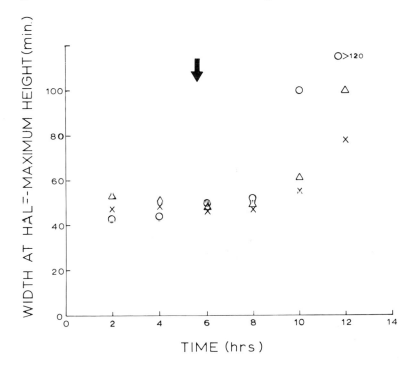

Fig. 3. - The half-width of the pulse at three positions, 10 (O), 20 (x) and 30 cm (Δ) from the fed leaf chamber. The region between the feeding chamber and the first counter was shaded at the time indicated by the arrow.

To illustrate the independence of variations in speed of translocation and the half-width of the pulse, data are given in Figure 3 for the half-width during an experiment in which the speed was also measured. Data from the speed are given elsewhere (Troughton 1975; Figures 8 and 9) and showed that the mean speed of translocation in the leaf increased from about 4 cm min^{-1} to 5.25 cm min^{-1} on imposition of a shading treatment between the feeding chamber and the first counter. It was also shown that the major change in speed was between the first two counters where the speed increased from 3.5 to 6.5 cm min^{-1}. The results shown in Figure 3 indicate that there were no pronounced changes in pulse shape during the period when the large changes in speed occurred. The plant was kept at high light for about 5 hours, during which time there were relatively small differences in half-width between the first three counters. The leaf segment between the fed region and the first counter was shaded at about 6 hours but there was relatively little change in pulse shape for the succeeding 4 hours, i.e. during the period that the speed showed large fluctuations. Late in the same day there was an increase in half-width on all counters which would be due to long term effects of the shading treatment.

DISCUSSION

The results presented here indicate the large variations which can occur in the half-width of the pulse of radioactive carbon fed to a maize leaf. It has been possible however, to localize the cause of some of these variations as being due to physiological events and conditions associated with the source region of the leaf.

Carbon-11 has been particularly useful in being able to clarify the respective roles of either the source region or events along the transport pathway in determining the shape of the pulse. The substantial change in pulse shape between that fed to the leaf as carbon dioxide and that appearing at the first counter, supports the view that pulse shape is primarily determined by the events of photosynthesis, carbon metabolism and loading in the source region of the leaf. Significant increases in width can occur with distance along the leaf but the magnitude of the effects in this study were considerably less than those associated with the source region.

Complementing this conclusion are the results which show the variation in pulse shape due to manipulating the light conditions in the source only. As shown here the half-width of the pulse decreased with increasing light level on the leaf. As has been reported elsewhere (Troughton et al. 1976) there is an immediate increase in half-width of the pulse when the source region of the leaf is shaded.

In comparing these results with those in other studies (Troughton 1975, Troughton et al. 1976) it is possible to show that the regulation of the speed of translocation is under the control of different parameters than those influencing half-width. For example, the speed of translocation has been shown to be approximately proportional to the light intensity whereas the half-width of the pulse shows an approximately exponential decline with increasing light. Similarly under conditions where the speed of translocation showed substantial variation, the half-width of the pulse was relatively unaffected. The working hypothesis that has emerged from these studies is that the shape of the pulse is primarily determined by the source region whereas the speed of translocation is a function of the whole leaf upstream from the measured region.

In general, these results provide evidence of a slow phloem loading process and the possibility of a bulk-flow mechanism in the sieve tubes. If it can be shown that the pulse-shape remains coherent over longer distances than established in this study, then the Münch theory of translocation becomes impellingly attractive.

ACKNOWLEDGEMENTS

The author appreciates the assistance of Mr. R. More, Mr. B.G. Currie and Mrs. K.A. Card during the course of these investigations.

REFERENCES

Canny, M.J. (1973). Phloem Translocation. Cambridge University Press, New York.

Fisher, D.B. (1970). Kinetics of C-14 translocation in soybean. 3. Theoretical considerations. *Plant Physiol.* *45*, 119-125.

More, R.D. and Troughton, J.H. (1973). Production of $^{11}CO_2$ for use in plant translocation studies. *Photosynthetica*, *7*, 271-274.

Troughton, J.H. (1975). Light level and the mean speed of translocation in *Zea mays* leaves. In: "Environmental and Biological Control of Photosynthesis" ed. by R. Marcelle. pp. 373-385. Dr. W. Junk b.v. The Hague.

Troughton, J.H., Chang, F.H. and Currie, B.G. (1974a). Estimates of a mean speed of translocation in leaves of *Oryza sativa* L. *Plant Science Letters*, *3*, 49-54.

Troughton, J.H., Moorby, J. and Currie, B.G. (1974b). Investigation of carbon transport in plants.1.The use of carbon-11 to estimate various parameters of the translocation process. *J. Exp. Bot. 25*, 684-694.

Troughton, J.H., Chang, F.H. and Currie, B.G. (1976). Relations between light level, sucrose concentration and translocation of carbon-11 in *Zea mays* leaves. *Plant Physiol.* (In Press).

Chapter 29

The Effect of Selected Sink Leaf Parameters on Translocation Rates

C.A. Swanson*, J. Hoddinott**, and J.W. Sij***

*Dept. of Botany, Ohio State Univ., Columbus, Ohio, U.S.A.: **Dept. of Botany, Univ. of British Columbia, Vancouver, B.C., Canada; ***Texas A & M Univ. Agric. Res. and Extension Center, Beaumont, Texas, U.S.A.

INTRODUCTION

In work currently in progress in our laboratory we are attempting to study further the respective roles of source and sink regions in controlling translocation rates and assimilate distribution patterns. The present paper reports particularly on our efforts to characterize more specifically the role of young, rapidly expanding leaves as transport sinks. Previous studies in this general problem area have boon reviewed recently by Wardlaw (1974).

METHODS

Bush bean plants (*Phaseolus vulgaris* L., cv. Black Valentine) were grown hydroponically under a 15-h photoperiod with a 24°C day temperature and 18°C night temperature. Light intensity at the primary leaf area was about 1200 ft-c. Plants were selected for experiment when the sink leaf (in most cases, the central leaflet of the first trifoliate leaf, counting upward) had reached the developmental stage required for the experiment. On the day before a run, the selected test plant was pruned to one source leaf and one sink leaflet. To eliminate alternative sinks as much as possible, the stem was heat-girdled 1 cm below the primary leaf node. Possible errors resulting from re-translocation between sinks were thus minimized.

The experimental system used in these studies permitted simultaneous measurements of net photosynthetic rates of the source leaf and translocation rates to the sink leaf, and has been previously described (Bailey 1975; Christy 1972; Coulson et al. 1972). In brief, the plants were mounted in the system shown schematically in Figure 1. At time 0,

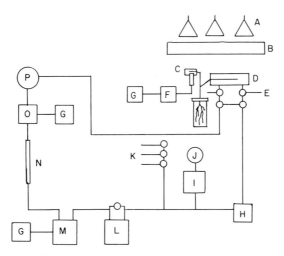

Fig. 1. - Schematic (abbreviated) of analytical system. A, Light bank; B, water shield; C, sink-leaf cuvette with G-M detector; D, source-leaf cuvette; E, air inlet; F, ratemeter; G, potentiometric recorder; H, water trap; I, syringe pump, motor-driven; J, elapsed running time indicator; K, calibration gas manifold; L, pump with by-pass control; M, Infrared gas analyzer; N, flowmeter; O, ion chamber electrometer; P, mixing chamber.

labelled CO_2 of known specific activity was admitted to the source-leaf cuvette and maintained at a constant concentration (usually 385 ± 6 µg ℓ^{-1}) for the duration of the run, usually 6 to 8 hours. Concurrently the influx of labelled translocate into the target leaf was monitored with a thin-window G-M detector connected to a ratemeter recorder. The absolute rate of carbon translocated, in µg $^{12}C + ^{14}C$ min^{-1}, was calculated from the observed accumulation rate of ^{14}C in the sink leaf, the counting efficiency, and the specific activity of the CO_2 in the source-leaf cuvette. Rate calculations of translocation were not started until 1.5 to 2 h after time 0 in order to permit isotopic equilibration of the transport molecules (mainly sucrose).

Net photosynthetic rates were calculated from the running time of the motor-driven syringe pump. The standing pressure in the syringe chamber increased 7.22 mm Hg per ml min^{-1} of the CO_2-demand rate at steady state, necessitating a PV correction factor that varied with the photosynthetic rate. If a rapid change in photosynthetic rate of the source leaf was induced, as for example, by a step-change in the light intensity, approximately 20 to 30 min were required for the syringe chamber pressure to reach a new equilibrium level, during which time calculations of the net photosynthetic rates were not possible.

To minimize changes in the source-detector geometry, the sink leaf was held in a special grid support mounted on the G-M detector. Usually, when all operating parameters were held constant, the translocation rate was also very constant. However, occasional spontaneous rate changes did occur. We tentatively attribute at least some of these changes to shifts in the sink-leaf position, perhaps due to nutational stresses developed in the subtending internodes.

Because of the close proximity of the sink leaf to the detector, water occasionally condensed on the detector end-window unless precautions were taken to force-ventilate the system. In earlier experiments, the window was ventilated by directing a jet of air on to its surface; in later experiments, a cuvette was designed which provided for laminar air flow both over and under the leaf. By using turn-over rates of 15 to 30 volumes per min in the sink-leaf cuvette, condensation was avoided even with relative humidities of the air-stream as high as 97%. The air was humidified by bubbling through 1M KCl.

RESULTS AND DISCUSSION

Sink-leaf light effect. - Table 1 presents the data of two series of experiments designed to ascertain the effect of sink-leaf irradiation on the rate of translocation (more accurately, the rate of apparent translocation) to the sink leaf. It should be kept clearly in mind that irradiance of the sink leaf was the only major variable in these experiments. Source-leaf irradiance was held constant, and as may be observed from Table 1, the rate of net photosynthesis during the time course of these experiments was very constant.

It is evident that illumination of the sink leaf significantly increased the rate of ^{14}C-accumulation in the sink leaf. The question is: does this increase represent a real increase in the rate of translocation? Several hypotheses to account for the observed increase may be readily suggested:

(1) Light increases the transpiration rate of the sink leaf. If the rate of water unloading from the sieve tubes to the surrounding mesophyll is thereby accelerated, the mass flow rate in the sieve tubes may also be accelerated. We attempted to test this hypothesis by comparing the translocation response to light under conditions of high relative humidity (95 to 97%) with earlier measurements carried out at ambient relative humidities, in the range of 30 to 50% (Table 1). In the low relative humidity series, the average

TABLE 1. Apparent translocation rate to the sink leaf in relation to sink leaf irradiance. Source leaf irradiated constantly.

R.H. in sink leaf cuvette		Relative translocation rate to sink leaf			Relative photosynthetic rate of source leaf		
		Sink leaf in light (Period 1)	Sink leaf in dark (Period 2)	Sink leaf in light (Period 3)	Sink leaf in light (Period 1)	Sink leaf in dark (Period 2)	Sink leaf in light (Period 3)
Ambient	a	100	60.0	107	100	99.1	99.0
	b		±6.0	±33.0		±2.7	±2.0
	c	(8)	(8)	(5)	(6)	(6)	(4)
95 to 97%	a	100	71.8	118	100	98.0	97.6
	b		±4.6	±15.7		±2.2	±2.6
	c	(6)	(6)	(5)	(6)	(6)	(5)

a. Rate of photosynthesis (or translocation) normalized to 100.

b. ± SD.

c. Number of observations.

translocation rate in the dark was about 60% of the rate in
the previous light period; in the high relative humidity
series, about 72%. The difference in these means is significant (>99% level of confidence), but the data base is not
substantial, and the two experimental series were carried out
several months apart. Nonetheless, present indications are
that an increase in transpiration rate of a sink will increase
the rate of translocation to that sink. We plan to follow
through on this problem by carrying out a series of experiments
holding sink-leaf irradiance constant and varying the relative
humidity, the converse of the high humidity - variable light
series.

   (2) Transpiration may decrease the water content of the
sink leaf, leading to a decrease in its absorber thickness
(mg cm^{-2}). A loss of 10% of the water content of the sink
leaf gives a calculated increase in count yield of over 70%,
more than adequate to account for the observed increase in
translocation. However, it is very unlikely that a decrease
in water content of this magnitude occurred, particularly
in the high relative humidity series. Furthermore, Mederski
and Alles (1968) have shown that in young soybean leaves subjected to increasing water stress, leaf area shrinkage may be
nearly proportional to water loss, with the result that the
absorber thickness remains nearly constant. On the basis of
the methodology employed in our studies (the near-saturation
humidity in the cuvette environment of the sink leaf) and
the leaf area shrinkage factor, we are inclined to rule out
any significant transpiration effect on the counting efficiency
per se.

   (3) In the light, metabolically produced $^{14}CO_2$ is re-
fixed in the sink leaf, necessitating a distinction between
net and total translocation in the same sense as between net
and total photosynthesis. There can be no doubt that this is
an important component but quantitation of this effect has
proven difficult, mainly because of difficulties we have had
in constructing a sink-leaf cuvette sufficiently leaf-proof to
permit accurate measurement of the $^{14}CO_2$ produced. In the
meantime we have carried out a series of isotope trapping
experiments involving the flushing of air mixtures containing
different concentrations of CO_2 (unlabelled) through the sink-
leaf cuvette. If the observed translocation rate (sink leaf
in light) in zero concentration of CO_2 is normalized to 100,
the rate in 350 µl l^{-1} was 94 ± 5 (n = 6), and in 910 µl l^{-1},
78 ± 15 (n = 4). These results are in the direction anticipated, on the basis that trapping efficiency should increase with
increasing CO_2 concentration. However, it was not possible to
over-ride the translocation response to light by isotope

trapping. In the high-humidity series given in Table 1, two
of the runs were carried out at 0 μl 1^{-1} CO_2 concentration,
two at 350 μl 1^{-1}, and two at 910 μl 1^{-1}. The averages for
the respective runs were 71, 72, and 72 (for this reason the
data were pooled for statistical purposes). It is evident
that the interpretation of the light effect remains perplex-
ing. Our next approach to this problem will be to measure
quantitatively $^{12}CO_2$ and $^{14}CO_2$ exchange in the sink leaves
in the light and dark. Turgeon and Webb (1975) have reported
that the *Cucurbita pepo* leaf blade is first capable of a
positive net CO_2 fixation when about 8% expanded. Fraser
and Bidwell's data (1974) indicate that positive net photosynthesis is first attained in the first trifoliate leaf of
bush bean about 2 days after the emergence of this leaf. In
the present experiments, the sink leaf averaged about 23% of
its final laminar length.

Developmental stage of sink leaf. - Figure 2 shows the
regression of translocation rate on sink-leaf size. Pending
further processing of data, the curve has been drawn only
visually; the dashed curve is strictly conjectural. A peak
import rate of 7.0 μg carbon min^{-1} (or 16.6 μg sucrose equivalent min^{-1}) is attained when the leaf is about 25% of its
final length. This ontogenetic parameter is essentially the

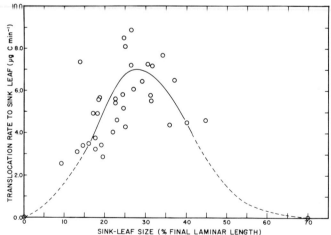

Fig. 2. - *Translocation rate to sink leaf in* μg
carbon (^{14}C + ^{12}C) *min^{-1} (source leaf)$^{-1}$. Data
have not been normalized to unit area of source
leaf.*

same as reported by Thrower (1962) for soybean and for sugar
beet by Fellows and Geiger (1974). Data reported in the

literature for sugar beet, squash, and soybean indicate that import ceases when the leaf attains 45 to 50% of its final size; it may be more accurate to constrain the curve for bush bean to intercept the abscissa at this point.

Figure 3 is a plot of the same data normalized to unit area of the sink leaf. The curve was computer fitted using an exponential regression program, and has a correlation coefficient of -0.94. The translocation rate to the sink leaf extrapolates to a limiting value of 4.1 µg carbon min^{-1} (or 9.8 µg sucrose equivalents min^{-1}). This intercept value may be a useful parameter for characterizing the sink-leaf system in comparative studies.

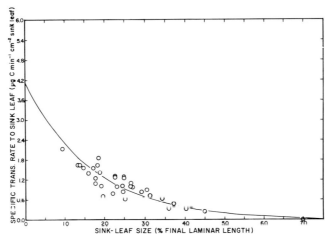

Fig.3. - Translocation rate to sink leaf in µg carbon min^{-1} cm^{-2} sink leaf (defined as the "specific translocation rate"). Y = 4.12e$^{-0.06x}$

Changes in assimilate demand. - Mullins (1970) reported that in bean plants the distribution of assimilate from a given primary leaf to the lateral leaflets of the first trifoliate leaf is highly asymmetric, about 90% of the assimilate imported by this sink being distributed to the near-leaflet ("near" in an orthostichous sense) and about 10% to the far-leaflet, in leaves from which the central leaflet has been removed. Since the near-leaflet sink with respect to one primary leaf is the far-leaflet sink with respect to the other primary leaf, the distribution patterns from the two primary leaves are opposite or complementary. If, then, one of the primary leaves is removed, the assimilate demand of its near-leaflet sink (the far-leaflet sink of the remaining primary leaf) should increase. How rapidly is this information transmitted to this source leaf and how rapidly does the distribution pattern compensate?

Bailey (1975) has recently reported on a series of studies designed to provide some partial answers to these questions. The general protocol was similar to that described in the section on methods except that double sink – double source test plants were used, the two lateral leaflets of the first trifoliate leaf serving as sinks and the two primary leaves as sources. One of the primary leaves was then supplied with labelled CO_2 under steady-state conditions, and the respective rates of translocation to the two sink leaves monitored. When the distribution ratio under these standard conditions had been determined (usually in the range of 6 : 1 to 10 : 1, near-to-far, under our conditions), the opposite primary source leaf was excised, and the time course in change of the distribution ratio followed. Figure 4 presents the data from a typical experiment. The initial distribution was 10.6 : 1 (5.3 µg **carbon** min^{-1} to the near-leaflet vs. 0.5 µg carbon min^{-1} to the far-leaflet). Immediately following

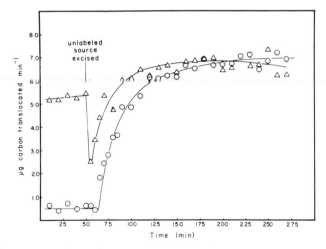

Fig. 4. - The time course of the translocation rate (µg carbon min^{-1}) from the labelled primary leaf to the near- and far-leaflets of the first trifolium leaf prior, and subsequent, to the excision of the opposite (unlabelled) primary leaf. Δ , near-leaflet; O , far-leaflet.

excision of the opposite source leaf, transport from the labelled source leaf to its near-leaflet sink was severely inhibited for a short period, then recovered rapidly to the pre-excision rate. Transport to the far-leaflet sink, after holding constant or declining slightly for a short lag period, also rapidly increased, attaining a final rate closely similar to the rate to the near-leaflet. Within 50 to 75 min, the distribution ratio was essentially 1 : 1. The lag response time for the initiation of compensated translocation was remarkably constant in these experiments, namely, 8.8 ± 1.6 min, based on 7 runs. The net photosynthetic rate of the source leaf remained constant despite the doubling of its export rate.

A few very preliminary experiments have shown that the distribution ratio of assimilate exported from a primary leaf is strongly controlled by the opposite primary leaf. A high rate of photosynthesis in the opposite source leaf is correlated with a high distribution ratio for the labelled source leaf. Darkening the opposite source decreases this ratio, the decrease increasing with the duration of darkening. In one run, 18 h of darkening was found to be physiologically equivalent to excision. The data suggest that control by the opposite source leaf is mediated by its rate of export.

These experiments tentatively suggest a methodology for investigating certain problems difficult to study by more conventional approaches. For example, we are interested in determining whether various atmospheric pollutants, such as SO_2, diminish translocation in proportion to photosynthesis. For these studies we propose using conventional steady-state labelling procedures, the SO_2 being administered to the opposite leaf in an open flow system. Maintaining SO_2 at a constant partial pressure in a closed loop system, as is required for the labelled CO_2, would present numerous and perhaps insurmountable complications.

ACKNOWLEDGEMENTS

This work was supported in part by U.S. National Science Foundation Grant No. GB40029.

REFERENCES

Bailey, B. Ann. (1975). A kinetic study of the partitioning of labelled translocate to double sinks in *Phaseolus vulgaris* L. M.Sc. thesis. Ohio State University, Columbus, Ohio.

Christy, A.L. (1972). Translocation kinetics in relation to source-leaf photosynthesis and carbohydrate concentrations in sugar beet. Ph.D. diss. Ohio State University, Columbus, Ohio.

Coulson, C.L., Christy, A.L., Cataldo, D.A. and Swanson, C.A. (1972). Carbohydrate translocation in sugar beet petioles in relation to petiolar respiration and adenosine 5'-triphosphate. *Plant Physiol. 49*, 919-923.

Fellows, R.J. and Geiger, D.R. (1974). Structural and physiological changes in sugar beet leaves during sink to
source conversion. *Plant Physiol. 54*, 877-885.

Fraser, D.E. and Bidwell, R.G.S. (1974). Photosynthesis and
photorespiration during the ontogeny of the bean plant.
Can. J. Bot. 52, 2561-2570.

Mederski, H.J. and Alles, W. (1968). Beta gauging leaf water
status: influence of changing leaf characteristics.
Plant Physiol. 43, 470-472.

Mullins, M.G. (1970). Transport of ^{14}C-assimilates in seedlings of *Phaseolus vulgaris* L. in relation to vascular
anatomy. *Ann. Bot. 34*, 889-896.

Thrower, S.L. (1962). Translocation of labelled assimilates
in the soybean. II. The pattern of translocation in
intact and defoliated plants. *Aust. J. Biol. Sci. 15*,
629-649.

Turgeon, R. and Webb, J.A. (1975). Leaf development and phloem
transport in *Cucurbita pepo*: carbon economy. *Planta
(Berl.) 123*, 53-62.

Wardlaw, I.F. (1974). Phloem transport: Physical Chemical
or Impossible. *Ann. Rev. Plant Physiol. 25*, 515-539.

CHAPTER 30

Translocation and the Diffusion Equation

J.B. Passioura

CSIRO, *Division of Plant Industry, Canberra, A.C.T., Australia.*

TRANSLOCATION MODELS

Of the many mechanisms which have been proposed for translocation, three imply that the flux of sugars in the phloem is proportional to their concentration gradient. These are:

(a) *The Canny-Phillips model.* - (Canny and Phillips 1963). This has, in each sieve tube, two streams of solution flowing in opposite directions. This model has been thoroughly discussed in the literature (Canny 1971, MacRobbie 1971) and we need not dwell on it further.

(b) *The reciprocating flow model.* - (Miller 1975). If the sugars in the phloem are in two pools, one stationary, the other moving to and fro, and if there is interchange between the pools, then there will be a net movement of sugars down the gradient of mean concentration. This type of movement has been explored theoretically by Farrell and Larsen (1973). From their results, we can conclude that the maximum value of the coefficient K is $\varepsilon U^2 \tau$ where ε is the proportion of solute in the stationary pool, U is the root mean square velocity of the solution (averaged over both pools), and τ is a time constant describing the rate at which the two pools will equilibrate. The value of K decreases rapidly with increasing ω (the frequency of the reciprocating flow) if $\omega\tau$ exceeds about 0.5. Thus, for given values of ε, U and ω, there is an optimum value for τ which will maximize K. Miller (1975) has suggested that reasonable values of ε, U, τ and ω may be 0.5, 0.03 cm s^{-1}, 30 s, and 0.01 s^{-1} respectively, thus giving a value of 0.015 cm^{2}s^{-1} for K, which is of the same order as the values given by Canny (1971) for some experimental systems. However, it is much below the value of 3 cm^{2}s^{-1} found by Passioura and Ashford (1974) for wheat roots, and it is hard to imagine reasonable values for the parameters which would give such a large K as this.

(c) *Münch pressure flow*. - This model is discussed in detail by Christy (this volume). It clearly predicts that sugars will move down their concentration gradient, but, because of the inverse relation between concentration and speed in this model, it seems unlikely, at first sight, that the flux would be proportional to the concentration gradient. Over the range of concentration which we would expect to find in the sieve tubes, however, this inverse relation between speed and concentration is closely counterbalanced by the increase of viscosity with concentration, so that the pressure gradient remains approximately constant, with the result that the flux of sucrose is very nearly proportional to its concentration gradient. These relations are illustrated in Fig. 1, where it is assumed that the hydrostatic pressure in the sieve tube is proportional to the osmotic pressure.

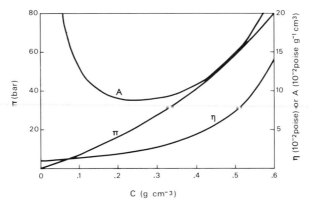

Fig. 1. - Osmotic pressure (π), viscosity (η), and A (= viscosity/sucrose concentration), as functions of sucrose concentration.

For Poiseuille flow, the pressure gradient (P') is proportional to speed (U) and viscosity. For a given flux of sucrose, S (g $cm^{-2}s^{-1}$), $U=S/C$, where C g cm^{-3} is the sucrose concentration. Thus $P' \propto \eta.S/C = S.A$. Since A is approximately constant in the range of sucrose concentration from 0.15 to 0.40, P' will also be approximately constant, and so will π'. Since π is almost linear with C over the same range, S will be very nearly proportional to the concentration gradient. Note that in this system a minimum in A implies a minimal loss of energy due to viscous drag. Hence in purely physico-chemical terms, the system is at its most efficient in the concentration range 0.15 to 0.40 gm cm^{-3} which, interestingly, is the range in which most measurements of sucrose concentration in the sieve tubes lie.

THE DIFFUSION EQUATION

If the flux of sugars is indeed proportional to their concentration gradient in the sieve tubes their movement may be described by the diffusion equation. In its simplest form, the diffusion equation is:

$$\frac{\partial C}{\partial t} = K \frac{\partial^2 C}{\partial x^2}$$

(1)

where C is concentration (g cm^{-3}); t is time (s); x is distance (cm); and K is a coefficient relating the rate of movement to the concentration gradient (cm^{2}s^{-1}).

To be able to describe a phenomenon by means of the diffusion equation is a great advantage, for hundreds of solultions of the equation are known and powerful generalizations can often be made about the relations between the variables. But it is important to realize that the equation has an infinite number of solutions and that any given distribution of C as a function of x can arise in an infinite number of ways. Thus it is fallacious to argue, as Canny (1971) often has, that because a particular distribution of C corresponds to one of the solutions of the equation, the behaviour of C can in general be described by the equation. The equation can be solved uniquely only if we specify the initial conditions, i.e. $C(x)$ at $t=0$, and the boundary conditions, i.e. the behaviour at $x=0$ and $x=L$ for $t>0$ where $0<x<L$ defines the space in which the equation applies. For example, the error function solution (Canny 1971) applies only if the space is semi-infinite (i.e. L is so large that any perturbation at $x=0$ does not reach $x=L$ during the time in which we are interested) and only if

$$t=0, \quad x>0, \quad C=C_i$$

(2)

and
$$t>0, \quad x=0, \quad C=C_o$$

(3)

where C_i and C_o are constants. If we are considering a translocation profile (Canny 1971) with $x=0$ being defined, say, as the junction between a leaf and its petiole, the boundary condition (3) is demonstrably not true, that is, there is no step change in C at $x=0$ (Mortimer 1965, Fisher 1970). Thus, in these circumstances, the fact that a profile can be described by the error function is evidence, if anything, that the diffusion equation does not apply.

In practice there are two ways in which we can test if translocation is described by the diffusion equation. One is to work with a steady state system, so that we have

$$F = K\frac{dC}{dx} = K\frac{\Delta C}{\Delta x}$$

(4)

where F, the rate of flow of solute, is proportional to the drop in concentration, ΔC, along a length of phloem, Δx. This is technically difficult and tedious and has been attempted only by Mason and Maskell (1928). The alternative is to work with a non-steady state system, where some of the technical difficulty and tedium can be removed by using radioactive tracers, although one must be wary of the fact that a tracer does not necessarily move in the same way as the bulk material. If we feed radiocarbon, say, to a leaf, we can follow the distribution of that carbon as a function of time and distance in the petiole of that leaf. The distribution of the tracer at any given time is called a translocation profile. To interpret any single translocation profile in terms of the diffusion equation, or indeed in terms of any transport equation, is a waste of time, as mentioned previously, unless we can specify the initial and boundary conditions. The initial condition is no problem, for at $t=0$ there is no tracer anywhere. The boundary condition at $x=L$ (i.e. the end of the petiole) is no problem providing the experiment is stopped before appreciable amounts of tracer arrive there. It is the condition at $x=0$ which presents difficulty. However, providing that we have a series of translocation profiles through time, we can specify it. We can define either the activity at $x=0$, or the flux of tracer at $x=0$, as functions of time. The average flux during the time interval between successive translocation profiles can be determined simply by measuring the total increase in activity in the petiole during the time interval. There are several sets of data in the literature which can be analysed in this way (e.g. Mortimer 1965, Fisher 1970, Moorby, quoted by Canny 1971). The resulting boundary conditions are complicated functions of time, so that appropriate analytical solutions of the diffusion equation are, in general, unobtainable. Recourse may however be made to numerical solutions obtained with the help of a computer. Such solutions show that the published translocation profiles are difficult to explain in terms of the diffusion equation; they are much better explained in terms of a model of bulk flow combined with lateral leakage (Passioura, unpublished) although even there the marriage is uneasy.

REFERENCES

Canny, M.J. (1971). Translocation: mechanism and kinetics. *Ann. Rev. Plant Physiol. 22*, 237-260.

Canny, M.J. and Phillips, O.M. (1963). Quantitative aspects of a theory of translocation. *Ann. Bot. 27*, 379-402.

Farrell, D.A. and Larsen, W.E. (1973). Effect of intra-aggregate diffusion on oscillatory flow dispersion in aggregated media. *Water Resour. Res. 9*, 185–193.

Fisher, D.B. (1970). Kinetics of ^{14}C translocation in soybean. I. Kinetics in the stem. *Plant Physiol. 45*, 107–113.

MacRobbie, E.A.C. (1971). Phloem translocation: Facts and mechanisms: a comparative survey. *Biol. Rev. 46*, 429–481.

Mason, T.G. and Maskell, E.J. (1928). Studies on the transport of carbohydrates in the cotton plant. *Ann. Bot. 42*, 571–636.

Miller, D.M. (1975). A further analysis of reciprocating flow in phloem tubes. *Can. J. Bot. 53*, 1149–1152.

Mortimer, D.C. (1965). Translocation of the products of photosynthesis in sugar beet petioles. *Can. J. Bot. 43*, 269–280.

Passioura, J.B. and Ashford, A.E. (1974). Rapid translocation in the phloem of wheat roots. *Aust. J. Plant Physiol. 1*, 521–527.

CHAPTER 31

Mathematical Models of Münch Pressure Flow: Basic Concepts and Assumptions

A.L. Christy

Research Department, Monsanto Agricultural Products Co., St. Louis, Mo. 63166, U.S.A.

Recently several papers have been published that describe mathematical models of the Münch pressure flow hypothesis of phloem transport. Eschrich et al. (1972) used a mathematical model to describe flow in tubular semi-permeable membranes and in a second paper (Young et al. 1973) extended this model to flow in sieve tubes. Christy and Ferrier (1973) independently derived a mathematical model of Münch pressure flow in the steady state and made sample calculations of flow in sieve tubes of sugar beet. Tyree and Dainty (1975) and Tyree et al. (1974) derived a similar steady state model that was employed in sample calculations of Münch pressure flow over long distances. Anderson (1974) derived yet another model of pressure flow based on standing gradient osmotic flow. Finally, Ferrier and Christy (1975) have obtained a time dependent solution of their model and applied it to cold inhibition of translocation and diurnal variation of water potential and loading rate in trees (Ferrier et al. 1975).

Since mathematical models of Münch pressure flow have been extensively discussed in the literature, I will only briefly describe our model to introduce the basic concepts of pressure flow and then consider some of the assumptions made in these models. Obviously, many of the assumptions made in deriving the mathematical models apply to the Münch pressure flow hypothesis.

BASIC CONCEPTS OF MÜNCH PRESSURE FLOW MODELS

The basic model consists of a single sieve tube divided into sieve tube elements by sieve plates (Fig. 1) and surrounded by a reservoir with a water potential (ψ_0) of -3 atm. The sieve tube is also divided into three regions of equal length: a source region, site of active sucrose load-

ing;   a path region; and a sink region, site of active
unloading (Christy and Ferrier 1973; Fig. 1).

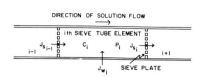

DIRECTION OF SOLUTION FLOW

Fig. 1. - Diagram of sieve tube element showing the computed variables and the relationship of the ith element to $i + 1$ and $i - 1$ elements (from Christy and Ferrier 1973)

The equation from irreversible thermodynamics for volume
flux, J ($cm^3cm^{-2}sec^{-1}$), across a membrane is

$$J = L_p (\Delta P - \sigma \Delta \Pi) \tag{1}$$

where P is hydrostatic pressure in atm, Π is osmotic pressure
in atm, Lp is the membrane conductivity in $cm^3cm^{-2}sec^{-1}atm^{-1}$,
and σ is the reflection coefficient for the solute. Since σ
for sucrose is assumed to equal 1.0 for the lateral membranes,
the flux of water from the reservoir into the ith sieve tube
element (Fig. 1) is given by

$$J_{w_i} = L_p (\psi_0 - P_i + C_i RT) \tag{2}$$

where ψ_0 is the water potential in the reservoir; P_i and C_i
are the hydrostatic pressure and sucrose concentration, respectively, in the ith sieve tube element; R is the gas constant; and T is the absolute temperature. The volume flux of
the solution down the tube, from the ith element to the $i + 1$
element (Fig. 1), is given by

$$J_{s_i} = L_s (P_i - P_{i+1}) \tag{3}$$

assuming that $\sigma = 0$ and L_s is the conductivity of the sieve
tube and plate.

Since water must be conserved, we have

$$J_{s_{i-1}} (1 - \alpha C_{i-1}) A_{s_{i-1}} + J_{w_i} A_{p_i} = J_{s_i} (1 - \alpha C_i) A_{s_i} \tag{4}$$

where αC is the fraction of solution volume occupied by sugar;
A_{s_i} is the sieve tube cross-sectional area in cm^2, and A_{p_i} is
the lateral membrane surface area in cm^2 of the ith sieve
tube element. By combining equations 2, 3 and 4, the hydrostatic pressure and the concentration in the ith element can
be calculated:

$$P_i = \frac{L_p A_{p_i}(\psi_0 + C_i RT) + L_s A_{s_{i-1}}(1 - \alpha C_{i-1})P_{i-1} + L_s A_{s_i}(1 - \alpha C_i)P_{i+1}}{L_p A_{p_i} + L_s A_{s_{i-1}}(1 - \alpha C_{i-1}) + L_s A_{s_i}(1 - \alpha C_i)} \tag{5}$$

$$C_i(t+\Delta t)=C_i(t)+\frac{(r_i+J_{s_{i-1}}C_{i-1}A_{s_{i-1}}-J_{s_i}C_iA_{s_i})\ \Delta t}{V_i} \qquad (6)$$

where r is the loading rate in μg sec^{-1}, t is time in sec.
and V is volume in cm^3.

A steady state solution of the resulting set
of equations was found by iterative numerical solution employ-
ing a Fortran program on an IBM 360 computer (Christy and
Ferrier 1973). For a discussion of a time dependent solution
of these equations see Ferrier and Christy (1975).

The results of a typical steady state solution of the
model are shown in Fig. 2. The osmotic and hydrostatic

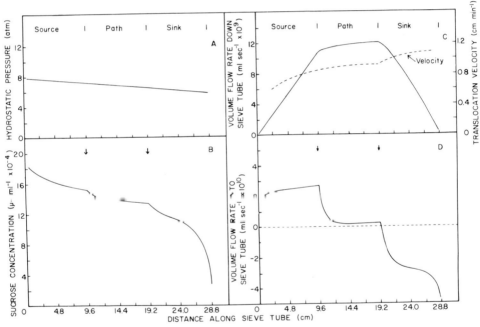

Fig. 2. - Results of a steady state solution, assuming
L_p = 5.0 x 10^{-7} cm sec^{-1}atm^{-1} and L_s = 10.2 cm
sec^{-1}atm^{-1}; D: positive values indicate flow into
the sieve tube and negative values indicate flow
out of the sieve tube (from Christy and Ferrier
1973).

pressure gradients calculated for the path region are 12.0
and 7.1 atm m^{-1}, respectively, and the velocity (where velocity = volume flow rate/A_{si}) at the center of the path is
0.9 cm min^{-1}. At the transition from loading in the source
to no loading along the path, and then to unloading in the

sink, marked changes occur in the concentration and volume flow rates into and along the sieve tube (Fig. 2, C and D). These changes would also be expected to occur in a plant depending on how sharp the transition is between the source, path, and sink and the unloading rate, if any, in the path. Changes in these parameters and in the osmotic and hydrostatic pressure gradients at the source, sink, and path transition zones will also depend on the extent of branching and changes in cross-sectional area of the sieve tube. For a more extensive discussion of sample calculations of the model the reader is referred to Christy and Ferrier (1973) and Tyree et al. (1974).

ASSUMPTIONS FOR MODELS OF MÜNCH PRESSURE FLOW

Several assumptions are necessary to develop a mathematical model of Münch pressure flow. In addition, a number of parameters are required to obtain sample calculations of these models. The following discussion is intended only as an introduction to these assumptions and parameters and is not an attempt to defend their validity.

To make representative calculations with pressure flow models it is necessary to have dimensions for sieve tube pores, sieve tube elements, and sieve tube cross-sectional areas. These dimensions are not readily available and may differ from one area of the plant to another (Housley 1974). Dimensions of the sieve tube used in model solutions are average observed values reported in the literature and are usually held uniform throughout the model due to the complexities of programming. The dimensions used in the models should give a good approximation of the events occurring in a functioning sieve tube.

Osmotically driven solution flow requires a differentially permeable membrane in the sieve tube. The ability to plasmolyze the contents of the sieve tube supports the existence of such a membrane. The conductivity of this membrane is extremely important in regulating translocation. However, conductivity values for sieve tube membranes are not presently available and those used in the models are estimates for other plant membranes. For this reason the effects of membrane conductivity on the pressure flow model have been extensively investigated (Christy and Ferrier 1973, Tyree et al. 1974).

Sieve tube conductivity (L_S) is another parameter needed for sample calculations of pressure flow models. Low conductivities may be a major obstacle to the acceptance of pressure flow as the mechanism of translocation. Sieve tube conductivity can be calculated from sieve tube dimensions using Poiseuille's equation (Horwitz 1958). Strictly speaking Poiseuille's equation does not apply to sieve tubes due to changes in momentum resulting from the influx and efflux of water along the sieve tube. Horwitz (1958) considered this problem and concluded that the errors would be small. Tyree et al. (1974) also suggested that additional resistance could result from bending of laminar flow lines to pass through sieve pores. However, it is not possible to estimate the magnitude of this effect.

The status of sieve pores is an important consideration when developing models of translocation. Sieve tube conductivity is calculated assuming that the pores are open (Christy and Ferrier 1973). The ultrastructure of a functional sieve tube remains a controversial area of translocation research and a number of papers have appeared demonstrating both plugged and unplugged pores. Fisher (1975) is the only investigator to date to report open pores in sieve tubes shown to be functional at the time of fixation.

Another problem for Münch pressure flow is the water potential gradient in the plant. Solution flow in the phloem from leaves to roots is against the water potential gradient and requires a greater osmotic pressure gradient to drive flow against this gradient. An evaluation of this problem using mathematical models (Tyree et al. 1974, Ferrier et al. 1975) have shown that water potential gradients in the range of those reported for plants may not pose a problem for Münch pressure flow.

Although there are several unknowns and assumptions in the pressure flow module, it is useful to carry out sample calculations with these models. These sample calculations have enabled us to study the operation of the pressure flow mechanism and evaluate the effects of various parameters on the models. These studies have led to the conclusion that the pressure flow hypothesis is consistent with observed properties of the translocation system including long distance translocation in trees (Christy and Ferrier 1973, Tyree et al. 1974, Ferrier et al. 1975).

REFERENCES

Anderson, W.P. (1974). The mechanisms of phloem translocation. In: Transport at the Cellular Level ed. by M.A. Sleigh, Society for Experimental Biology Symposium No. XXVIII. pp. 63-86.

Christy, A.L. and Ferrier, J.M. (1973). A mathematical treatment of Münch's pressure-flow hypothesis of phloem translocation. *Plant Physiol.* 52, 531-538.

Eschrich, W., Evert, R.F. and Young, J.H. (1972). Solution flow in tubular semipermeable membranes. *Planta (Berl.)* 107, 279-300.

Ferrier, J.M. and Christy, A.L. (1975). Time-dependent behavior of a mathematical model for Münch translocation. *Plant Physiol.* *55*, 511-514.

Ferrier, J.M., Tyree, M.T. and Christy, A.L. (1975). The theoretical time-dependent behavior of a Münch pressure-flow system: The effect of sinusoidal time variation in sucrose loading and water potential. *Can. J. Bot. 53*, 1120-1127.

Fisher, D.B. (1975). Structure of functional soybean sieve elements. *Plant Physiol. 56*, 555-569.

Horwitz, L. (1958). Some simplified mathematical treatments of translocation in plants. *Plant Physiol. 33*, 81-93.

Housley, T.L. (1974). An evaluation of the pressure flow hypothesis of phloem transport. Ph.D. Thesis, University of Georgia.

Tyree, M.T., Christy, A.L. and Ferrier, J.M. (1974). A simpler iterative steady state solution of Münch pressure-flow systems applied to long and short translocation paths. *Plant Physiol. 54*, 589-600.

Tyree, M.T. and Dainty, J. (1975). Mechanism of phloem transport: Theoretical considerations. In: Transport in the Phloem. Encyclopedia of Plant Physiology (N.S.) ed. by M.H. Zimmermann and J.A. Milburn, Springer-Verlag, Heidelberg.

Young, J.H., Evert, R.F. and Eschrich, W. (1973). On the volume-flow mechanism of phloem transport. *Planta (Berl.) 113*, 355-366.

LONG DISTANCE TRANSPORT - SUMMARY AND DISCUSSION

*Adapted from presentations given by the Session Chairmen :
B.E.S. Gunning, C.A. Swanson and D.B. Fisher*

It is probably true to say that a knowledge of cell
structure is just as important in understanding long distance
transport as it is in short distance transfer processes. However, although far from being finalized the problem of structure relating to function now appears to be capable of being
resolved and a good indication of this was given in three
remarkable displays of technical wizardry. A display of
serial sections sequences by M.H. Zimmermann encompassing
long stretches of vascular networks, the uses of a range of
variations with the technique of freeze substitution by
D.B. Fisher and the exploitation of snap cooling in order to
examine the ultrastructure of sieve elements by G.P. Dempsey
et al. Also T.P. O'Brien stressed the general problems in-
herent in the use of chemical fixatives in producing E.M.
pictures of sieve tubes, problems that arise even with the
best non-coagulent fixatives. One crucial criterion emphasised
by several speakers was the need to know whether the sieve
elements were actively transporting prior to fixation, as it
is counter productive to examine sieve elements that may have
been wrecked and then to attribute a function to the bits of
wreckage. There was general agreement between the chemical
fixers and the freeze substituters, that the plasma membrane
is present in a sieve tube and the sieve elements are a part
of the symplast, with the main questions centering round P-
protein. The groups of filaments, or skeins described by
G.P. Dempsey have features in common with Thaine's strands,
but are not enclosed by a membrane. The general consensus,
based largely on freeze substitution techniques, would appear
to be that 'P-protein' is parietal in a working sieve element and only filling the sieve pores if there is damage.

The longevity of sieve elements, a topic raised by
M.H. Zimmermann is a problem in cell biology. How does an
enucleate cell, which is connected only by plasmodesmata to
a nucleate neighbour retain its plasma membrane and other
constituents for up to a year? T.P. O'Brien made the suggestion that protophloem, in which sieve elements are not associated with companion cells, may only be of use in unloading
and their short life may be the consequence of the lack of
a companion cell, even though there are plasmodesmatal connections to other cells.

On the basis of ultrastructure and osmotic pressures
the companion cells would appear to be the site of a step up in

assimilate concentration from the surrounding mesophyll. The
characteristic branched plasmodesmata that lead from the companion cell to the sieve element have a high frequency per
unit area of cell wall and are widespread taxonomically. However the question of evolution was raised by T.P. O'Brien and
from the work of R.F. Evert and his associates it would appear
that the branching of these plasmodesmata is not found below
the gymnosperms, however the characteristic high frequency of
the connections goes right down through the pteridophytes, and
we have to get to the leptoids of the bryophytes before this
disappears. Also relevant to sieve tube loading is another
structural feature, the continuity of the endoplasmic reticulum through the plasmodesmata from the companion cell to the
sieve tube. The possibility then arises that there is loading at the sub-cellular as well as the cellular level.

Little was said about the structural features of the unloading sites in the phloem, and indeed little seems to be
known about the unloading process, even though it might well
be the focus of control for the flow of carbon within the
plant. Plasmodesmata do seem to occur throughout the vascular
system and there could well be passive unloading in growing
tissue with continuous symplastic connections right from the
leaf to the root cap. However I.R. Cowan pointed out that
such passive unloading might inefficiently dissipate the
potential energy which the plant has stored in the concentrated phloem sap. There was fairly general agreement that
longitudinal transport in the phloem was unlikely to limit
the movement of carbon within the plant and that therefore
the control of this movement is probably exerted during the
loading or unloading of the phloem. Current models of translocation tend to take the flux of sugars as an independent
variable and then explore the concentrations, velocities and
pressures of the phloem sap within this constraint. Future
developments of these models may take the flux as a dependent
variable and thus focus our attention on those parts of the
system which may be controlling the movement of the sugars.

A fairly reliable picture of the material being transported through the phloem may be obtained by collecting
phloem exudates, particularly when this is done in combination with radioisotope tracer experiments. Of the collection
procedures probably the aphid stylet is the most reliable,
and least subject to artifacts because of the highly localized nature of the puncture and the ability of the stylet to
minimize contamination from adjacent cells. However transport fluids vary in composition and concentration with many
factors such as age of plants, nutritional status, season,
presence or absence of disease and last, but not least, the

collection site itself on the plant. Longitudinal concentration gradients, for different solutes, arise through selected exchange between the transport stream and the bordering walls and protoplasts, involving both reversible and irreversible uptake. Perhaps more thought needs to be given to the manner in which the plant maintains two transport fluids in close association, with the phloem and xylem differing widely in concentration and relative balance of solutes.

It would appear that present methodology will allow the investigation of phloem translocation during the unsteady state. This capability is an important key to understanding how translocation responds, for example to an osmotic change, or during the redistribution of translocate by new channels. It is also an important approach to the problem of how transport responds to active control mechanisms. Clearly the greater the number of parameters measured in studies using radio-isotopes, the better the interpretation. Unfortunately our capabilities are limited, and in some cases even to get a single set of values requires some very rough measurements.

Water potential is one particularly important area, that has not received as much attention as it should, particularly as we are concerned with an osmotic system. When treatments can be expected to lead to changes in water status it is important to measure these changes and translocation physiologists must be concerned with the wonders of stomatal physiology, water movement and resistance.

A valuable aspect of translocation is the premium it places on biological insight into integration and organization at the whole plant level. In attempting to understand such a complex system on a quantitative basis, modelling must inevitably play an increasingly important role and it is important that we deal with the unsteady state behaviour of the system, as was the case with the Christy-Ferrier model.

INTEGRATING SYSTEMS

CHAPTER 32

The Control of Water Movement Through Plants

J. B. Passioura

CSIRO, Division of Plant Industry, Canberra, A.C.T., Australia.

INTRODUCTION

Plants have several ways of influencing the rate at which water flows through them. The systems involved help a plant in times of water stress, enabling it to survive or even prosper in its normal habitat if it is wild, or to give reasonable yields if it is cultivated.

Environmental stresses operate over a wide range of time scales, particularly water stress, where at one extreme moving clouds can change the radiation load in seconds, and, at the other extreme, the supply of water in the soil may take weeks or even months to run down. Likewise, the processes in plants which influence water stress operate over a wide range of time scales, from minutes for stomatal behaviour, to days or even weeks for leaf shedding and root growth. The purpose of this paper is to discuss the control of water flow through plants in relation to hypothetical strategies which the plants might have for coping with their environments over a range of time scales. The diversity of plants and their environments is enormous. To prevent the discussion being too diffuse, therefore, its scope is restricted to mesophytes, particularly annual crops. Many of the stratagems adopted by xerophytes have been discussed by Slatyer (1964) and by Morrow and Mooney (1974).

PROCESSES

Table 1 lists physiological processes which affect plant water relations together with time scales at which they operate. Of these, root growth influences the supply of water, leaf growth, shedding, and movement, influence the demand for water, and stomatal movement acts as a valve whose setting can be influenced by both past and present conditions.

TABLE 1. Physiological Processes Affecting Plant Water
Relations.

Process	Time Scale
Root growth	days
Leaf growth (area, thickness)	"
Leaf shedding or senescing	"
Conditioning (adaptation at cellular level)	"
Leaf movement (photonasty, rolling)	hours
Changes in hydraulic resistance	hours, minutes
Stomatal movement	hours, minutes

    Table 2 lists environmental processes which affect plant
water relations. The time scales, in general, apply to the
field, and may be one or two orders of magnitude smaller in
controlled environments. For example, a large plant in a
small pot may deplete its water supply in less than a day,
whereas a similar plant in the field may take weeks to deplete
the soil of water. Similarly, the change from night to day
or from high to low humidity may take seconds in a controlled
environment cabinet compared with an hour or more in the field.
Such rapid changes do not allow long-term adaptive responses
in plants to manifest themselves, with the consequence that
controlled-environment research on water relations has concentrated largely on rapid processes like stomatal movement,
changes in hydraulic resistance of roots, and short-term leaf
expansion (e.g. Hsiao 1973, Kuiper 1972). These short-term
processes, in addition to being more obvious in the laboratory,
have the added attraction of seeming more mysterious and
therefore more exciting than something as prosaic as, say,
the slow changes in leaf area which one can observe in the
field. Yet, it is the control of leaf area and morphology
which is often the most powerful means a mesophytic plant
has for influencing its fate if it is subject to long-term
water stress of the sort it might experience in the field.
    To control transpiration is to control the proportion
of incoming energy which is dissipated as latent heat. A
plant with a large leaf area index (LAI) can keep this proportion low by keeping its leaves parallel to the sun's rays,
a trick which requires special machinery (Begg and Torssell
1974), or by closing its stomates, thus increasing the temperature of its leaves. The latter solution is an unsatis-
factory one in the long term, for to have a large area of hot
green leaves losing water through their cuticles while barely
photosynthesising, though still respiring, must lead to major
inefficiencies in water use. A better solution is to control
the leaf area.

TABLE 2. Environmental Processes Affecting Plant Water Relations.

Process	Time Scale
Run-down in soil water	weeks, days
Diurnal evaporative demand	hours
Seasonal evaporative demand	months, weeks
Rain, irrigation	hours to seconds
Cloud movement	hours to seconds

The effect of leaf area on transpiration rate is obvious in an isolated plant, but it is also large in a canopy if the LAI is small. To a good approximation for our purposes, the evaporation rate from a freely transpiring crop canopy is proportional to LAI up to 2.5, above which evaporation occurs at potential rate (Ritchie 1974). Below 2.5, some of the incoming energy is dissipated as sensible heat from the surface of the soil, which is usually dry if the crop's growth is being limited by water. To vary leaf area takes time, but given that time, a plant has an efficient way of controlling its long-term transpiration rate.

GOALS AND STRATEGIES

We may usefully view a plant as an intricate control system in which responses to stress are strategies directed towards achieving certain goals. The most important of these goals, and the one about which there would be least debate, is genetic survival. We can be much less confident about what the lesser goals might be, but simplistic guesses about them, such as those made below, serve as a useful framework for discussing interactions between the processes listed in Tables 1 and 2. Strategies concerned with achieving a goal at a given time scale usually imply lesser goals at smaller time scales; there is a hierarchy of goals which roughly correlates with time scale. The examples discussed below start with the greater goal and follow some of its implications through lesser goals. Most of the examples relate to crops, particularly cereals, and two main goals are discussed. One is to maximise dry weight, given a limited supply of water, the other is to maximize grain yield given a limited total dry weight, that is, to maximize what Harper and Ogden (1970) call reproductive effort.

Maximizing reproductive effort.- If water is limiting, the grain yield of cereals depends on the amount of water which the plants use between anthesis and maturity (Nix and Fitzpatrick 1969, Passioura 1976). To maximize grain yield

in these circumstances, the plants must maximize the amount
of water used between anthesis and maturity. If the water
supply comes mainly from current rainfall, there is little
room to manoeuvre, but if the supply comes mainly from water
stored in the soil, plants which conserve such water for use
after anthesis will produce more grain than those which do
not (Passioura 1972). How then can a plant restrict its
early water use? The most effective way is to restrict leaf
area, but to do this whether there is a good supply of water
or not is a poor solution, for a plant with this strategy is
unable to respond to good conditions. A better strategy is
to restrict leaf area only if the topsoil is dry, so that
only where the plant is relying on stored water does it
conserve water. The root system of the temperate cereals is
so arranged that it can neatly exert this type of control
(Passioura 1974). It is a dual root system, and comprises
(1) the seminal roots, which arise from the seed and usually
penetrate deeply into the subsoil, and (2) the nodal roots,
which arise from the crown and tend to be restricted to the
topsoil. In a good season, the topsoil is frequently wet,
the nodal roots provide an ample water supply to the shoot,
and the plant grows fast. In a poor season, the topsoil is
chronically dry, the nodal roots develop poorly, if at all,
and the plant must rely on its seminal roots to extract water
from the subsoil. Now the hydraulic resistance to vertical
flow through the basal part of the seminal root system is
very large, for virtually all of the water collected in the
subsoil has to be carried by only one xylem vessel in each
main root through the dry topsoil to the shoot. The plant
seems unable to get water fast enough in these circumstances
and it responds by slowing its growth rate (Passioura 1972).
Yet it is interesting to note that such a plant does not
show the usual symptoms of water stress. Even in the extreme
case where a plant is forced to rely on only one seminal
root to extract water from the subsoil, its midday leaf water
potential is indistinguishable from that of a plant with a
full complement of roots, and so is its stomatal resistance
(Passioura, unpublished). The only obvious difference between the plants in the first few weeks of growth is in their
size, and it is through size that they control their water
use.

The strategy of conserving water for later use seems
appropriate for isolated plants, which have no competitors,
or for essentially pure stands like crops. Where there is
competition for water, however, such a strategy seems in-
appropriate. A conservative plant sharing a water supply
with a spendthrift one is at a disadvantage, for the water

it saves will be used by its competitor. A better strategy
would be to grow as fast as possible, thus using water as fast
as possible, until the water supply becomes so low that little
further growth can be made, then to hang on, keeping stomates
closed most of the time, slowly shedding leaves while waiting
for rain. This type of behaviour seems to occur in many crops
(Ritchie 1974) even though they are pure stands, and may
reflect the fact that the plants have been selected, at least
in part, for their ability to compete with weeds. Plants
showing this behaviour keep their stomates open even when
their leaf water potential is very low, for example, Jordan
and Ritchie (1971) found that cotton grown in the field kept
its stomates open at a leaf water potential as low as -27 bars,
while that grown in a glasshouse closed its stomates sharply
at -16 bar. The difference was presumably due to a difference
in the rate of onset of water stress; the plants in the field
may have had more time in which to make an osmotic adjustment
of the type discussed by Turner (1975).

Maximizing dry weight of shoot. - If a plant is to maximize the size of its shoot given the constraint of a limited
water supply, its leaves must photosynthesize with as little
loss of water as possible, and it must invest as little
assimilate in the roots as it can.

A suitable strategy for achieving a low transpiration
ratio depends on the time scale, the pattern of water supply,
and other features of the environment. Slatyer (1964) has
discussed features relevant to xerophytes. The main strategy
for mesophytes is to close their stomates during periods of
high evaporative demand, or to time their development so that
they avoid such periods. If the time scale is seasonal, and
the plants are relying on stored water, then they will produce
most dry matter if they grow as fast as possible during
periods of low evaporative demand. For example, in a mediterr-
anean climate, there is no point in an annual plant saving its
water in the spring, for the rate of inflation is such that
its savings would be almost without value by summer.

If the plants are relying on current rainfall, then the
time scale is hours or days rather than weeks, and a low
transpiration ratio may presumably be achieved by midday
closure of the stomates. However, there is little direct
evidence that such a strategem influences transpiration ratio
in the field.

While stomatal control of transpiration is an option
available to all higher plants, some plants have special
machinery for effecting morphological changes in response to
water stress. The nastic movements of sunflower and of
Townsville stylo (Begg and Torssell 1974) presumably allow

the plants to lower their transpiration ratios.

Another morphological feature which may influence transpiration ratio is thickness of leaves. There is some evidence that the photosynthetic rate of a leaf may be positively correlated with its thickness or its specific leaf weight (Khan and Tsunoda 1970, Pearce et al. 1969). Since the amount of water lost by a leaf is, in general, proportional to its area, and if its photosynthesis depends more on its weight than on its area, then a thick leaf should be more efficient in its use of water than a thin one. In a crop canopy with a large LAI such an effect would be swamped. But if the LAI were small, so that transpiration was limited partly by leaf area, then a plant which invested its photosynthate in producing thick leaves rather than large ones would produce more dry matter when supplied with a given amount of water. There is no firm evidence that this is so, but it is suggestive that, of two varieties of wheat with which I have been working, one always produces about 20% more dry matter than the other when supplied with a given amount of water, and has leaves which are about 30% thicker under similar conditions.

Having thick rather than large leaves is, however, a conservative trait, and, as before, is likely to be a useful strategy only for isolated plants, or plants in pure stands. Where there is competition for a limited water supply, the leafy plant is likely to do better.

The above discussion has implicitly assumed that the plants have a fixed and well-defined supply of water. In real situations, however, the supply is usually not well defined, and can be increased by the growth of the root system. Much has been written about the ratio of root to shoot, and the way in which it responds to temperature, nutrition and light (Troughton 1974). The ratio, as we might expect, is increased if nutrition is poor, and decreased if light is poor. Little has been written about the effect of water status on the ratio, but what there is suggests, again as we might expect, that a plant diverts more of its assimilate to its roots if water is hard to get (Troughton 1970, Passioura and Ashford 1974). That is, the plant influences its long-term water status by the way it partitions its assimilate.

CONCLUSION

Much of the preceding discussion has been speculative and much may appear to be only tenuously connected to the title of this paper. But the theme has been the control of water flow through plants in the broadest sense - from the soil surrounding the roots, to the air surrounding the leaves, of plants which are responding to their environments at an ontogenetic time scale. The emphasis has been on slow processes because these seem to have the most direct influence on a plant's achieving its most important goal, namely, genetic prosperity. This is not to say that the quicker processes are unimportant. On the contrary, for it is the integration of the quick processes which produce the slow ones. Indeed, the way in which physiological processes occurring at a time scale of minutes or hours interact to influence the development of a plant over days or weeks is probably the least understood, yet possibly the most important area there is in plant physiology. If it is the least understood it is because our imaginations have been stirred more by the successes of the biochemist than by the problems of the ecologist.

REFERENCES

Begg, J.E. and Torssell, B.W.R. (1974). Diaphotonastic and parahelionastic leaf movements in *Stylosanthes humilis* H.B.K. (Townsville Stylo). In "Mechanisms of Regulation of Plant Growth" ed. by R.L. Bieleski, A.R. Ferguson, and M.M. Cresswell. The Royal Society of New Zealand, Wellington. pp.277-283.

Harper, J.L. and Ogden, J. (1970). The reproductive strategy of higher plants. I. *J. Ecol. 58*, 681-698.

Hsaio, T.C. (1973). Plant responses to water stress. *Ann. Rev. Plant Physiol. 24*, 519-570.

Jordan, W.R., and Ritchie, J.T. (1971). Influence of soil water stress on evaporation, root absorption, and internal water status of cotton. *Plant Physiol. 48*, 783-788.

Khan, M.A. and Tsunoda, S. (1970). Leaf photosynthesis and transpiration under different levels of air flow rate and light intensity in cultivated wheat species and its wild relatives. *Jap. J. Breed. 20*, 305-314.

Kuiper, P.J.C. (1972). Water transport across membranes. *Ann. Rev. Plant Physiol. 23*, 157-172.

Morrow, P.H. and Mooney, H.A. (1974). Drought adaptations in two Californian evergreen sclerophylls. *Oecologia 15*, 205-222.

Nix, H.A. and Fitzpatrick, E.A. (1969). An index of crop water stress related to wheat and grain sorghum yield. *Agric. Met. 6*, 321-337.

Passioura, J.B. (1972). The effect of root geometry on the yield of wheat growing on stored water. *Aust. J. Agric. Res. 23*, 745-752.

Passioura, J.B. (1974). The effect of root geometry on the
     water relations of temperate cereals (wheat, barley,
     oats). In "Structure and Function of Primary Root
     Tissues" ed. J. Kolek. Veda, Bratislava. pp. 357-363.
Passioura, J.B. (1976). Physiology of grain yield in wheat
     growing on stored water. *Aust. J. Plant Physiol. 3,*
     (in press).
Passioura, J.B. and Ashford, A.E. (1974). Rapid translocation
     in the phloem of wheat roots. *Aust. J. Plant Physiol. 1,*
     521-527.
Pearce, R.B., Carlson, G.E., Barnes, D.K., Hart, R.H. and
     Hanson, C.H. (1969). Specific leaf weight and photosyn-
     thesis in alfalfa. *Crop Sci.9,* 423-426.
Ritchie, J.T. (1974). Atmospheric and soil water influences
     on the plant water balance. *Agric. Met. 14,* 183-198.
Slatyer, R.O. (1964). Efficiency of water utilization by arid
     zone vegetation. *Annals Arid Zone 3,* 1-12.
Troughton, A. (1970). Grass roots. Annual Report Welsh Plant
     Breeding Station. Aberystwyth. pp. 87-99.
Troughton, A. (1974). The growth and function of the root in
     relation to the shoot. In "Structure and Function of
     Primary Root Tissues" ed. J. Kolek. Veda, Bratislava.
     pp. 153-164.
Turner, N.C. (1975). Concurrent comparisons of stomatal
     behaviour, water status, and evaporation of maize in soil
     at high or low water potential. *Plant Physiol. 55,*
     932-936.

Chapter 33

Assimilate Partitioning: Cause and Effect

I.F. Wardlaw
CSIRO Division of Plant Industry, Canberra, Australia

INTRODUCTION

The result of genetic variation, or a change in the environment is of interest to the plant breeder and agronomist, but often we do not fully understand the physiological basis of the changes observed. One problem has been the failure to distinguish cause and effect in the interactions between the supply of metabolites (source), the growing parts (sink) and the vascular system (translocation). An excellent example of this can be seen in the association, obtained in some plants, between low photosynthetic rates and slow growth. Whether this is a causal relationship is difficult to establish and this particular problem has been discussed in some detail in this volume by A.Luuumann et al.

Interactions of course occur also between plants in crop situations and at the cellular and biochemical level, but the discussion that follows is centered on the broad interactions between photosynthesis, translocation and growth, particularly in relation to environmental responses and the control of partitioning of dry matter in plants.

EXPERIMENTAL GUIDELINES

Artificial manipulation of assimilate supply, the transport system and rate of growth are necessary to provide a set of guidelines for the control and function of each part of the system, which can then be matched against the whole plant response.

VARIATIONS IN ASSIMILATE SUPPLY

Variation in supply is generally achieved in three distinct ways, by defoliation, shading, or a change in the ambient CO_2 level. The end result may not necessarily be similar, but the effect can be followed, both through dry weight changes and by observing the pattern of movement of ^{14}C-labelled assimilates.

With a reduction in assimilate supply two general observations may be made. Firstly, there will be a change in the partitioning of dry material resulting from a polarizing of metabolites to the dominant centres of growth (sinks). Secondly, there will be less assimilate stored along the pathway of movement. Both these attributes can be seen in experiments on wheat, where shading the culm between the penultimate leaf and the ear resulted in a marked increase in the movement of ^{14}C-labelled assimilates from that leaf to the grain (Marshall and Wardlaw 1973), yet despite the increased total movement of ^{14}C there was less retention of this along the transport pathway. However removing leaves from the lower part of the vegetative shoot of either Darnel (*Lolium temulentum*) (Evans and Wardlaw 1964), or soybean (Thrower 1962), resulted in a relatively small, although significant stimulation of movement of ^{14}C from the upper leaves to the roots. The latter are only poor competitors in comparison with the expanding leaves! One outcome of this apparent difference in 'sink' capacity is a reduction in the root: shoot ratio with decreasing light intensity (Welbank et al. 1973).

Table 1 illustrates the effect of supply on assimilate distribution in two cereals, *Sorghum* and wheat, during grain development. With *Sorghum* (Table 1a) variation in the assimilate supply by altering light intensity or CO_2 levels had relatively little effect on grain dry weight, with most of the compensatory response occurring in the roots. Studies on the distribution of ^{14}C-labelled flag leaf assimilates in wheat, at two light levels (Table 1b), also demonstrated the dominant role of the grain as a sink following anthesis. However the patterns of ^{14}C-accumulation suggest that grain growth in the latter case was maintained under low light more as a result of reduced storage in the stem than from competition with the roots.

Caution must be used to interpret experiments, where assimilate supply is varied by light, as this may also have a direct effect on the pattern of plant development. For example, work by Starck (1973) on radish showed that despite an overall reduction in plant dry weight in response to shading, the petiole of the leaf was actually heavier under low light conditions. This developmental effect of light was clearly demonstrated by the pattern of distribution of ^{14}C-photosynthate.

TABLE 1. The Effect of Supply on Assimilate Distribution

(a) Dry weight increase (g) in *Sorghum* plants during grain filling (from Fischer and Wilson 1975).

Plant Part	Natural Light		CO_2 conc. of air (μl 1^{-1})		
	100%	72%	400	300	250
Grain	69	67	68	68	67
Stem and leaf	6	2	5	4	5
Other	4	4	5	5	2
Roots	5	-1	17	8	3
Total	84	72	95	85	77

(b) Distribution of ^{14}C-photosynthate (%) from the flag leaf of wheat (from Wardlaw 1970).

Irradiance (400–700 nm)	Plant Part				
	Ear	Peduncle	Flag	Lower Stem	Roots and Crown
96 Wm^{-2}	49	11	11	19	10
14 Wm^{-2}	71	5	14	3	7

VARIATIONS IN THE TRANSPORT PATHWAY

The influence of the path of transport on assimilate distribution is associated with variation in the carrying capacity, the ability to transfer metabolites laterally to adjacent non-conducting tissue, the exchange between individual vascular elements, or bundles and the pattern of vascular connections within the plant.

The relation between assimilate distribution and phyllotaxic patterns and the restricted movement between vascular bundles, has been observed in many plant systems (Wardlaw 1968). This buffering of movement between traces can be seen to some extent in the node and adjacent stem tissue of wheat, where altering the assimilate requirement of the ear has only a marginal effect on the transport of ^{14}C-labelled assimilates out of the flag leaf, one of the main sources of ear photosynthate (Wardlaw 1965).

The path of movement can be restricted experimentally, as in the classical work of Mason and Maskell (1928) on cotton, where they either restricted the area of bark through which translocation could occur, or varied the length of an already restricted connecting zone of bark. The reduction in transport was much less than the reduction in the cross-sectional area of bark and this compensation in transport could largely be accounted for by a change in the concentration gradient of sucrose across the restricted zone. Similarly the transport of carbohydrate through different lengths of bark strip was also associated with a change in sucrose gradient.

More recently it was shown by Fischer and Wilson (1975) that cutting half the peduncle of *Sorghum*, below the ear, had no significant effect on grain development, indicating that compensation in the movement of assimilates through the remaining vascular tissue was adequate to restore the original level of supply. A similar response has been observed in wheat (Wardlaw and Moncur 1976), thus there is no evidence in these two cereals that the vascular system of the peduncle limits the supply of assimilates available for grain development.

VARIATIONS IN DEMAND

Increased demand by a specific growth centre will change the pattern of assimilate distribution, decrease temporary storage along the transport pathway and may increase the speed of movement of assimilates through the vascular system to that centre (Thrower 1962; Wardlaw 1965, 1974a; Moorby, Troughton and Currie 1974).

Figure 1 summarises a series of experiments on wheat (Wardlaw and Moncur 1976), in which the speed of movement of ^{14}C-labelled flag leaf assimilates, through the peduncle to the ear, was compared with the rate of import of dry matter into the ear. Import (demand) was varied either by removing grains from selected spikelets, or by preventing ear photosynthesis with 3-(3,4-dichlorophenyl)-1,1-dimethylurea (DCMU). As the demand increased so did the speed of movement. An interesting feature of these results was that, in every experiment, it was found that doubling the rate of import of dry matter by the ear, more than doubled the speed of movement of ^{14}C-assimilates through the peduncle. Thus it follows that the concentration of assimilates being transported was lower under high demand conditions and this in turn is consistent with the reduction in lateral movement and reduced storage of assimilates along the path of transport when the demand is high. The effect of demand on storage was evident in an earlier experiment on wheat (Wardlaw 1965), where although the amount of ^{14}C-labelled flag leaf photosynthate reaching the ear was reduced by 80% with the removal of 2/3 of the grain the accumulation of ^{14}C in the peduncle was more than doubled.

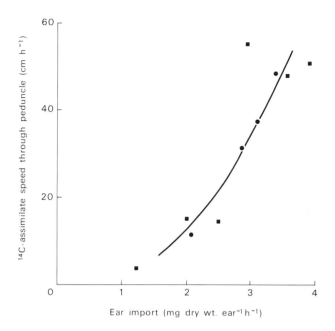

Fig. 1. - *The relation between the rate of import of dry matter by the ear of wheat and the speed of* ^{14}C-*photosynthate movement through the peduncle. (Adapted from Wardlaw and Moncur 1976).*

RESPONSE PATTERNS

GENETIC

Space will not allow more than a short comment on cause and effect associated with genetic changes in plant growth. The control of partitioning and the interaction between organs competing for a limited supply of substrate is of considerable importance to the crop physiologist interested in the yield of cereals. For example the greater ear development of the dwarf wheats, in comparison with many taller varieties, may be a response associated with reduced competition by the stem (Wardlaw 1975), but there is little direct evidence to substantiate this suggestion. Another example of current interest is the apparent association between high growth rates, rapid translocation, efficient vein loading and high photosynthetic rates in C_4- compared with C_3-species (Hofstra and Nelson 1969; Lush and Evans 1974). Any one, or all of these processes could be limiting in the C_3-species, but again there is little experimental evidence available on which to interpret the possible interactions.

ENVIRONMENTAL

The main emphasis in this brief presentation is on plant responses to temperature and water stress. The effect of varying light levels has been discussed to a limited extent earlier and to this information can be added the observation by Moorby, Troughton and Currie (1974) that the speed of movement of ^{11}C-labelled assimilates may be enhanced in response to irradiance, in a species where the demand for assimilates is high. The papers by Pate and Loneragan in this volume deal with the movement of nutrients, and only a few additional comments will be made here.

Temperature. - Figure 2 illustrates the response to temperature in wheat during grain development (data from Wardlaw 1974b) of net flag leaf photosynthesis, vein loading (^{14}C-export), translocation (^{14}C-movement) and growth (based on the rate of accumulation of ^{14}C in the grain). For comparative purposes all values have been expressed as percent of the maximum rate obtained.

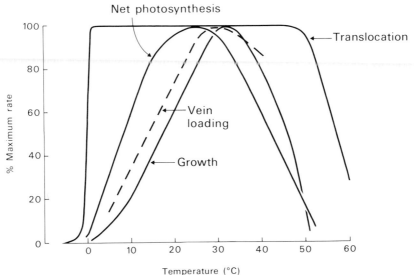

Fig. 2. - The relative effect of temperature on net photosynthesis, vein loading (^{14}C-export), translocation (the movement of ^{14}C-photosynthate through a 10 cm length of peduncle), and growth (the rate of ^{14}C import by the grain) in wheat. (Adapted from Wardlaw 1974b).

Initially translocation through the peduncle was not altered by a 10 cm long jacket ranging in temperature from 1°

to 50°C. However, although high rates of transport were maintained even after three days treatment at 1°C, transport did deteriorate with time at 50°C. This ability to transport at low temperatures is common to many species (Geiger 1969). Growth was more sensitive to temperature than photosynthesis or vein loading in the range 15 to 25°C. However below 15°C the combined effect of reduced photosynthesis and reduced vein loading, may result in the reduced availability of photosynthate becoming a factor in the temperature response. A comparison of these results can be made with earlier experiments on wheat (Wardlaw 1970), where air temperatures around the whole plant were held at either 15/10°C or 27/22°C. At the higher temperatures there was greater movement of ^{14}C-photosynthate to the ear and very much reduced storage in the stem, as well as having a small effect on movement to the root. These results, taken together with the individual responses presented in Figure 2 and the earlier observations on the effect of sink size on transport, indicate that the dominant effect of temperature is directly on grain growth with a consequent change in the availability of dry matter for storage and root growth.

Water stress. - A sugar cane crop is often allowed to dry out towards maturity, with a resulting rise in the sugar content of the stem at harvest, and based on an analysis of ^{14}C-movement in sugar cane, Hartt (1967) concluded that moisture stress affected translocation more severely than photosynthesis. However a real alternative that was not adequately considered here, is the possibility that the observed changes were the consequence of reduced expansion growth in the leaves and roots.

The effect of water stress on assimilate partitioning in *Lolium temulentum* (Wardlaw 1969) showed that there was a small, but significant, increase in the dry weight of the base of the mature leaf sheath (12% after 2 days stress) and continued accumulation of dry matter in the base of the young expanding leaf and the tip of the root, the weight of the basal 4 cm of leaf increasing from 6.4 to 8.5 mg, as the extension rate decreased from 2.6 to 0.4 mm h^{-1} with stress. These data suggest that the initial response to stress was a reduction in extension growth and not translocation. Other experiments clearly showed that photosynthesis was reduced in response to water stress, well after the reduction in extension growth was observed. The change in pattern of distribution of ^{14}C-assimilates from the uppermost mature leaf of *Lolium*, also indicated an enhancement of carbohydrate accumulation in the leaf sheath under stress, with a parallel reduction in movement to the roots and tillers.

Experiments in which the expansion of isolated root and leaf sections floated on various concentrations of mannitol was followed for a 6 hour period (Figure 3), indicated that the additional dry matter accumulating at the root tip, and in the growing leaf base under water stress, was osmotically active. For comparison the effect of a reduced supply of photosynthate was obtained by defoliating plants 4 days prior to sampling. Expansion on mannitol was reduced by this last treatment and this result, together with the sequences observed in the intact plant suggest that changes in photosynthetic rate and the availability of carbohydrate were not in any way associated with the water stress effects on growth in these short term experiments.

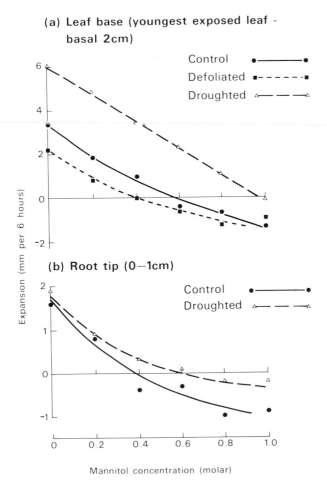

(a) Leaf base (youngest exposed leaf - basal 2cm)

Control ●———●
Defoliated ■-----■
Droughted △—— —△

(b) Root tip (0—1cm)

Control ●———●
Droughted △—— —△

Expansion (mm per 6 hours)

Mannitol concentration (molar)

Fig. 3. - The effect of mannitol concentration on the expansion of the growing leaf base (a) and root tip (b) isolated from control, droughted and defoliated plants. Each result is the mean of 3 plants. Growth rates prior to isolation were controls 2.2 mm h^{-1}, defoliated 1.7 mm h^{-1}, and droughted 0.6 mm h^{-1}.

Nutrition. - Lee, Whittle and Dyer (1966) indicated that boron deficiency resulted in a lower velocity of assimilate transport in sunflower and Hartt (1969) reported that potassium deficiency also reduced the velocity of assimilate movement in sugar cane. The observations, discussed earlier, that variations in growth are reflected in the speed of movement of assimilates are an indication that the effects of both potassium and boron deficiency are directly acting through growth and only indirectly on translocation. This is in agreement with the conclusion reached earlier by Neales (1959) on the effect of boron, based on the measurement of sugar levels in boron deficient plants. However the role of nutrients in translocation needs further study, particularly potassium, which could well be active in the water relations of the phloem as well as the leaf.

CONCLUSION

The data available for wheat and other C_3-species indicate that the transport system is stable under a wide range of environmental stress conditions and has a carrying capacity in excess of that required to maintain normal growth. However the latter conclusion may need to be qualified when we have a fuller understanding of restrictions on transport in situations such as the shoot apex and the developing grain of cereals (cf. Jenner in this volume). Thus assimilate partitioning will be regulated both by assimilate supply and demand, with this interaction depending on both environmental factors and the stage of plant development.

Insensitivity to low temperature and the change of speed of photosynthate movement in relation to demand, which does not appear to be associated with a hormonal control system (Wardlaw and Moncur 1976), is not compatible with a mechanism of transport, such as electro-osmosis or contractile microtubules, activated by metabolic processes along the pathway.

REFERENCES

Evans, L.T. and Wardlaw, I.F. (1964). Inflorescence initiation in *Lolium temulentum* L. IV. Translocation of the floral stimulus in relation to that of assimilates. *Aust. J. Biol. Sci. 17*, 1-9.

Fischer, K.S. and Wilson, G.L. (1975). Studies of grain production in *Sorghum bicolor* (L. Moench). III. The relative importance of assimilate supply, grain growth capacity and transport system. *Aust. J. Agric. Res. 26*, 11-23.

Geiger, D.R. (1969). Chilling and translocation inhibition. *Ohio J. Sci. 69*, 356-366.

Hartt, C.E. (1967). Effect of moisture supply upon transloc-
    ation and storage of ^{14}C in sugarcane. *Plant Physiol. 42,*
    338-346.
Hartt, C.E. (1969). Effect of potassium deficiency upon
    translocation of ^{14}C in attached blades and entire plants
    of sugarcane. *Plant Physiol. 44,* 1461-1469.
Hofstra, G. and Nelson, C.D. (1969). A comparative study of
    translocation of assimilated ^{14}C from leaves of different
    species. *Planta (Berl.) 88,* 103-112.
Lee, K.W., Whittle, C.M. and Dyer, H.J. (1966). Boron defic-
    iciency and translocation profiles in sunflower. *Physiol.
    Plant. 19,* 919-924.
Lush, W.M. and Evans, L.T. (1974). Translocation of photo-
    synthetic assimilate from grass leaves, as influenced by
    environment and species. *Aust. J. Plant Physiol. 1,*
    417-431.
Marshall, C. and Wardlaw, I.F. (1973). A comparative study
    of the distribution and speed of movement of ^{14}C assim-
    ilates and foliar-applied ^{32}P-labelled phosphate in
    wheat. *Aust. J. Biol. Sci. 26,* 1-13.
Mason, T.G. and Maskell, E.J. (1928). Studies on the trans-
    port of carbohydrates in the cotton plant II. The
    factors determining the rate and the direction of move-
    ment of sugars. *Ann. Bot. 42,* 571-636.
Moorby, J., Troughton, J.H. and Currie, B.G. (1974). Invest-
    igations of carbon transport in plants II. The effects
    of light and darkness and sink activity on translocation.
    J. Exp. Bot. 25, 937-944.
Neales, T.F. (1959). Effect of boron supply on the sugars
    soluble in 80% ethanol, in flax seedlings. *Nature 183,*
    483.
Starck, Z. (1973). The effect of shading during growth on
    the subsequent distribution of ^{14}C-assimilates in
    Raphanus sativus. Bull. Acad. Polon. Sci. 21, 309-314.
Thrower, S.L. (1962). Translocation of labelled assimilates
    in the soybean II. The pattern of translocation in
    intact and defoliated plants. *Aust. J. Biol. Sci. 15,*
    629-649.
Wardlaw, I.F. (1965). The velocity and pattern of assimilate
    translocation in wheat plants during grain development.
    Aust. J. Biol. Sci. 18, 269-281.
Wardlaw, I.F. (1968). The control and pattern of movement of
    carbohydrates in plants. *Bot. Rev. 34,* 79-105.
Wardlaw, I.F. (1969). The effect of water stress on trans-
    location in relation to photosynthesis and growth II.
    Effect during leaf development in *Lolium temulentum* L.
    Aust. J. Biol. Sci. 22, 1-16.

Wardlaw, I.F. (1970). The early stages of grain development
    in wheat: Response to light and temperature in a single
    variety. *Aust. J. Biol. Sci. 23*, 765-774.
Wardlaw, I.F. (1974a). Phloem transport: Physical, chemical
    or impossible. *Ann. Rev. Plant Physiol. 25*, 515-539.
Wardlaw, I.F. (1974b). Temperature control of translocation.
    In "Mechanisms of regulation of plant growth" eds.
    R.L. Bieleski, A.R. Ferguson, and M.M. Cresswell. *Roy.
    Soc. N.Z. Bull. 12*, 533-538.
Wardlaw, I.F. (1975). The physiology and development of temp-
    erate cereals. In "Australian field crops. 1. Wheat and
    other temperate cereals", pp.58-98 eds. A. Lazenby and
    E.M. Matheson, Angus and Robertson, Sydney.
Wardlaw, I.F. and Moncur, L. (1976). Source, sink and hormo-
    nal control of translocation in wheat. *Planta (Berl.)
    128*, 93-100.
Welbank, P.J., Gibb, M.J., Taylor, P.J. and Williams, E.D.
    (1973). Root growth of cereal crops. In Rothamsted Exp.
    Stat. Rep. Part 2, 26-66.

CHAPTER 34

Effects of Sink Size, Geometry and Distance from Source on the Distribution of Assimilates in Wheat

M.G. Cook and L.T. Evans
CSIRO Division of Plant Industry, Canberra, Australia

INTRODUCTION

Shifts in the partitioning of photosynthetic assimilates among the organs of plants have played a major role in their adaptation and evolution, both in nature and in agriculture. Labelled isotopes offer an easy way to determine patterns of assimilate distribution, but the extent of our empirical knowledge is by no means matched by our understanding of the principles involved.

So vital and complex a process must depend on many interacting factors. Distance and the pattern of vascular connections between source and sink are important. So too are environmental conditions; in water-stressed plants, for example, shoot growth is restricted and root growth enhanced. The relative size and geometric arrangement of competing sinks may also be important, as suggested by the evolution of fewer, larger, more compact organs in the course of domestication of many crop plants (Evans 1976).

In order to examine the effect of relative sink size on the distribution of assimilates we need systems in which competing sinks identical in every way except in size can be set up equidistant and with equivalent pathways from the source of assimilates. The only previous experiment which meets these criteria is that of Peel and Ho (1970), in which small and large colonies of aphids were established on stem cuttings of willow equidistant from a leafy shoot which was exposed to $^{14}CO_2$. Although the weight of honeydew produced by the competing colonies was proportional to their size, its specific activity was far higher in the larger colonies, even when they were spread over a proportionally greater area. Such a bias in favour of larger sinks could be an extremely important determinant of crop productivity. Earlier experiments by Rawson and Evans (1971) comparing 6 cultivars of wheat had shown that the rate of increase in kernel weight tended to be greater the more grains there were per ear, which might reflect a similar phenomenon. We therefore set out to analyse an experimental system in wheat similar to that of Peel and Ho.

THE EXPERIMENTAL SYSTEM

Plants of cv. Sonora were grown singly in pots under a daylength of 16 h in a glasshouse controlled at 21°C by day and 16°C at night. They were supplied with nutrient solution and water daily. Plants in which the ears of the first and second tillers were comparable and began anthesis within one day of each other were selected for experiment. Earlier work with cv. Sonora grown under the same conditions had provided a fairly complete picture of the course of grain growth and the supplies of assimilate available for it (Evans and Rawson 1970; Rawson and Evans 1971).

Plants were cut back to the main stem and the first two tillers. $^{14}CO_2$ was fed only to the flag leaf of the main stem, 11 days after anthesis when the grains were in their linear phase of growth. All other leaves on the plant were removed at least 2 days earlier, as well as the ear on the main stem which would normally receive most of the assimilates from the flag leaf at that stage. Since the two tillers had been deprived of leaves, their ears received assimilates from the main stem flag leaf, about 100 cm distant from each of them. The numbers and arrangement of grains in these two competing ears were varied 2 days before exposure to $^{14}CO_2$ by removal of grains, the total number of grains in the two ears being kept constant. This could be done without injury to the ears, and care was taken to leave grains of approximately equal stage and size in the competing tiller ears, so that they differed only in number.

SOME PROBLEMS

a) *Equivalence of tillers.* - Although the ears of the first and second tillers were approximately equidistant from the main stem flag leaf, that on the first tiller was slightly more advanced, often beginning anthesis one day earlier, and therefore having slightly larger grains during the experimental period. However, in all experiments with competing ears of different grain number, the ear with the larger number was assigned to tiller 1 in half of the replicates (usually totalling 8-12), and to tiller 2 in the other half.

b) *Ear photosynthesis.* - The ears of cv. Sonora are awned and their photosynthesis contributes about 30% of the assimilate requirements of intact ears 11 days after anthesis. Where grain numbers are severely reduced, ear photosynthesis

might therefore supply much of the required assimilate.
Bremner and Rawson (1972) found photosynthate fixed by individual florets not to move to other spikelets, and only to more
distal florets within the same spikelet. Consequently, in ears
with only a few basal grains left, we did not expect assimilates from other spikelets and more distal sterilized florets
to contribute to the requirements of the remaining grains.
However, an experiment in which only the uppermost or lowest
one third of intact or partially degrained ears were exposed
to $^{14}CO_2$ revealed substantial movement of labelled assimilates
from those parts of the ear to all others, not only in partially degrained ears but also in intact ears. Movement to the
central grains was much greater when all others were removed.

Consequently, attempts were made in later experiments to
eliminate ear photosynthesis by spraying with 5 x 10^{-5}M DCMU,
found to be very effective for this purpose in earlier experiments by King et al. (1967). The effectiveness of such sprays
was monitored by infra-red gas analysis, and was unfortunately
neither as complete nor as prolonged as in previous experiments, possibly because the treatments were applied earlier
while the ears were still developing.

c) *Stem reserves*. - Another possible source of assimilates, apart from the flag leaf of the main stem, is the reserve of mobilizable carbohydrate stored in the culms. This
usually contributes less than 10% to final grain weight, and
is not fully mobilized even under quite severe stress (Rawson
and Evans 1971). In many of our experiments we tried to
deplete these reserves substantially well before the competition treatments by removing all tiller leaves as early as 7
days before anthesis, wrapping stems, and repeatedly spraying
the tiller ears with DCMU. The effect of these pretreatments
on stem weights and extractable carbohydrates was monitored
in several experiments. They were substantially reduced, but
not eliminated. Nevertheless, such treatments reduced grain
size to a considerable degree.

Thus, although the main stem flag leaf was not the sole
source of assimilates to the two ears in our experimental
plants, it is likely that they were actively competing for the
labelled assimilates from it in many of our experiments.

RESULTS

a) *Effect of relative size of competing sinks*. - Three
experiments examined the effect of varying grain numbers in
the two ears. The results of two of them are given in Fig. 1.
In all experiments, whether or not the ears were sprayed with
DCMU, the specific activity of the grains in the ear with more
kernels was several times greater than of those in the ear
with fewer kernels, and the ratio of their specific activities

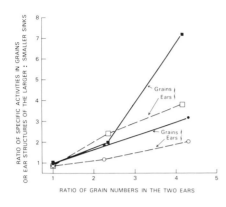

Fig. 1. - Ratio of the specific activity in grains (solid symbols) and structures (open symbols) of the two ears as influenced by the relative number of grains in the two competing ears. Grains were removed and tillers defoliated 2 days before exposure of the flag to $^{14}CO_2$, and grains were harvested 2 days after exposure. Ears were sprayed with DCMU 2-3 days before and 7 days after anthesis (Exp 3,□ and Exp 4,○).

increased with increase in the grain number ratio. The bias in ^{14}C distribution in favour of the ears with more grains was sustained when comparisons were restricted to grains in identical positions in the two ears, and was found also in cv. Kalyansona as well as in cv. Sonora. The ratio of the specific activities of the ear structures without grains also increased with increase in the grain number ratio, and in other experiments this was found to happen also in the tiller stems.

b) *Time trends and effects of geometry.* - In one experiment, $^{14}CO_2$ fixation took place at various intervals after degraining. Exposure to $^{14}CO_2$ immediately after degraining resulted in no bias in favour of the larger sink when grains were harvested 2 days later, but exposures 2-6 days after degraining all resulted in a fairly constant bias in favour of the larger sink.

Three experiments examined the effect of duration between $^{14}CO_2$ fixation and harvest on the bias in ^{14}C partitioning between the two ears. In all of them, a significant bias in favour of the grains in the larger sink was apparent within one day of exposure to $^{14}CO_2$, became most marked after 2-3 days, and then declined a little, as shown in Fig. 2. Clearly, the bias in favour of the larger sink is not an evanescent phenomenon.

Competition between ears with the same number of grains arranged in a variety of patterns was examined in two experiments. No significant bias in favour of any arrangement was found, but the more compact arrangements appeared to have a slight advantage.

c) *Effects of defoliation and inhibition of ear photosynthesis.* - A substantial bias in favour of the larger sink was found in all early experiments whether or not the tillers were defoliated early and the ears sprayed with DCMU. The results given in Fig. 2 show that the bias in ^{14}C distribution

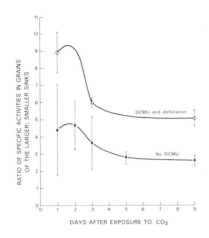

RATIO OF SPECIFIC ACTIVITIES IN GRAINS OF THE LARGER: SMALLER SINKS

DAYS AFTER EXPOSURE TO CO₂

DCMU and defoliation

No DCMU

Fig. 2. - Change with time after exposure to $^{14}CO_2$ in the ratio of specific activity in the grains of ears with 24 kernels compared with those with only 6 (Experiment 11). In one group of plants the tillers were defoliated and the ears sprayed with DCMU 2 days before anthesis (open circles). No DCMU was applied to the other group, and tillers were not defoliated until degraining (closed circles).

in favour of the larger sink was enhanced by treatments reducing reserves and ear photosynthesis before anthesis. Measurement of extractable carbohydrates showed that these treatments greatly reduced the stem reserves. In spite of this, the rate of growth per grain over the period from ^{14}C exposure to harvest appeared to be the same whether there were many or few grains in the ear, and whether reserves were diminished or not. Presumably ear photosynthesis was sufficient, even after the early DCMU spray, to make up the needs of the grains in the smaller sink.

More frequent spraying of the ears with DCMU was therefore used in two experiments, the results of which are given in Table 1. As is evident from the grain weights at harvest, the earlier the defoliation and the more DCMU sprays applied, the more adverse was the effect on early grain growth. Coupled with this trend there was an increase in the total ^{14}C activity assimilated by the flag leaves (of more than 30% in experiment 13) and in the proportion exported by the main stem to the tillers (from 31 to 61% in the same experiment). This feedback effect of increased demand on photosynthetic rate and assimilate export (cf. King et al. 1967), indicates that the plants were increasingly stressed for assimilate supply. At the intermediate levels of stress, the bias in specific activity in favour of the larger sink was marked but under more extreme stress and more complete inhibition of ear photosynthesis during ^{14}C distribution the bias was considerably reduced.

d) *Interaction of relative distance and size of sinks.* - The system used in the experiments described above was modified in two experiments by the removal of the entire main stem, and the replacement of its flag leaf by one of the top

TABLE 1. Effect of tiller defoliation and inhibition of ear photosynthesis on the bias in ^{14}C distribution in favour of the larger sink

Experiment	DCMU sprays(days from anthesis)	Defoliation	Specific activity ratio	Dry matter(mg) per grain	
				18 grains	5 grains
13	0	A+9	5.82±0.68	11.6	13.8
	A-2	A-2	6.45±0.50	9.3	11.6
	A-2,+2,+6,+10	A-2	1.56±0.11	7.2	9.1
				20 grains	5 grains
14	A-2	A-2	8.63±0.80	8.4	10.2
	A-2,+2,+6,+8,+10	A-2	1.32±0.10	6.2	7.4
	A+8,+10	A-2	4.29±0.64	8.2	10.1
	A-2,+2,+6,+8,+10	A+9	1.40±0.08	6.8	8.6

three leaves on the first tiller.   In this way various sink size ratios could be combined with various distances from the source leaf.   Results of the two experiments are combined in Fig. 3.   The relative distance of the two competing sinks from the source leaf is clearly the dominant influence, with a marked bias in favour of the closer sink even though the minimum distance from the source was more than 50 cm.   The bias in favour of a larger sink could not overcome the handicap of greater distance, but is clearly expressed whether the larger sink is closer to or more distant from the source leaf.

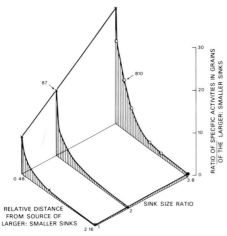

Fig. 3. - Interrelations between relative sink size and relative distance of the two ears from the source leaf on the ratio of specific activities in the grains of the larger and smaller sinks. (Results of experiments 7 and 10 combined.)

DISCUSSION

Three features of these experiments accord with those of Peel and Ho (1970). (i) There was a consistent and pronounced bias in favour of the larger sink with regard to its specific activity, in spite of similar rates of exudation or grain growth. (ii) The relative sizes of the competing sinks were more important than their geometry. (iii) Other sinks on the pathway to the dominant sink also profited from the bias in distribution, whether small colonies of aphids or the stem and structure of the ear with more grains.

Important as relative size is, relative distance from the source is even more so, and might be by far the dominant factor in other systems where path lengths less than 50 cm could be explored. The great advantage of proximity probably means that assimilates from ear photosynthesis or from mobilization of reserves in the peduncle will be called upon to support kernel growth in the more heavily degrained ears before those from the far more distant flag leaf of the main stem. Doing so contributes to the bias in favour of the larger sink for labelled assimilates from the flag leaf. We cannot be sure how much of the bias in specific activity measured in the grains is due to compensation by reserves and ear assimilates, but taken as a whole our results suggest that size is an important factor in the competition between sinks for assimilates.

ACKNOWLEDGEMENTS

We are most grateful to Kati Bretz for expert technical assistance with these experiments.

REFERENCES

Bremner, P.M. and Rawson, H.M. (1972). Fixation of $^{14}CO_2$ by flowering and non-flowering glumes of the wheat ear, and the pattern of transport of label to individual grains. *Aust. J. Biol. Sci. 25*, 921-930.

Evans, L.T. (1976). - Physiological adaptation to performance as crop plants. *Proc. Roy. Soc. London* (In Press).

Evans, L.T. and Rawson, H.M. (1970). - Photosynthesis and respiration by the flag leaf and components of the ear during grain development in wheat. *Aust. J. Biol. Sci. 23*, 245-254.

King, R.W., Wardlaw, I.F. and Evans, L.T. (1967). - Effect of assimilate utilization on photosynthetic rate in wheat. *Planta (Berl.) 77*, 261-276.

Peel, A.J. and Ho, L.C. (1970). - Colony size of *Tuberolachnus salignus* (Gmelin) in relation to mass transport of ^{14}C-

labelled assimilates from the leaves in willow. *Physiol. Plant. 23*, 1033–1038.

Rawson, H.M. and Evans, L.T. (1971). – The contribution of stem reserves to grain development in a range of wheat cultivars of different height. *Aust. J. Agric. Res. 22*, 851–863.

CHAPTER 35

Sink Effects on Stomatal Physiology and Photosynthesis

P.E. Kriedemann*, B.R. Loveys**, J.V. Possingham**
and M. Satoh***

*CSIRO, Division of Horticultural Research, Merbein, Victoria,
**CSIRO, Division of Horticultural Research, Adelaide, S.A.,
***Sericultural Experiment Station, Suginami-Ku, Tokyo, Japan

INTRODUCTION

Photosynthesis is subject to both environmental constraints and internal regulation which takes the form of source/sink feedback control over CO_2 assimilation. Photosynthetic responses to these sinks often embody changes in both stomatal and internal resistances to gas exchange, and have been widely documented. In general, a leaf's rate of CO_2 assimilation shows a positive response to increased demand for photosynthate, or vice versa, and the question arises as to how this feedback control is mediated. Do centers of growth (or other importing organs) have some subtle means of indicating their requirement to photosynthetic tissues, or are photosynthetic responses a direct consequence of substrate withdrawal from source leaves with associated "mass-action" effects?

Both views have gained qualified support in the published literature including the possible involvement of growth regulators. Short term positive responses (minutes or hours) have already been attributed to auxins (Bidwell et al. 1968), while gibberellins and cytokinins have been implicated in the longer term developmental changes (days). These include photosynthetic rejuvenation of mature or even senescing leaves following treatments such as decapitation or fruit removal in either herbaceous or woody perennial test plants (Wareing et al. 1968, Sweet and Wareing 1966), and altered stomatal behaviour following photoperiodic flower induction (Meidner 1970).

The present paper documents stomatal and photosynthetic responses to such manipulative treatments in three types of test material (mulberry, capsicum and grapevine) and implicates a further category of growth regulators in source/sink feedback effects, viz. the naturally occurring substances abscisic acid, and its close metabolic relative, phaseic acid.

MATERIALS AND METHODS

Test plants of mulberry (*Morus alba*, L.) and grapevine (*Vitis vinefera*, L. cv. Cabernet Sauvignon) were raised from rooted cuttings in 12 cm containers of potting mixture under greenhouse conditions (steam heated-evaporatively cooled). *Capsicum annuum* L. was raised under the same conditions from seed.

Young mulberry trees (24 to 28 leaves) were ranked according to size and then allocated to one of three groups:- 1. Control (intact plant); 2. Decapitated between nodes 12 and 14 (from base) with axillary buds retained and allowed to grow; 3. Decapitated as above but with axillary buds removed.

Photosynthetic activity of the uppermost remaining leaf, and a second leaf between nodes 9 and 11 (third position down), was then measured for single individuals on six occasions over the ensuing thirty days. Samples were also taken at intervals for determination of specific leaf weight, chlorophyll, total sugars and starch content.

Capsicum plants 8-10 weeks old and bearing 3 or 4 maturing fruits were also ranked according to size and distributed between the four treatments shown in Fig. 3. Photosynthetic and stomatal responses were followed for the next ten days. In related experiments on regulation of photosynthesis in capsicum, ^{14}C-labelled fruits from parent plants exposed to 100µc $^{14}CO_2$ for 24 hours (4 days previously) were grafted into the cut stump of a previously decapitated but unlabelled plant. Photosynthetic response to the initial decapitation and to the additional labelled sink were measured over 6 days, and ^{14}C-labelled metabolites were extracted from these same leaves on recipient plants. One further experiment was concerned with photosynthetic responses and associated anatomical changes in mature lower leaves on decapitated capsicum plants. Gas exchange was measured on 7 occasions over 16 days and the same leaves were sampled on 3 occasions for determination of mesophyll cell size, chloroplast number per cell, and chlorophyll content (see Possingham and Smith 1972, for methods).

Test plants of grapevine were well established at the time of experimentation (12 weeks after planting) and had grapes 2-4 mm in diameter. Plants were selected for uniform shoot growth and bunch development. Lateral buds were excised and the single shoot trimmed to 6 leaves. In cinctured (girdled) vines, a 2 mm collar of internodal tissue (cortex) was removed below the bunch. Plants were transferred to growth cabinets 7 days before the experiment; temperature was 25/22°C (day/night) and quantum flux density above the crop was $550\mu E/m^2/s$.

Physiological responses. - Stomatal resistances of leaves adjacent to reproductive nodes were measured on vines inside the growth cabinet using a diffusion porometer (Lambda Model Li60). For other material, both photosynthesis and transpiration were measured under laboratory conditions (incorporating control over temperature, light intensity, humidity and CO_2 concentration) using infrared analysis for both CO_2 and H_2O vapour exchange.

Oxygen evolution by leaf slices was measured with a pair of plexiglass cuvettes fitted with Clark-type oxygen electrodes. Tissue slices $860\mu m$ x 7 mm were cut from leaves of spinach plants (*Spinacia oleracea*) (3 weeks old) grown in nutrient solution under controlled conditions. Thirty slices (65 mg fresh weight and containing approximately $80\mu g$ chlorophyll) were transferred to each of two cuvettes containing 4 ml of 50 mM PIPES buffer at pH 5.6. One cuvette contained $12\mu g/ml$ phaseic acid. Slices were stirred magnetically for one hour under incandescent light ($140Wm^{-2}$ total energy inside cuvette) and then deoxygenated with a gentle stream of N_2 before adding $25\mu l$ of 1M $KHCO_3$ to instigate O_2 evolution. Photosynthetic activity in the two cuvettes was followed concurrently on a 2 channel recorder and O_2 evolution rates (derived from slopes) were expressed as p mol O_2/s (see Table 5).

Abscisic acid (ABA) and phaseic acid (PA) were identified and assayed in capsicum and grapevine leaves after methanol extraction and purification by solvent partitioning followed by t.l.c. and g.l.c. using an electron-capture detector (see Loveys and Kriedemann 1974).

Carbohydrate determinations were based on oven dried (60°C) material extracted with 80% ethanol and centrifuged to yield a supernatant (soluble sugars) and a pellet containing starch. This residue was then subjected to hydrolysis (α amylase), and sugars in this hydrolysate and in the original supernatant were measured by phenol/sulphuric acid hydrolysis and subsequent colourimetic assay (see Dubois et al. 1956).

Leaf chlorophyll content (area or fresh weight basis) was based on absorbance at 652 nm of 80% acetone extracts after the method of Bruinsma (1961).

RESULTS

Mulberry. - The time course of physiological responses of leaf 12-14 in mulberry plants following decapitation with

403

or without axillary bud removal is shown in Fig. 1 and Table
1. Leaf photosynthesis (Fig. 1) increased $c.40\%$ within 8 days
of decapitation with a continuing increase where axillary buds
had been removed. This same pattern of photosynthetic response, including the distinction between decapitation plus
bud and decapitation minus bud, has now become apparent in 4
separate experiments, 2 in Japan (Satoh and Hazama 1971, Satoh
and Ohyama 1971) and 2 in Australia. Parallel measurements
of gas exchange on lower leaves showed similar treatment
effects to those in Fig. 1 but a more protracted time course.

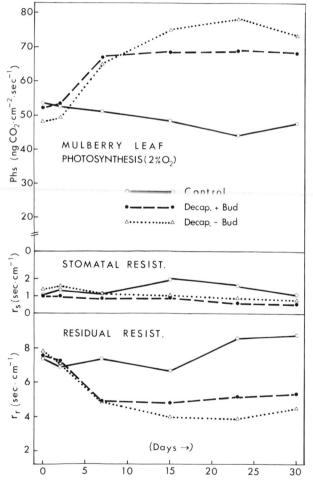

*Fig. 1 - Photosyn-
thetic and stomatal
response to manipu-
lative treatments
in mulberry. De-
capitation, com-
bined with bud re-
moval or retention,
was performed on
day 0 subsequent
to initial labora-
tory measurement of
gas exchange.*

Values derived for gaseous diffusive resistances on these
same leaves (Fig. 1) show that stomatal resistance was uni-
formly low for all treatments throughout the experiment, so
that photosynthetic changes shown above can be attributed to

internal processes, i.e. a drop in residual resistance.

TABLE 1. Leaf response to decapitation and bud removal in mulberry trees (*Morus alba* L.)

Time Elapsed (days)	Treatment	Chlorophyll Content (μg cm^{-2})	Sugars (mg glucose equiv. dm^{-2})	Starch
0	Control Decap.-buds }	47.2	70.4	93.2
8	Control	50.9	66.4	91.5
	Decap.-buds	68.6	63.7	45.4
17	Control	46.2	65.6	63.0
	Decap.-buds	90.5	70.3	37.9
38	Control	52.2	59.0	78.5
	Decap.-buds	75.4	88.6	51.8

Photosynthetic responses following decapitation were associated with substantial increases in chlorophyll content which had virtually doubled on a leaf area basis in the de-capitated -buds treatment after 17 days (see Table 1) so that photosynthesis per unit chlorophyll remained virtually un-changed. Nevertheless, photosynthetic performance was im-proved and the light response curves for CO_2 fixation ($2\%O_2$) ᴜʜᴏᴡɴ ɪɴ ꜰɪɢ. 2 ɪᴍᴘʟʏ ᴛʜᴀᴛ ʟᴇᴀᴠᴇꜱ ᴏɴ ᴅᴇᴄᴀᴘɪᴛᴀᴛᴇᴅ ᴍᴏᴅᴅᴇ ᴘʟᴀɴᴛꜱ achieved both greater photosynthetic capacity at light satur-ation and higher photochemical efficiency (compare initial slopes under light limited conditions). This photosynthetic superiority of leaves on decapitated -buds plants compared with controls, correlates to some degree with lower levels of starch (see Table 1, righthand column) but levels of soluble sugars do not bear any consistent association with rates of assimilation.

Capsicum. - Photosynthetic responses to decapitation are again evident in Fig. 3 which shows the time course of changes in stomatal resistance and CO_2 fixation by capsicum leaves following certain manipulative treatments, (see Fig. 3 legend). Mature leaves on intact plants (treatment 1) show the gradual decline in assimilation rate and associated rise in stomatal resistance characteristic of leaves approaching senescence, and this pattern is accentuated if fruits are re-moved (treatment 2). Presumably, developing fruits on intact plants help to sustain the photosynthetic activity of adjac-ent leaves. If instead of fruit removal, the entire upper region is excised (young leaves plus fruits), then remaining

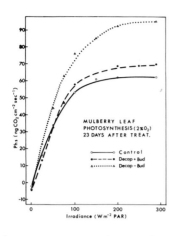

Fig. 2 - Light response curves for photosynthesis in mulberry leaves following decapitation. Regression analysis performed on the first 5 points of Control *and* Decapitation - buds *showed a significant difference between slopes* (t_6 = 4.83)

leaves on capsicum plants appear to be rejuvenated as reported above in Fig. 1 for mulberry. The temporary increase in stomatal resistance following such decapitation in capsicum (treatment 3) is eliminated if fruits are retained (see treatment 4 in Fig. 3) so that the stomatal resistance now remains low while photosynthesis shows a progressive increase.

Photosynthetic response to developing fruits was also evident in related experiments on capsicum where fruits excised from plants previously fed $^{14}CO_2$ were grafted into the cut stumps of non-labelled recipient plants. It was hoped that any compound acting as a "sink signal" would have become labelled while that fruit was still attached to the parent plant, and that such labelled material would be detectable in the recipient plants after grafting. Results from one such experiment are summarized in Table 2. The additional sink has led to a marked increase in photosynthesis due primarily to a reduction in residual resistance. In addition, small amounts of ^{14}C-labelled metabolites were detected in these same leaves. Following methanol extraction of the leaf tissue, solvent partitioning into 4 fractions and subsequent paper chromatography, discrete zones of ^{14}C-activity were apparent in each leaf fraction which corresponded to zones of similar R_f on chromatograms of the ^{14}C-fruit extract. ^{14}C-labelled sugars were abundant in the fruit, but barely detectable in leaves. Presumably, some radioactive metabolites (regulatory compounds?) had moved from the donor fruit into the recipient plant, but quantities were insufficient for further investigation.

The regreening and photosynthetic rejuvenation which followed decapitation in *Capsicum* (shown previously for treatment 3 in Fig. 3) was accompanied by an increase in specific leaf weight with an associated increase in chlorophyll content

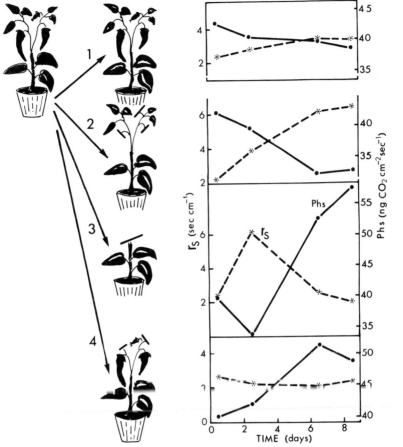

Fig. 3 - *Time course of stomatal (r_s) and photo-*
synthetic response (Phs) of mature leaves
on Capsicum annuum L. following manipulative
treatments: 1. Control (intact); 2. Fruit
removed (all leaves retained); 3. Decapitated
(old leaves retained); 4. Partial defoliation
(young and expanding leaves removed and mature
leaves retained). (All values shown are mean
data from 3 replicates)

from 28 to $62\mu g/cm^2$. Since photosynthetic activity per unit
chlorophyll showed no general increase, the faster assimilation per unit leaf area following decapitation can probably
be attributed to a greater depth of chlorophyll containing
tissue, and more specifically to increases in both cell size
and chloroplast number per cell as outlined below.

407

TABLE 2. Photosynthetic response to fruit grafting in
Capsicum annuum L.

Time Elapsed (days)	Treatment	Photosynthesis (ng CO_2 $cm^{-2}sec^{-1}$)	Stomatal Resistance (sec cm^{-1})	Residual Resistance (sec cm^{-1})
0	Intact (then decapitated)	43.9	1.6	–
1	decapitated	57.5	1.8	13.3
4	decapitated* (then grafted ^{14}C-fruit)	66.1	2.8	–
6	" "	98.9	1.9	8.3

*
Control plants received a small piece of peduncle from donor
plants on day 4, but photosynthesis remained close to the
original 66 ng CO_2 $cm^{-2}sec^{-1}$.

Anatomical changes were followed in a parallel experiment
on *Capsicum* plants where photosynthesis reached 92 ng CO_2/cm^2/
s within 10 days of decapitation compared with 55 (same units)
on intact control plants. Subsequent measurements of gas exchange, up to 17 days after decapitation, revealed no further
increase. Photosynthetic response to decapitation was associated with an increase in cell size and in chloroplast number
per cell. For example: at day 10 the cell area of control
and decapitated plants was 23 and 35 mm^2 x 10^{-4} respectively,
while chloroplast number per cell was 48 and 83 respectively.
By day 17, decapitated plants had shown a further increase in
cell area to 41 mm^2 x 10^{-4} while chloroplast number per cell
had reached 106. The net result of this cell enlargement,
plus greater numbers of chloroplasts per cell, was a substantial increase in chlorophyll content per unit leaf area. Total
chlorophyll (a + b) content after 10 days was 32.5 compared
with 53.5 $\mu g/cm^2$ in control and decapitated plants respectively.

One physiological consequence of this increase in pigment
content is that photosynthesis, if expressed on a chlorophyll
basis, remains virtually unchanged over the entire period.
Assimilation numbers were 1.69 and 1.72 ng CO_2/μg chlr./s for
control and decapitated plants respectively on day 10, and
compare with 1.71 (same units) for decapitated plants on day
17. Clearly, the increased rates of CO_2 assimilation (leaf
area basis) referred to above, can be attributed to formation

of additional chloroplasts following decapitation.

Fruits modify both the time course of leaf response to decapitation (compare treatments 3 and 4 in Fig. 3) as well as final values (compare treatments 1 and 2). Such photosynthetic and stomatal differences as shown in Fig. 3 after 8 days were confirmed in a third capsicum experiment (Table 3) where fruit removal led to a reduction in photosynthesis that was largely attributable to stomatal factors. This increase in stomatal resistance following fruit removal was associated with greater levels of endogenous inhibitors in the foliage of de-fruited plants (see ABA and PA data in Table 3). Since ABA levels need only double to elicit stomatal closure under other circumstances (e.g. moisture stress or exogenous supply to turgid leaves, Kriedemann et al. 1972) the present data would be regarded as physiologically significant.

TABLE 3. Leaf response to fruit removal in *Capsicum annuum* L. (Values shown are mean data from 3 replicates taken 7 days after fruit removed)

Treatment	Photosynthesis (ng CO_2 $cm^{-2}sec^{-1}$)	Stomatal Resistance (sec cm^{-1})	Residual Resistance (sec cm^{-1})	ABA[*]	PA[**] (µg kg^{-1}F wt)
Fruit Present	62.8	1.57	4.05	11.9	7.2
Fruit Removed	48.6	3.46	4.69	26.1	11.5

[*]Abscisic acid   [**]Phaseic acid   (treatment differences were statistically significant at the 1% level of probability)

Grapevine. - Apparent correlations between stomatal response to fruit removal and endogenous levels of ABA and PA as shown above, were also demonstrated with small grapevines - (see Table 4). Within 4 days of removing bunches from either intact or cinctured (girdled) plants, there was a substantial increase in leaf resistance, inhibitor level, and starch content. Fruit removal in combination with stem cincturing had the most pronounced effect. Starch levels doubled, stomatal resistance increased 5-fold and inhibitor levels showed a general increase, especially PA, which increased 5-fold compared with foliage on intact plants.

Phaseic acid effects. - In related vine experiments reported by Loveys and Kriedemann (1974, and literature cited) increased stomatal resistance due to fruit removal with or without cincturing, had always been accompanied by statistically significant and proportional reductions in photosynthesis; these were in turn correlated with increased PA. It was of interest therefore to know whether PA might have some direct effect on the photosynthetic apparatus which would in turn influence stomatal resistance and lead to the type of correlation between stomatal resistance and PA shown in Tables 3 and 4. Accordingly, some native PA extracted from plant material was applied to spinach leaf slices and caused a direct inhibition of photosynthesis as shown in Table 5. Non-stomatal inhibition of photosynthesis in excised leaves has been previously demonstrated (Kriedemann et al. 1975) and the present data (Table 5) add confirmation to these observations. Oxygen evolution showed 72% inhibition after 1 hour. By contrast, pretreatment with ABA (10^{-5}M) for 1 hour has no effect on O_2 evolution by this leaf slice system (Kriedemann et al. 1975).

TABLE 4.   Vine leaf response to sink removal after 4 days

Treatment	Stomatal Resistance (s cm^{-1})	Relative Leaf Water content* (%) (arcsin %)		Starch (% dry wt)	ABA (ng g^{-1} Fr. wt)	PA
Stem Intact						
Fruit present	1.41	92.8	2.67	19.9	67	16
Fruit removed	2.10	92.8	2.67	26.8	55	13
Stem Cinctured						
Fruit present	2.60	93.5	2.51	32.1	99	39
Fruit removed	7.14	94.2	2.38	40.0	110	75
LSD (P ≥ .05)	1.23		0.41	3.10	**	**

*Mean values from 6 replicate determinations;   **All replicates were bulked and analysis performed on duplicate subsamples.

At light saturation (660 W/m^2) O_2 evolution by control slices was equivalent to 40.8 ng CO_2/cm^2/s and compared favourably with the gas exchange rate of intact leaves, viz. 80-100 ng CO_2/cm^2/s.

TABLE 5. Phaseic acid effects on photosynthetic oxygen evolution by spinach leaf slices

| Time (h) & Treatment* | Rate (p mol s^{-1}) | | % Inhibition |
	Control	Phaseic Acid	
1 $\quad$ (140 Wm^{-2})	555	157	72
1.5 $\quad$ (660 Wm^{-2} saturating light)	1683	357	79
5.0 $\quad$ (140 Wm^{-2})	762	0	100
(Buffer removed, slices washed and resuspended in H_2O)			
5.5 $\quad$ (140 Wm^{-2}) Fresh Buffer No PA	480	15	96
7.0 $\quad$ (140 Wm^{-2}) Fresh Buffer No PA	393	7.3	98
22.5 $\quad$ (140 Wm^{-2}) Fresh Buffer No PA	160	0	100
22.6 $\quad$ Dark Respiration	155	115	25

*Tissue was transferred to cuvettes at time zero and incubated in buffer alone (50 mM PIPES at pH 5.6) or in the presence of PA (12 µg/ml) for 1 hour in light (140 Wm^{-2}, total energy)

Leaf slice photosynthesis (Table 5) was still totally inhibited 22.5 h after the initial treatment with PA. At that stage, control slices were still showing almost 30% of their original activity. Despite this total inhibition of photosynthesis due to PA, the respiratory activity of these treated slices held at 75% of control rates (bottom line Table 5). This effect helps to identify PA as a specific inhibitor of photosynthesis rather than a general metabolic poison.

DISCUSSION

Leaf gas exchange is clearly responsive to source-sink relationships where any form of modification such as organ removal or partial defoliation may influence both stomatal physiology and photosynthesis. The precise nature of this feedback system is obscure. Conceivably assimilate level could contribute towards control by processes analogous to mass action. In biochemical terms, key enzymes in photosynthetic pathways would be subject to end product inhibition while additional physiological processes could operate at the "whole chloroplast" level. For example, plastids might tend to "round up" due to accumulation of starch plus osmotically-active assimilate, and this change in their morphology would reduce their surface volume ratio with associated effects on both light interception and substrate exchange. In general, higher levels of leaf carbohydrate do appear to be associated with lower rates of assimilation and *vice versa* but such correlation does not necessarily imply causation (e.g. see Neales and Incoll 1968). Notwithstanding problems of compartmentation, there are notable exceptions, where higher levels of leaf sucrose accompany increased rates of photosynthesis (Thorne and Koller 1974). Conversely, photosynthetic rate and starch content may fall simultaneously following treatment with growth retardants (Marcelle et al. 1974). Moreover, the abbreviated time course of stomatal and photosynthetic response to manipulative treatments often contrasts with a more gradual adjustment of leaf carbohydrate. Humphries (1963) for example, demonstrated that primary leaves of *Phaseolus vulgaris* showed no variation in net assimilation rate despite steady accumulation of photosynthate. Consequently, some additional form of control is called for, and growth regulatory compounds could be instrumental.

Some developmental changes following photoperiodic induction or decapitation which enhance photosynthesis, have been attributed to altered distribution of growth substances including root-derived cytokinins (Beever and Woolhouse 1974, Meidner 1970), but anatomical changes that accompany such rejuvenation must be recognized. Photosynthetic response to decapitation in capsicum (see Results section) can be attributed to a greater depth of assimilatory tissue and the associated increase in chloroplast numbers per cell. Root-derived cytokinins may well have initiated these anatomical changes but in photosynthetic terms, their action was probably indirect.

On the other hand, short term responses that result in close coupling between source and sink, as manifest in *Phaseolus vulgaris*, have been attributed to auxin (Bidwell et al. 1968), and the apparent migration of certain ^{14}C-labelled metabolites from ^{14}C-sink to ^{12}C-source leaves in the present capsicum grafting experiment supports this possibility. In such cases, the sink's synthesis of growth substances might well fulfill the dual role of influencing assimilate distribution as well as promoting increased photosynthesis by source leaves.

Acting in concert with these substances that elicit short term positive responses (auxin) or others which foster a resurgence of synthetic events (cytokinins) we now envisage a further category of naturally occurring compounds that participate in the regulation of stomatal opening and photosynthetic fixation of CO_2. These would include ABA and PA and their effects would superimpose over the more progressive adjustments in rates of gas exchange already attributed to assimilate concentration within the photosynthetic apparatus. The data presented have implicated ABA and PA in source-sink feedback effects, and imply that fine control over gas exchange could be exerted by their combined action.

ACKNOWLEDGEMENTS

We are indebted to Ms. Edith Törökfalvy and Mr. Henk van Dijk for expert technical assistance, and to the Australian Government for financial support (Japan Science and Technology Award to Mitsumasa Satoh).

REFERENCES

Beever, J.E. and Woolhouse, H.W. (1974). Increased cytokinin export from the roots of *Perilla frutescens* following disbudding or floral induction. In: "Mechanisms of Regulation of Plant Growth" eds. R.L. Bieleski, A.R. Ferguson and M.M. Cresswell, Bull. 12, The Roy. Soc. N.Z. Wellington, pp. 681-686.

Bidwell, R.G.S., Levin, W.B. and Tamas, I.A. (1968). The effects of auxin on photosynthesis and respiration. In: "Biochemistry and Physiology of Plant Growth Substances" eds. F. Wightman and G. Setterfield, The Runge Press Ltd., Ottawa, Canada, pp. 361-376.

Bruinsma, J. (1961). A comment on the spectrophotometric determination of chlorophyll. *Biochim. Biophys. Acta 52*, 576-578.

Dubois, M., Gilles, K.A., Hamilton, J.K., Rebeis, P.A. and Smith, F. (1956). Colorimetric methods for determination of sugars and related substances. *Anal. Chem. 28*, 350-356.

Humphries, E.C. (1963). Dependence of net assimilation rate on root growth of isolated leaves. *Ann. Bot. 27*, 175-183.

Kriedemann, P.E., Loveys, B.R. and Downton, W.J.S. (1975). Internal control of stomatal physiology and photosynthesis. II. Photosynthetic responses to phaseic acid. *Aust. J. Plant Physiol. 2*, 553-567.

Kriedemann, P.E., Loveys, B.R., Fuller, G.L. and Leopold, A.C. (1972). Abscisic acid and stomatal regulation. *Plant Physiol. 49*, 842–847.

Loveys, B.R. and Kriedemann, P.E. (1974). Internal control of stomatal physiology and photosynthesis. I. Stomatal regulation and associated changes in endogenous levels of abscisic and phaseic acids. *Aust. J. Plant Physiol. 1*, 407–415.

Marcelle, R., Clijsters, H., Oben, G., Bronchart, R. and Michel, J.M. (1974). Effects of CCC and GA on photosynthesis of primary bean leaves. In: "Plant Growth Substances 1973" ed. Y. Sumiki, Hirokawa Publishing Co. Inc. Tokyo, pp. 1169–1174.

Meidner, H. (1970). Effects of photoperiodic induction and debudding in *Xanthium pennsylvanicum* and of partial defoliation in *Phaseolus vulgaris* on rates of net photosynthesis and stomatal conductances. *J. Exp. Bot. 21*, 164–169.

Neales, T.F. and Incoll, L.D. (1968). The control of leaf photosynthesis rate by the level of assimilate concentration in the leaf: A review of the hypothesis. *Bot. Rev. 34*, 107–125.

Possingham, J.V. and Smith, Jennifer W. (1972). Factors affecting chloroplast replication in spinach. *J. Exp. Bot. 23*, 1050–1059.

Satoh, M. and Hazama, K. (1971). Studies on photosynthesis and translocation of photosynthate in mulberry trees. I. Photosynthetic rate of remained leaves after shoot pruning. *Proc. Crop Sci. Soc. Japan 40*, 7–11.

Satoh, M. and Ohyama, K. (1971). Studies on photosynthesis and translocation of photosynthate in mulberry trees. II. The effect of shoot pruning and lateral buds removal on the photosynthetic rate of remained leaves. *Proc. Crop Sci. Soc. Japan 40*, 525–530.

Sweet, G.B. and Wareing, P.F. (1966). Role of plant growth in regulating photosynthesis. *Nature 210*, 77–79.

Thorne, J.H. and Koller, H.R. (1974). Influence of assimilate demand on photosynthesis, duffusive resistances, translocation and cabohydrate levels of soybean leaves. *Plant Physiol. 54*, 201–207.

Wareing, P.F., Khalifa, M.M. and Treharne, K.J. (1968). Rate-limiting processes in photosynthesis at saturating light intensities. *Nature 220*, 453–457.

CHAPTER 36

Implications for Plant Growth of the Transport of Regulatory Compounds in Phloem and Xylem

R.W. King

CSIRO, Division of Plant Industry, Canberra, A.C.T. Australia

INTRODUCTION

Over 30 years ago Chibnall (1939) suggested that roots synthesized compounds which were essential for shoot growth. It was implied that these factors were transported over considerable distances between sites of production and action. Basipetal transport from the shoot apex of factors controlling lateral bud development was also well documented at about this time (see Phillips 1975). However, most attention was focussed, subsequently, on regulation of growth by substances produced close to the responsive tissue and for which transport from cell-to-cell could be studied in excised tissue pieces. With the exception of the floral stimulus for which long distance transport has been accepted for some 40 years (see Evans 1971), only recently has interest been renewed in the question of vascular transport of growth substances.

GROWTH REGULATORS IN THE PHLOEM AND XYLEM

All the known classes of growth substances, apart from ethylene have been found in xylem (see Table 1) and phloem (see Table 2). Additional reports of hormones in phloem or xylem are provided in the review by Crafts and Crisp (1971).

A wide range of species, both woody and herbaceous, are covered by reports of hormones in phloem and xylem sap (see Crafts and Crisp 1971) and this implies some physiological significance of their vascular transport. In fact, even at the apparently low content of gibberellin in xylem sap in grapes (Table 1), Skene (1967) calculated that enough GA could pass into leaves by transpiration to account for their total content of gibberellin. Unfortunately such calculations may introduce many errors including problems of technique of collection (see Pate this volume); exudate sap

TABLE 1. Content of hormones in xylem sap

Compound	Concentration µg/l	Environmental response	Reference
Abscisic Acid	10 - 50		Lenton et al. 1968
	0.05 → 1.6*	low root temp.	Atkin et al. 1973
	<5 → 40	water stress	Hoad 1975
	40	spring sap flow	Harrison and Saunders 1975
Auxin	3*		Sheldrake 1973
	3		Hall and Medlow 1974
Cytokinin	9 - 17*		Kende 1965
	3 → 30*	plant age	Sitton et al. 1967
	8 → 1*	low root temp.	Atkin et al. 1973
	5 → 150*	fruit develop-ment	Beever and Woolhouse 1973
	40 → 180*	photoperiod	Henson and Wareing 1974
Gibberellin	25*		Carr et al. 1964
	0.07 - 0.3*		Skene 1967
	0.01 - 0.02*		Luckwill and Whyte 1968
	20 → 2*	low root temp.	Atkin et al. 1973

* Identification based on "bioassay"
Environmentally increased content indicated by →

concentrations and fluxes may be at variance with normal transpiration fluxes; inhibitors and promotors may interfere with a bioassay on unpurified samples; and the transport form of a compound may not have been the form assayed. Cytokinins for instance may be transported in the phloem as the physiologically less active nucleotide and this may explain the high levels of activity reported by Vonk (1974) for *Yucca*. By contrast, auxin if transported in phloem as the conjugate IAA-aspartate (Hoad et al. 1971) might be lost during sample purification.

TABLE 2.   Content of hormones in phloem sap

Compound	Concentration µg/l	Environ- mental response	Reference
Abscisic Acid	100 - 300		Lenton et al. 1968
	60 → 1500	Water stress	Hoad 1973
Auxin	10*		Hall and Baker 1972
	4		Hall and Medlow 1974
Cytokinins	10*		Hall and Baker 1972
	27 → 184*	Photoperiod, flowering	Phillips and Cleland 1972
	520*		Vonk 1974
Gibberellin	2*		Hall and Baker 1972

* and → as in Table 1.

MOVEMENT OF GROWTH SUBSTANCES IN PHLOEM AND XYLEM

Not only are growth substances present in vascular fluid but they are transported over long distances. Two types of experiments have confirmed this supposition. Firstly, when radioactive growth regulators are applied elsewhere in the plant they soon appear in phloem and xylem exudates (see Eschrich 1968). Secondly, radioactive growth regulators move in an essentially identical pattern to the flow of assimilate or water in phloem or xylem, respectively. For example, growth regulators applied to roots or to the base of cut stems move with the transpiration stream into terminal regions such as the lobes of leaves (Couillerot and Bonnemain 1975) or awns and grain of barley ears (see Michael and Seiler-Kelbitsch 1972). In the phloem, radioactively labelled growth sub- stances move from leaves at speeds similar to those of assimilate (Bonnemain 1971, Morris et al. 1973, Goldsmith et. al. 1974). Also autoradiographic examination has shown that activity is localized over sieve tubes (Goldsmith et al. 1974). Moreover, the pattern of movement of labelled growth substances is similar to that of assimilate exported via the phloem to various actively growing regions or sinks (McComb 1964; Goldsmith et al. 1974). Whether at lower, endogenous, concen-

trations growth substances are transported so simply from
leaves in the phloem or from roots in the xylem is, however,
a question yet to be answered. Similar criticism applies to
evidence of the ready lateral transport of applied growth substances from xylem to phloem (Bowen and Wareing 1969) and
across the bark into phloem of willow (Hoad et al. 1971). In
fact, too ready an interchange and circulation of growth substances within xylem and phloem poses considerable difficulties
for any simple theory of correlative control of plant growth
by compounds transported over long distances in vascular
tissue. In this regard it would be valuable to have evidence
that a compound accumulates in a receptor region and controls
growth there, whether it is exogenously or endogenously supplied.
Attention must also be focussed on identification of the site(s)
of synthesis of the various growth substances. In general,
such evidence is scant although, if synthesis does not occur
throughout the plant, then it appears to be most important at
sites close to the region of physiological response to a growth
regulator. To best highlight these points, subsequently, five
aspects of hormonal control of growth are discussed: (1) cell-
to-cell and vascular transport of auxin; (2) environment
responses of transport; (3) sites of synthesis and transport
requirements; (4) movement of cytokinins in xylem and, (5)
movement of the floral stimulus in the phloem.

Phloem transport of floral stimulus. - The simplicity
with which growth of a plant can be controlled by unidirect-
ional channelling of the transport of a regulatory factor in
the phloem is illustrated from studies on the movement of the
floral stimulus. Despite its elusive nature, there is little
doubt that this stimulus is produced in leaves and moves to
the apex in the phloem (see summary in Evans 1971). Living
tissue is required and lateral transport is minimal.

Also, in *Pharbitis nil*, for example, the speed of movement of floral stimulus (24-37 cm/h) may be as rapid as
assimilate (33-37 cm/h) (King et al. 1968). Furthermore, in
Perilla, when an induced source-leaf is grafted to a vegetative receptor plant the pattern of flowering matches the
pattern of supply of photosynthetic assimilate from the donor
leaf (Fig. 1a,b). Moreover, a non-induced leaf prevents
flowering of specific shoots by providing an alternative
supply of stimulus-free assimilate with the resultant reduction of assimilate import from the donor leaf (Fig. 1b,c).

One problem of transport of the floral stimulus has
arisen in studies on the speed of translocation. Although in
the short-day plant *Pharbitis* the floral stimulus moves with
much the same velocity as assimilate (see above), in *Lolium*
a long-day plant the velocity of assimilate movement (77-105
cm/h) is obviously different from that of floral stimulus

(2 cm/h) (Evans and Wardlaw 1966). Maybe the stimuli differ between long- and short-day plants, the transport paths could differ or species differences in phloem characteristics such as in lateral loss or the chromatograph properties of the sieve tube could give rise to such disparate results.

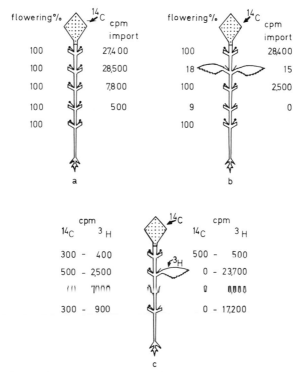

Fig. 1. - *Flowering response and assimilate import by axillary shoots of* Perilla *induced to flower by a grafted donor leaf (◈) as affected by the presence of non-induced leaves. Flowering shoots (↘); vegetative shoots (＼). (After King and Zeevaart 1973).*

Xylem transport of cytokinin. - It was almost 20 years after Chibnall's (1939) report of the control of leaf senescence by roots that Richmond and Lang (1957) reported that kinetin delayed senescence of detached leaves. However, despite the many reports which followed of cytokinins in xylem sap (see Table 1), the evidence is still only circumstantial that cytokinins are synthesized in roots and that this source is important for growth of intact plants.

The control of senescence by root produced factors has been the most widely studied response (see Woolhouse 1974).

The findings of Wareing et al. (1968) illustrate one such example of the effect in whole plants of various combinations of roots and shoots on two indicators of senescence .... leaf photosynthesis and the activity of a photosynthetic enzyme. They found that photosynthesis was reduced by root removal but that this effect could be prevented by applying kinetin to the leaf. Photosynthesis was also enhanced by defoliation, a cytokinin sparing treatment which would allow more of a limited root pool of cytokinin to pass to the remaining leaves. The response of a senescent leaf to removal of other leaves, may, in fact, be extremely rapid. For example, Callow reported increases in synthesis of chloroplast ribosomal RNA within 12 h of removing other leaves (see Woolhouse 1974). Over the following two days the chlorophyll content and photosynthetic rate also increased dramatically.

Other root controlled responses which can be mimicked by cytokinin application include axillary shoot growth and shoot form - whether plagio- or diageotropic (see Phillips 1975), or the ability of growing shoots to flower or continue vegetative growth (Migniac 1971). Reduced growth under unfavourable environmental conditions such as low temperature, salinity stress and root flooding may also result from a decrease in cytokinin and gibberellin production by roots (see Carr and Roid 1968; Atkin et al. 1973). A decrease in the cytokinin concentration in root exudate might also be involved in the senescence which occurs at late stages of fruit development of annual plants (see Sitton et al. 1967; Beever and Woolhouse 1973).

Whether xylem transport of growth substances from root to shoot is normally important for shoot growth is, however, far from settled. Especially puzzling is the evidence that reduced root growth associated with early stages of flower development (see Woolhouse 1974) or with growth retardant treatment (Skene 1970) is associated with an increased flow of cytokinin from roots. The converse could have been predicted. Certainly low temperature treatment or flooding depresses root exudation of cytokinins (Table 1 and Carr and Reid 1968).

Another difficulty is that not only cytokinins but also gibberellins, abscisic acid, and auxin may be transported from roots in the xylem (see Table 1) and all these compounds have been implicated in the control of leaf senescence in one or other species (see Woolhouse 1974).

SYNTHESIS SITES AND TRANSPORT REQUIREMENTS

The simple model presented above of production in roots of cytokinins and their movement in the xylem to shoots provides an apparently tidy example of correlative control of plant growth by a transported hormone. However, the root system may not be the sole site of cytokinin synthesis....if at all. Shoots, young fruits, seeds (see Kende 1971) and leaves (Hewett and Wareing 1973) may also synthesize cytokinins. A similar diversity of sites of synthesis is suggested for auxin, abscisic acid and gibberellins (see references in Wightman and Setterfield 1968, Milborrow 1974). Thus, rather than suggesting their import in xylem or phloem, *in situ* synthesis of hormones may explain much of the evidence of variation in the level of a growth substance during growth of an organ, or under stress conditions.

Accumulation of abscisic acid in wheat grains 25 to 40 days after anthesis provides one such example (King 1976). Vascular import of this abscisic acid is not essential since the isolated grain can synthesize abscisic acid from added mevalonic acid (Milborrow and Robinson 1973). In fact import in the xylem is unlikely since the major transpiring organs of the wheat ear, the awns, contain only about 5% of the total AbA in the whole ear and yet, together with the rachis, they account for 32% of the dry weight of the ear (King 1976). Phloem import also seems unlikely since the timing of maximum grain growth and, hence, of maximum assimilate import is apparently earlier (5-10 days) than the time of the rapid increase in abscisic acid.

There is also a considerable *in situ* component in the pool of cytokinins and gibberellins in developing pea seeds. In excised cultured pea pods seed growth continues and the content of gibberellins may increase 300 fold in 10 days (Baldev et al. 1965) while cytokinin activity increases 7 fold in 12 days (Hahn et al. 1974). Clearly allowance must be made for contributions by localized sources of growth substances when assessing physiological significance of vascular sources of a compound.

Although it may be of little value to calculate fluxes of a growth substance into a shoot without also measuring *in situ* synthesis, by contrast, the extent to which variation in transpiration rate influences shoot growth may provide some indication of the importance of xylem fluxes of growth substances. At present there is conflict in the available data. On the one hand O'Leary and Knecht (1971) found that maintaining bean plants at a relative humidity (RH) of 95-100% over 3 weeks reduced water use by 70% compared to plants maintained at 70-75% RH. Nevertheless growth and yield were not altered significantly. Possibly there was a compensating increase in the concentration of hormones in the transpiration stream. Alternatively, growth could have been maintained even with a three-fold decrease in the delivery of hormones

to the shoots. However, in barley there appears to be a relationship between grain cytokinin, ear transpiration and
grain growth. For example Michael and Seiler-Kelbitsch (1972)
have reported that grain growth in barley cultivars correlates
with the endogenous level of cytokinin in the grain. Moreover,
and most significantly, reduced cytokinin production in water-
logged roots reduced both growth and the cytokinin content of
grain. Similarly, removing the awn, the primary site of
water loss from the ear, reduced both growth and the cytokinin content of the grain. Clearly if sap-borne hormones play
a role in the regulation of shoot metabolism and growth it
would be of considerable value to obtain further information
on responses to changes in transpiration.

ENVIRONMENT AND TRANSPORT RESPONSES

Transport is clearly implicated where growth of an organ
is influenced by environmental treatments imposed elsewhere
on the plant. Indeed, to establish a direct relationship between a growth substance and growth this approach of manipulating the environment is, in theory, most likely to succeed.

Slowing of shoot growth when roots are exposed to flooding or to low temperatures provides one example. In the xylem
sap there is an associated decrease in the level of growth
promotors such as gibberellins and cytokinins and an increase
in the inhibitor, abscisic acid (see Carr and Reid 1968;
Atkin et al. 1973, Table 1). However these data, as such, are
somewhat inconclusive. It has not been shown that the changes
in hormones transported from roots preceded growth responses.
Furthermore, despite their magnitude (5 to 30 fold) there is
no indication that such changes need be of importance for
growth.

Growth of stressed plants provides another suggestive
illustration of a relationship to the level of a growth substance....abscisic acid. From studies on responses of wilted
detached organs (see Milborrow 1974) it is known that the
leaves are a major site of stress-induced synthesis of abscisic
acid. Thus the finding in *Lolium temulentum* that a brief 8
hour water stress leads to dramatic increase in the level of
abscisic acid in the leaf some 8-12 hours before there is an
increase at the apex (Fig. 2) implies transport from leaf to
apex of abscisic acid. Furthermore, the increase in the apex
occurs well after the plants have been rewatered and regained
full turgor at which time the content of abscisic acid in the
leaf has also dropped. Thus it is unlikely that stress was
acting directly on the apex. Stress-imposed inhibition of
flowering in *Lolium* can therefore be related to the timing of

arrival of abscisic acid at the apex which in turn correlates
with the time when flowering is most sensitive to applied
abscisic acid (King and Evans unpublished).

Obviously in this work with *Lolium* a more detailed analysis of localized synthesis of abscisic acid and of its transport is required. However the findings of Hoad (1973, 1975)
are relevant. He found a 5-13 fold increase in 24 hour in the
content of abscisic acid in the leaves and shoot tip of water
stressed plants of *Ricinus* (see Table 3). The large increase
in shoot tips only occurred if they were allowed to wilt on
the plant. Thus, abscisic acid produced in the leaves was
probably transported to the shoot tip. It is significant that
Hoad also found increases in the content of abscisic acid in
phloem sap of *Ricinus* (Table 3) and this may have led subsequently to an increase in the xylem apparently by recirculation
from the phloem (Hoad 1975).

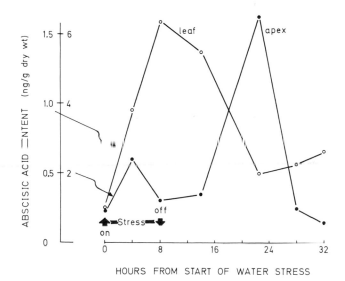

Fig. 2. - Increase in the abscisic acid content of
the leaf and apex of Lolium *plants exposed*
to an 8 h water stress using polyethylene
glycol. (King and Evans unpublished).

Unfortunately, the evidence summarized above fails to
establish final causality in the control of growth. Brief
stress applied locally to root or shoot should be further
examined to develop a clear time sequence of changes to confirm hormone production in a source, its transport, accumulation
and control of receptor organ growth.

Photoperiod is another environmental treatment which influences the levels of transported hormones (Tables 1 and 2).
On exposure to long days leaves apparently export more gibberellins in the phloem (Hoad and Bowen 1968). In short days
there may be an increase in the phloem content of cytokinin
(Phillips and Cleland 1972) and of an inhibitor which may be
abscisic acid (Hoad 1967). The level of cytokinin in root
sap, however, may decline dramatically on transferring plants
to short days (Henson and Wareing 1974). As yet there is no
evidence that these responses are of importance to plant growth
although it was hoped that the increase in inhibitor content
in phloem sap would explain the induction of dormancy on exposure of woody species to short days. The rapid decline in
cytokinin content of xylem sap of the short day plant *Xanthium
strumarium* is also not likely to be of significance to flowering which can occur in the absence of roots (Chailakyan 1958).

TABLE 3.   The effect of a 24 h water stress on the content
of abscisic acid in leaves, phloem and shoot tips
of *Ricinus* (after Hoad 1973)

Tissue	Abscisic acid content		
	Control	Stressed	Increase
Leaf (ng/gm fresh wt)	31	410	13.2
Phloem (ng/ml sap)	105	1019	9.7
Shoot tip attached (ng/gm fresh wt)	131	830	6.3
Shoot tip detached (ng/gm fresh wt)	197	344	1.7

CELL-TO-CELL AND VASCULAR TRANSPORT OF AUXIN

As well as their transport in phloem or xylem some growth
substances and, in particular auxin, move from cell-to-cell
in living non-vascular tissue (see references in Wightman and
Setterfield 1968). The speed of movement is considerably
slower than that of phloem transport of assimilate (1-2 cm/h
vs. 20-200 cm/h). Also, movement may be polar down the stem
from the young growing leaves of the shoot tip.... a transport
pattern not expected of movement in xylem or phloem.
Recently Phillips (1975) has reviewed the importance of
polar, cell-to-cell movement of auxin in the correlative control of plant growth and especially in apical dominance. Of
particular significance is the fact that transport is in living cells since stem girdling blocks transmission of the
correlative signal in apical dominance and allows lateral bud
outgrowth (see Phillips 1975). Moreover this transport is

not in living phloem. For instance, Morris and Kadir (1972)
found that in pea only an insignificant amount of radioactivity was recovered from the honeydew of aphids feeding on
plants which were transporting auxin basipetally from the
shoot tip. Aphid honeydew did, however, contain auxin if it
had been applied to a leaf. In addition, auxin transport from
the shoot tip could be blocked with 2,3,5-triiodobenzoic acid
(TIBA) (Morris et al. 1973) yet this compound had no effect on
phloem transport of auxin or assimilate (Morris et al. 1973;
Goldsmith et al. 1974). Such TIBA treatment also results in
the release of lateral buds from apical dominance (Kuse 1953)
the inhibition of auxin-directed transport of metabolites
(Davies and Wareing 1965) and alters root development (McDavid
et al. 1972). There can be little doubt, therefore, that
cell-to-cell transport of auxin in non-vascular tissue is of
paramount significance in the correlative control of growth
in plants.

Possibly the auxin found in xylem and phloem (Tables 1
and 2) is unimportant in the correlative control of plant
growth. However, without analysis of fluxes into receptor
regions and some knowledge of the dynamic balance of auxin at
such a site, this latter conclusion is too extreme. Nevertheless, it is a suggestion favoured by the evidence of Morris
and Kadir (1972) of a considerable isolation and minimal lateral exchange between vascular and non-vascular transport
paths for auxin. This is despite the fact that applied auxin
may move laterally between xylom and phloem (Bowen and Wareing
1969) and from bark to phloem (Hoad et al. 1971).

Further examination of these distinctive pathways of vascular and non-vascular auxin transport is urgently required
particularly with regard to the form and quantities of hormones
transported in all pathways. In addition, since TIBA can be
transported throughout the plant (Ohki and McBride 1973), it
interferes with apical growth (White and Hillman 1972) and it
can prevent loading of auxin into phloem (see Phillips 1975),
other methods of blocking cell-to-cell movement should be
utilized. Of particular value might be low temperature or
anaerobiosis to block auxin (Gregory and Hancock 1955) but not
assimilate transport (see Wardlaw, this volume).

CONCLUDING COMMENTS

Although there is much suggestive evidence it is not yet
possible to conclude that vascular transport of the known
plant growth substances is indisputably involved in the control of plant growth. The floral stimulus, by contrast, is
clearly transported in the phloem from leaf to shoot apex
where it triggers a dramatic morphological change. This

contrast between floral stimulus and the four major classes
of growth substances is, however, instructive in itself. Not
only is the site of synthesis of floral stimulus localized,
there is a precise environmental control over its production
and there is evidence of minimal lateral transfer and recirculation of the stimulus during its transport in the phloem
(Fig. 1). Clearly identification of source and sink is as
important as the analysis of vascular sap.

Accepting the inevitable, however, most compounds will
be synthesized in many organs depending on their stage of
development. Moreover, recycling will occur in xylem and
phloem. Therefore the best approach available might be to
utilize labelled precursors of the compound in question. Such
studies would be less likely to suffer from the criticism
which arises when use is made of exogenous application of a
growth substance.

When synthesis occurs in many sites it is also a problem
to estimate the significance of each site even when transported hormones can be monitored. In this regard the use of
specific inhibitors of hormone synthesis would be of value.
Autonomy of synthesis of a compound *in situ* could also be examined more thoroughly using organ culture techniques such
as applied by Baldev et al. (1965) and Hahn et al. (1974) to
the growth of excised pea seeds. If, however, the contribution of hormones produced elsewhere in the plant was minute
but continuous as in the possible "drip feeding" of cytokinins into leaves, as suggested by Carr and Reid (1968),
then no culture technique would necessarily provide the
correct answers. Moreover, not only the rate of supply but
its timing could be important. If for example cytokinins
and gibberellins in the xylem play a role in regulation of
senescence this may be of significance only for mature organs.
The young leaf may be able to synthesize these compounds itself (see above).

By contrast to cytokinins and gibberellins, for auxin
it might be reasonable to suggest that its transport in the
phloem and xylem has no significance for growth. Of the
studies discussed above on the movement of radioactive auxin
in phloem none reported changes in growth of receptor organs.
It is clear, nevertheless, that polar transport of auxin from
the shoot tip is of considerable importance in the control
of growth. There also appears to be anatomical isolation of
the polar from the vascular pathway (see Morris and Kadir
1972; Morris et al. 1973) and it would be interesting to
know why this is so. One serious problem with polar auxin
movement is, however, the lack of a defined receptor. Although bud growth may be under the control of apical dominance

it does not seem that auxin moves in quantity into inhibited buds (Phillips 1975).

Unfortunately the apparent ready recirculation of growth substances between xylem and phloem and their possible transport in an additional non-vascular path make it hard to see how a compound would not eventually reach the "forbidden fruit". Measurements of fluxes coupled with treatments such as blocking cell-to-cell pathways or varying transpirational demand nevertheless offer considerable potential in assessing the significance of transport. Further understanding of the role of hormone transport in the control of growth will not come, however, from examination of the transport path alone but from studies of both sources and sinks. Then it may be possible to explain how plant growth regulators control the flow of metabolites and hence control the growth of a plant.

REFERENCES

Atkin, R.K., Barton, G.E. and Robinson, D.K. (1973). Effect of root growing temperature on growth substances in xylem exudate of *Zea mays*. *J. Exp. Bot. 24*, 475-487.

Baldev, B., Lang, A. and Agatep, A.O. (1965). Gibberellin production in pea seeds developing in excised pods: effect of growth retardant AMO-1618. *Science 147*, 155-157.

Beever, J.E. and Woolhouse, H.W. (1973). Increased cytokinin from root systems of *Perilla frutescens* and flower and fruit development. *Nature 246*, 31-32.

Bonnemain, J.L. (1971). Transport et distribution des traceurs après application de AIA-2-^{14}C sur les feuilles de *Vicia faba*. *C.R. Acad. Sci. Paris 273*, 1699-1702.

Bowen, M.R. and Wareing, P.F. (1969). The interchange of ^{14}C kinetin and ^{14}C gibberellic acid between the bark and xylem of willow. *Planta (Berl.) 89*, 108-125.

Carr, D.J. and Reid, D.M. (1968). The physiological significance of the synthesis of hormones in roots and their export to the shoot system. In: The biochemistry and physiology of plant growth substances. ed. F. Wightman and G. Setterfield, The Runge Press, Ottawa, Canada. pp. 1169-1185.

Carr, D.J., Reid, D.M. and Skene, K.G.M. (1964). Supply of gibberellins from the root to the shoot. *Planta (Berl.) 63*, 382-392.

Chibnall, A.C. (1939). Protein metabolism in the plants. Yale Univ. Press. New Haven. pp. 255-266.

Chailakyan, M.Kh. (1958). The photoperiodic receptivity of isolated leaves. *Soviet Plant Physiol. 118*, 9-12.

Couillerot, J.P. and Bonnemain, J.L. (1975). Transport et devenir des molécules marquées après l'application d'acide gibbérellique - ^{14}C sur les jeunes feuilles de tomate. *C.R. Acad. Sci. Paris 28*,1453-1456.

Crafts, A.S. and Crisp, C.E. (1971). Phloem transport in plants. W.H. Freeman, San Francisco.

Davies, C.R. and Wareing, P.F. (1965). Auxin-directed transport of radio-phosphorus in stems. *Planta (Berl.) 65*, 139-156.

Eschrich, W. (1968). Translokation radioactiv markierter indolyl-3-essigsaure in Siebröhren von *Vicia faba*. *Planta (Berl.) 78*, 144-157.

Evans, L.T. (1971). Flower induction and the florigen concept. *Ann. Rev. Plant Physiol. 22*, 365-394.

Evans, L.T. and Wardlaw, I.F. (1966). Independent translocation of ^{14}C-labelled assimilates and of the floral stimulus in *Lolium temulentum*. *Planta (Berl.) 68*, 310-326.

Goldsmith, M.H.M., Cataldo, D.A., Karn, J., Brenneman, T. and Trip, P. (1974). The rapid non-polar transport of auxin in the phloem of intact *Coleus* plants. *Planta (Berl.) 116*, 301-317.

Gregory, F.G. and Hancock, C.R. (1955). The rate of transport of natural auxin in woody shoots. *Ann. 19*, 451-465.

Hahn, H., de Zacks, R. and Kende, H. (1974). Cytokinin formation in pea seeds. *Naturwissenschaften 61*, 170.

Hall, S.M. and Baker, D.A. (1972). The chemical composition of *Ricinus* phloem exudate. *Planta (Berl.) 106*, 131-140.

Hall, S.M. and Medlow, G.C. (1974). Identification of IAA in phloem and root pressure saps of *Ricinus communis* L. by mass spectrometry. *Planta (Berl.) 119*, 257-261.

Harrison, M.A. and Saunders, P.F. (1975). The abscisic acid content of dormant birch buds. *Planta (Berl.) 123*, 291-298.

Henson, I.E. and Wareing, P.F. (1974). Cytokinins in *Xanthium strumarium*: a rapid response to short day treatment. *Physiol. Plant. 32*, 185-187.

Hewett, E.W. and Wareing, P.F. (1973). Cytokinins in *Populus* x *robusta* (Schneid): light effects on endogenous levels. *Planta (Berl.) 114*, 119-129.

Hoad, G.V. (1967). (+)-abscisin II, (+)-dormin in phloem exudate of willow. *Life Sci. 6*, 1113-1118.

Hoad, G.V. (1973). Effect of moisture stress on abscisic acid levels in *Ricinus communis* L. with particular reference to phloem exudate. *Planta (Berl.) 113*, 367-372.

Hoad, G.V. (1975). Effect of osmotic stress on abscisic acid
    levels in xylem sap of sunflower (*Helianthus annuus* L.).
    Planta (Berl.) 124, 25-29.
Hoad, G.V. and Bowen, M.R. (1968). Evidence for gibberellin-
    like substances in phloem exudate of higher plants.
    Planta (Berl.) 82, 22-32.
Hoad, G.V., Hillman, S.K. and Wareing, P.F. (1971). Studies
    on the movement of indole auxins in willow (*Salix
    viminalis* L.). *Planta (Berl.) 99,* 73-88.
Hocking, T.J., Hillman, J.R. and Wilkins, M.B. (1972). Move-
    ment of abscisic acid in *Phaseolus vulgaris* plants.
    Nature 235, 124-125.
Kende, H. (1965). Kinetin-like factors in the root exudate
    of sunflowers. *Proc. Nat. Acad. Sci. 53,* 1302-1307.
Kende, H. (1971). The cytokinins. *Int. Rev. Cytol. 31,* 301-
    338.
King R.W. (1976). Abscisic acid in developing wheat grains
    and its relationship to grain growth and maturation.
    Planta (Berl.) (in press).
King, R.W. and Zeevaart, J.A.D. (1973). Floral stimulus
    movement in *Perilla* and flower inhibition caused by non-
    induced leaves. *Plant Physiol. 51,* 727-738.
King, R.W., Evans, L.T. and Wardlaw, I.F. (1968). Transloca-
    tion of the floral stimulus in *Pharbitis nil* in relation
    to that of assimilates. *Z. Pflanzenphysiol. 59,* 377-388.
Kuoo, C. (1953). Effect of 2,3,5 tri indobenzoic acid on the
    growth of lateral bud and on tropism of petiole. *Mem.
    Coll. Sci. Univ. Kyoto Ser.B. 20,* 3-15.
Lenton, J.R., Bowen, M.R. and Saunders, P.F. (1968). Detect-
    ion of abscisic acid in xylem sap of willow by gas-
    liquid chromatography. *Nature 220,* 86.
Luckwill, L.W. and Whyte, P. (1968). Hormones in the xylem
    sap of apple trees. In: *Plant Growth Regulators Soc.
    Chem. Ind. (Lond.) Monogr. 31,* 87-101.
McCombe, A.J. (1964). The stability and movement of gibber-
    ellic acid in pea seedlings. *Ann. Bot. 28,* 669-687.
McDavid, C.R., Sagar, G.R. and Marshall, C. (1972). The
    effect of auxin from the shoot on root development in
    Pisum sativum L. *New Phytol. 71,* 1027-1032.
Michael, G. and Seiler-Kelbitsch, H. (1972). Cytokinin con-
    tent and kernel size in barley grains as affected by
    environmental and genetic factors. *Crop Sci. 12,* 162-
    165.
Migniac, E. (1971). Influence des racines sur le développe-
    ment végétatif on floral des bourgeons cotylédonaires
    chez le *Scrofularia arguta:* role possible des cytokin-
    ines. *Physiol. Plant 25,* 234-239.

Milborrow, B.V. (1974). The chemistry and physiology of abscisic acid. *Ann. Rev. Plant Physiol. 25*, 259–307.

Milborrow, B.V. and Robinson, D.R. (1973). Factors affecting the biosynthesis of abscisic acid. *J. Exp. Bot. 24*, 537–548.

Morris, D.A. and Kadir, G.O. (1972). Pathways of auxin transport in the intact pea seedling (*Pisum sativum* L.). *Planta (Berl.) 107*, 171–182.

Morris, D.A., Kadir, G.O. and Berry, A.J. (1973). Auxin transport in intact pea seedlings (*Pisum sativum* L.): inhibition of transport by 2,3,5-triiodobenzoic acid. *Planta (Berl.) 110*, 173–182.

Ohki, K. and McBride, L.J. (1973). Deposition, retention and translocation of 2,3,5-triiodobenzoic acid applied to soybeans. *Crop Sci. 13*, 23–26.

O'Leary, J.W. and Knecht, G.W. (1971). The effect of relative humidity on growth, yield and water consumption of bean plants. *J. Amer. Soc. Hort. Sci. 96*, 263–265.

Phillips, I.D.J. (1975). Apical dominance. *Ann. Rev. Plant Physiol. 26*, 341–367.

Phillips, D.A. and Cleland, C.F. (1972). Cytokinin activity from phloem sap of *Xanthium strumarium* L. *Planta (Berl.) 102*, 173–178.

Richmond, A.E. and Lang, A. (1957). Effect of kinetin on protein content and survival of detached *Xanthium* leaves. *Science 125*, 650–651.

Sheldrake, A.R. (1973). Do coleoptile tips produce auxin? *New Phytol. 72*, 433–447.

Sitton, D., Itai, C. and Kende, H. (1967). Decreased cytokinin production in the roots as a factor in shoot senescence. *Planta (Berl.) 73*, 296–300.

Skene, K.G.M. (1967). Gibberellin-like substances in root exudate of *Vitis vinifera*. *Planta (Berl.) 74*, 250–262.

Skene, K.G.M. (1970). The relationship between the effects of CCC on root growth and cytokinins in the bleeding sap of *Vitis vinifera* L. *J. Exp. Bot. 21*, 418–431.

Vonk, C.R. (1974). Studies on phloem exudation from *Yucca flaccida* Haw. XIII. Evidence for the occurrence of a cytokinin nucleotide in the exudate. *Acta. Bot. Neerl. 23*, 541–548.

Wareing, P.F., Khalifa, M.M. and Treharne, K.J. (1968). Rate-limiting processes in photosynthesis at saturating light intensities. *Nature 220*, 453–457.

White, J.C. and Hillman, J.R. (1972). On the use of Morphactin and triiodobenzoic acid in apical dominance studies. *Planta (Berl.) 107*, 257–260.

Wightman, F.G. and Setterfield, G. (1968). Biochemistry and Physiology of Plant Growth Substances. The Runge Press Ltd. Ottawa.

Woolhouse, H.W. (1974). Longevity and senescence in plants. *Sci. Prog. Oxf.* 61, 123-147.

Chapter 37

Hormone-directed Transport of Metabolites

J.W. Patrick
Department of Biological Sciences, University of Newcastle,
N.S.W. 2308, Australia

INTRODUCTION

The concept that plant hormones may influence metabolite transport was introduced by Went (1939) in his "nutrient – diversion theory" to explain apical dominance. Subsequently, the concept has been offered as an explanation for apical control of leaf senescence (Wareing and Seth 1967) and patterns of photosynthate movement within plants (Thrower 1967). Some recent reviews covering the phenomenon of hormone-directed transport are those by Moorby (1968), Peel (1974) and Phillips (1975).

In this review, the term "hormone-directed transport" has been used to describe the observed affects of plant hormones on metabolite mobilization that are considered to involve long-distance transport in the phloem. Effects of hormones on short-distance (cell to cell) transfer will be discussed only in relation to events involving long-distance transport. Furthermore, hormonal regulation of metabolite transport in leaves *per se* (c.f. Mothes and Engelbrecht 1961) will not be considered.

The first part of the review summarizes some of the evidence that purports to demonstrate that plant hormones influence metabolite transport and describes what are considered to be the more pertinent features of the phenomenon. The remainder of the review forms an attempt to analyse the possible ways in which hormones may act on metabolite transport.

RELATIONSHIP BETWEEN PLANT HORMONES AND METABOLITE MOVEMENT

INTACT PLANT OBSERVATIONS

Studies exploring the relationship between endogenous hormonal levels of and dry matter accretion by sinks, such as fruits, have yielded little evidence to suggest that these two factors are closely related (Nitsch 1970). Even in those cases where a close correlation appears to exist (c.f. Eeuwens and Schwabe 1975), closer scrutiny, using the more reliable techniques of combined gas chromatography and mass spectrometry, suggests that these apparent correlations found by using biological assays may prove to be oversimplifications of both the qualitative and quantitative hormonal content of the tissues (e.g. Frydman et al. 1974). However further work in this area may profit from considerations of hormonal compartmentalization within the sink tissues, the relative metabolite-mobilizing potencies of the various hormone fractions and changes in the degree of competition for assimilates during plant ontogeny.

Further support for a hormonal influence on metabolite transport comes from studies with plant pathogens. Localized infection of plants with fungi or parasitic higher plants has been found to result in a substantial diversion of photosynthates to the site of infection and, in the case of parasitic fungi, this is associated with an increased production of plant hormones at the site of infection (Smith et al. 1969).

APPLICATION OF HORMONES TO INTACT PLANTS

An approach that has been utilized to examine the relationship between plant hormones and metabolite mobilization is to supplement endogenous levels of hormones with an exogenous source. Thus, spraying the foliage of *Vitis vinifera* cuttings with gibberellic acid (GA_3) or benzyladenine (BA) stimulated ^{14}C-photosynthate transport to the shoot tips (Shindy and Weaver 1967) whilst BA treatment of roots enhanced transport of ^{14}C-photosynthate to these organs (Shindy, Kliewer and Weaver 1973). Similar effects on metabolite transport have been found following the treatment of fruit with an exogenous hormonal source (Kriedeman 1968; Weaver et al. 1969). These studies lend support to the contention that metabolite transport to a sink is under some form of hormonal control. However, they do not distinguish between whether the applied hormone is the active mobilizing agent or transport is responding to induced changes (qualitative and/or quantitative) in the endogenous hormone content.

MODIFIED PLANT SYSTEMS - DECAPITATED SEEDLINGS AND STEM SEGMENTS

A more definitive approach in exploring the relationship between plant hormones and metabolite transport has been to employ modified plant systems in which the endogenous source of hormone is removed and replaced by an exogenous source. The now classical system is the decapitated seedling, in which the shoot apex and attendant growing regions are removed and the cut stem surface treated with a hormone supplied as either a dispersion in lanolin paste or dissolved in aqueous solution. Under these conditions, metabolite transport to the site of hormone application has been found to be enhanced in a wide range of plant material including both herbaceous and woody species (see for example Davies and Wareing 1965; Bowen and Wareing 1971; Hatch and Powell 1971). Furthermore, the amounts of metabolites transported appears to be hormone-concentration dependent (Patrick and Woolley 1973; Johnstone, Patrick and Wareing, unpublished data). Studies with stem segments prepared from the same plant species utilized as decapitated seedlings have also been found to exhibit hormone-directed transport (Patrick and Wareing 1973; Johnstone personal communication).

Together, the above observations provide some evidence that there is a causal relationship between plant hormones and metabolite transport. Assuming that stem segments would have a relatively poor capacity to synthesize hormones, it would seem that applied hormones are the active mobilizing agents.

CERTAIN CHARACTERISTICS OF HORMONE-DIRECTED TRANSPORT

RELATIVE EFFECTIVENESS OF THE VARIOUS HORMONE CLASSES

The effectiveness of a particular hormone will depend, at least to some extent, on the physiological state of the treated plant or plant part. Furthermore, since it is unlikely that the hormone will affect all processes governing metabolite flux between source and sink, then in order for the hormone effect on transport to be observed it must act on the process rate-limiting transport (see below). Therefore, failure of a plant hormone to influence metabolite movement should be examined in terms of the above possibilities before accepting its impotency as a metabolite mobilizing agent.

The stimulatory effects of cytokinins, gibberellins and auxins applied to intact plants on metabolite transport have been referred to earlier (Shindy and Weaver 1967; Shindy et al. 1973; Weaver et al. 1969). Transport studies with decapitated seedlings have established that indol-3yl-acetic acid (IAA) and a range of synthetic auxins stimulate movement of metabolites to the hormone-treated stem stump (c.f. Bowen and Wareing 1971). For these experimental systems, the effectiveness of cytokinins and gibberellins is less clear. Early work with these compounds suggested they did not influence metabolite transport (Davies and Wareing 1965; Seth and Wareing 1967). However, recent work has demonstrated the stimulatory action of both cytokinins and gibberellins on metabolite transport is a number of plant species (c.f. Hew et al. 1967; Lepp and Peel 1970; Mullins 1970; Hatch and Powell 1971). The discrepancy with earlier work may have been

caused by the poor release of cytokinins and gibberellins from
aqueous lanolin pastes compared to that from aqueous solutions
(Johnstone, Patrick and Wareing, unpublished data). In decapitated seedlings of *Phaseolus vulgaris* ethylene was found to
promote whilst abscisic acid inhibited transport (Mullins 1970).
Thus, it would seem that all classes of plant hormones are
capable of affecting metabolite movement in stems.

INTERACTIONS BETWEEN THE VARIOUS HORMONE CLASSES
Studies with decapitated seedlings of *Phaseolus vulgaris*
showed that cytokinins and gibberellins act synergistically
with auxin in stimulating metabolite transport in stem stumps
(Seth and Wareing 1967; Mullins 1970). These data may suggest
that all the hormones act on the same transport process. This
contrasts with the additive effects of these hormones on
transport in decapitated stems of *Malus sylverstris* which
could be interpreted as action of hormones on different
transport processes (Hatch and Powell 1971). This discrepancy
between the two sets of data clearly indicates that further
information is required to better understand these hormonal
interactions.

POLARITY OF HORMONE-DIRECTED TRANSPORT
For stems, it is now well established that hormones stimulate acropetal metabolite transport (see earlier discussion).
Basipetal metabolite transport is stimulated by IAA (Patrick
and Wareing 1973; Altman and Wareing 1975) and also by
gibberellins and cytokinins (Johnstone, personal communication).
Since metabolite flow in these studies was essentailly
uni-directional between a single source and a hormone-treated
region, the data provide little information on the directional
influence of hormones. This possibility has been explored
using stem segments taken from *Phaseolus vulgaris* seedlings.
Application of sucrose solution mid-way along each segment
showed that ^{14}C-metabolite movement was directed to the
points of IAA application at either cut end at the expense of
transport in the other direction (Table 1). Moreover, there
was some indication that IAA, applied at both cut ends, may
enhance acropetal and basipetal transport simultaneously
(Table 1). Confirmatory evidence for the latter conclusion
was obtained using stem segments which were supplied with
sucrose solutions at both cut ends, one of which contained
[14]C-sucrose. Mobilization of ^{14}C-metabolites to IAA applied
mid-way along the internode was stimulated from both sucrose
sources. These data contrast with those of Lepp and Peel
(1970) who found that both IAA and kinetin repelled rather
than attracted metabolites when these hormones were applied

TABLE 1. - Effect of IAA applied in lanolin pastes (100 ppm) to "acropetal" and "basipetal" cut ends of *Phaseolus vulgaris* stem segments (11 cm in length) on the transport of ^{14}C-metabolites supplied as 0.1 m ^{14}C-sucrose solutions applied to the mid-points of the segments. Transport assayed as radioactivity (cpm) accumulated in the distal 1.5 cm lengths of segments treated with lanolin at both ends (lanolin), IAA at the "acropetal" end (IAA_a); IAA at the "basipetal" end (IAA_b) and IAA at both ends (IAA_{a+b}).

Direction of ^{14}C-transport		Hormone Treatment			
		Lanolin	IAA_a	IAA_b	IAA_{a+b}
Acropetal a ↑ ±IAA		371	1672	147	626
		±98*	±210	±28	±137
↓ ^{14}C					
Basipetal b ↓ ±IAA		1627	590	3243	1824
		±229	±143	±355	±361

*Standard error of the mean. Ten replicates per treatment.

to *Salix* spp. bark strips. At this stage it is difficult to reconcile these two sets of data but they may reflect alterations in the physiology of the system caused by removal of the bark from the wood, as hormonal attraction has been observed in other moody stems (c.f. Davian and Wareing 1965; Hatch and Powell 1971).

MECHANISMS OF HORMONE-DIRECTED TRANSPORT
POSSIBLE SITES OF ACTION

SIMPLE FLOW MODEL

The mechanism by which hormones exert their effect on metabolite transport remains unclear (c.f. Mullins 1970; Bowen and Wareing 1971) and may involve either indirect and/or direct effects on transport processes. Possible indirect effects of hormones on metabolite transport could include hormone-induced changes in (i) the capacity of assimilatory tissues to synthesize metabolites available for export (source strength), (ii) the capacity of importing tissues to accumulate metabolites unloaded from the phloem (sink strength), (iii) the relative ability of other sink regions to compete for metabolites (mobilizing ability of competing sinks), (iv) artifacts of the experimental systems. Plant hormones may act directly on phloem transport by affecting (v) sieve-tube loading mechanisms, (vi) sieve-tube unloading mechanisms, (vii) longitudinal transfer within the sieve-tubes.

The transport of metabolites from source to sink depends on a number of processes (i.e. source activity, sieve-tube loading, longitudinal transfer, sieve-tube unloading, sink activity) arranged in series and any one of these processes may be rate-limiting. The mobilizing ability of a sink may be considered as the summed capacities of these processes between the reference sink and its source relative to those of other competing sinks. Thus, the distribution of metabolites exported from a source will depend on the relative mobilizing abilities of the various sink regions.

Given these conditions, a hormone will affect metabolite transport if it acts on the process rate-limiting movement between the reference sink and its source and/or by directly affecting the mobilizing abilities of competing sinks. Hormonal affects on longitudinal transfer, sieve-tube unloading mechanisms and sink strength will result in directed transport whereas affects on source strength and sieve-tube loading mechanisms will only alter metabolite flux but have no directional influence.

INDIRECT HORMONAL EFFECTS ON METABOLITE TRANSPORT

Source strength. - Short-term changes in source strength of a mature leaf would be entirely attributable to changes in its assimilatory rate (i.e. source activity). All classes of plant hormones appear capable of affecting net photosynthetic rates of leaves (Kriedemann et al., this volume) but affects on metabolite export have yet to be examined. However, transfer of shoot-produced hormones to mature leaves would seem to be minor compared to that for root-produced hormones (Goldsmith 1969). Thus, the mobilizing ability of root sinks may depend on hormonal control of photosynthesis but this would not seem possible for hormones produced in shoot sinks. The latter proposition is consistent with the observation that IAA and GA$_3$ applied to decapitated stems of *Glycine max* seedlings stimulated transport without affecting leaf photosynthetic rates (Hew et al. 1967).

Sink strength. - The term "sink strength" is used here to describe the potential capacity of ground tissues to accumulate metabolites and may be further defined as the product of "sink activity" and "sink size" (c.f. Wareing and Patrick 1975). Both these components of sink strength are amenable to experimental measurement. "Sink activity" is regarded as the potential rate of metabolite uptake per unit weight of sink tissue per unit time. For carbohydrates, sink activity may be assayed by incubating sink tissues in ^{14}C-sucrose solutions and determining rates of ^{14}C uptake (c.f. Patrick and Wareing 1970). Dry weight measure of sink regions provide estimates of sink size.

The main pathway of metabolite transfer from the phloem to the ground tissues appears to be via the apoplast (Glasziou and Gayler 1972) and hence two semi-permeable membrane boundaries (phloem and ground tissues) must be crossed. Under these conditions, sink strength would influence metabolite efflux from the phloem by depleting apoplast metabolite levels. It is considered that sink activity would depend upon the rates at which endogenous metabolite pools are depleted and/or upon the activities of cellular nutrient-uptake mechanisms. Both of these factors are susceptible to hormone action.

Circumstantial evidence implicating hormone-induced depletion of metabolite levels in influencing transport has been found in decapitated seedlings in *Phaseolus vulgaris* in which enhanced transport was associated with increased protein and nucleic acid synthesis (Mullins 1970). Whilst it is probable that these changes in protein and nucleic acid synthesis may have led to concomitant changes in sink strength, this was not tested.

Removal of metabolites from the apoplast may also be regulated by the activities of cellular nutrient-uptake mechanisms. For instance, rates of sucrose accumulation by ground tissues of *Saccharum* spp. stems were found to be controlled by the activity of free-space invertase and the levels of this enzyme may be under hormonal control in elongating stems (Glasziou and Gayler 1972). An interesting feature of this type of sink is that it provides the opportunity for hormones to influence the selective nature of the uptake mechanisms (c.f. Ilan et al. 1971). Indeed, the observation that auxins stimulated the transport of ^{14}C-sucrose but not ^{32}P in decapitated stems of *Coleus* spp. and *Helianthus annus* may reflect the operation of such a hormonally-controlled metabolite-selective sink (Bowen and Wareing 1971).

Mobilizing ability of competing sinks. - It is possible that hormones synthesized in one sink region may influence metabolite movement to that sink by acting on the mobilizing ability of competing sinks. Thus, GA applied to *Solanum tuberosum* plants after tuberization resulted in a marked shift in the distribution of ^{14}C-photosynthates from the tubers to the shoots and this effect was considered to be caused by a GA-induced reduction in the sink strength of the tubers (Booth and Lovell 1972).

Artifacts of the experimental systems. - It is not feasible to deal with experimental systems individually so a few generalizations will be made that apply equally to all.

All promotory hormone classes are known to have senescence-delaying capabilities (Sacher 1973) and it is possible that they act by merely maintaining cellular activities whilst the control declines (Mullins 1970; Patrick and Wareing 1970). However, the observation that decapitated stems of *Phaseolus vulgaris* seedlings treated with plain lanolin for 3 days retained their short-term responsiveness to auxin in terms of enhanced metabolite transport makes this proposition less tenable (Patrick and Wareing 1973).

The high concentrations of hormone applied in lanolin pastes (e.g. 1000 ppm) has caused some concern as to whether they are physiological (Mullins 1970). However, the high concentrations required would seem to be more a function of hormone release from the paste (Patrick and Woolley 1973).

Severing the phloem by either decapitation or leaf removal will lead to some damage of the transport channel through pressure release and consequent surging of sieve-tube contents. Moreover, wound responses are inevitable, including the production of callose around the sieve-pores. Although the phloem appears to retain some of its functional capacity (Patrick and Wareing 1973), it is possible that damage repair may allow increased transport and that the rate of repair is under hormonal control.

There are numerous other possible artifacts introduced by manipulative treatments, some of which have been explored by Patrick and Wareing (1973).

DIRECT HORMONAL EFFECTS ON METABOLITE TRANSPORT

Enhanced transport:independent of source and sink strengths. - In order to decisively demonstrate a direct effect of hormones on a phloem transport process, it is desirable to establish that hormone-induced transport does not depend on either changes in source and sink strength or on the mobilizing abilities of competing sinks.

It is proposed that *Phaseolus vulgaris* seedlings, treated with IAA under conditions described by Patrick and Wareing (1973), provide experimental systems that meet the specifications described above (see Patrick and Wareing 1970,1973, 1976). The following discussion largely refers to data obtained from studies with these experimental systems.

Sieve-tube loading mechanism. - Direct measures of sieve-tube loading are lacking for the *Phaseolus vulgaris* systems but estimates of net ^{14}C-export from the primary leaves were found to be unaffected by IAA treatment (Patrick and Wareing 1973). However, in so far as sieve-tube loading is an energy-dependent transfer mechanism across semi-permeable boundaries (Wardlaw 1974), it is feasible that the process could be influenced by plant hormones. Some evidence for such a possibility was found for bark strips of *Salix* spp. treated with IAA and kinetin (Lepp and Peel 1970). Whether similar hormonal affects on sieve-tube loading may occur in leaves remains to be elucidated.

Sieve-tube unloading mechanisms. - Hormonal stimulation
of metabolite transport will inevitably lead to increased rates
of sieve-tube unloading. The question is whether these increased rates are a cause or a result of enhanced metabolite
transport. Accepting that hormones (within short-term treatments of mature stems) do not alter the sink strength of the
treated tissues, then there are a number of pieces of circumstantial evidence that may be interpreted to indicate that hormone-directed transport is mediated through effects on sieve-
tube unloading mechanisms. For instance, the observation that
auxins of varying mobilities are equally effective as mobilizing agents is consistent with a localized hormonal action
(Bowen and Wareing 1971).

Based on the assumption that transfer of metabolites from
the phloem to stem ground tissues takes place via the apoplast
(c.f. Glasziou and Gayler 1972), then more direct estimates
of sieve-tube unloading may be obtained from measures of
efflux rates of transported metabolites into the tissue free-
space. Using such an approach, it has been found that the rates
of photosynthate efflux into the free-space of *Vicia faba*
stems was increased by parasitization by *Cuscuta* spp. suggesting some form of influence (possibly hormonal) of the parasite
on phloem-unloading mechanisms of the host plant (Wolswinkel
1974). Preliminary results obtained from ^{14}C-sucrose transport studies in *Phaseolus vulgaris* stem segments have shown
that whilst the absolute rate of ^{14}C-efflux into the stem free-
space was enhanced by IAA, the relative rate was less than the
controls (Table 2). These data indicate that even if sieve-
tube unloading mechanisms are stimulated by IAA, they alone
cannot fully account for IAA-induced transport in the *Phaseolus
vulgaris* systems.

Longitudinal transfer in sieve-tubes. - Because of the
present lack of understanding of the mechanisms responsible
for longitudinal transport in sieve-tubes (Wardlaw 1974),
evidence for a hormonal affect on this process must at best
be circumstantial. The observation that sieve-tube loading
and unloading do not appear to fully account for hormone-
induced transport provides some tentative evidence that hormones could be acting on longitudinal transfer directly.

An approach pioneered by Davies and Wareing (1965) to
distinguish localized (sink and sieve-tube unloading) from
remote (longitudinal transfer and sieve-tube loading) effects
was to prevent the basipetal movement of IAA down the
decapitated internode by applying the auxin-transport inhibitor,
2, 3, 5 - triiodobenzoic acid (TIBA), in lanolin as a ring
around the stem stump. Under these conditions, TIBA inhibited

TABLE 2. The effect of IAA on the efflux of acropetally transported ^{14}C-metabolites into the free space of 9 hour-treated
stem segments. The upper 3 cm of segments 6 cm in length was
assayed.

Hormone Treatment	Rate of ^{14}C Mobilization (cpm/h)	Rate of ^{14}C Efflux into Free-Space (cpm/h)	Percentage of ^{14}C Mobilized Effluxed into Free Space
Lanolin	472± 61	247± 31	68.4±7.4
IAA	2607±484	909±174	27.5±5.6

IAA-induced metabolite transport and on this evidence it was
proposed that IAA acted on longitudinal transfer by being
present along the entire length of the transport channel
(Davies and Wareing 1965). However, it is possible that TIBA
interfered with other processes (c.f. Milthorpe and Moorby
1969) and moreover, the TIBA concentration employed may have
been phytotoxic (Mullins 1970). Indeed, with regard to the
latter possibility, application of more physiological TIBA
concentrations (10^{-4} and 10^{-3}M) to decapitated stems was found
to stimulate metabolite transport whilst inhibiting basipetal
IAA transport (Mullins 1970). These data are difficult to
reconcile with those of Davies and Wareing (1965) but our
experience has shown that care must be taken to work with sub-
optimal levels of IAA for metabolite transport otherwise TIBA
treatment stimulates metabolite transport possibly by reducing
tissue levels of IAA. The observation that acropetal ^{14}C-
photosynthate transport is unaffected whilst basipetal IAA
movement is inhibited in stems of intact plants treated with
TIBA (Morris et al. 1973) does not necessarily provide evidence to exclude auxin action remote from its site of synthesis
as the movement of other hormones would not be impaired by
such a treatment.

Attracted by the potential usefulness of such an approach,
we have carried out further studies with TIBA and other inhibitors of polar IAA transport. Using stem segments a 0.5 per
cent concentration of TIBA in lanolin was found to inhibit
acropetal IAA-induced metabolite transport but have no effect
on IAA-induced basipetal metabolite transport (Table 3).
Thus, at this concentration, the TIBA effect on transport
would not seem to result from a generalized effect such as
phytotoxicity or alteration of sink activity. Furthermore,
application of IAA below the TIBA ring restored normal IAA-
indiced transport (Table 4) suggesting that TIBA was acting

to block the action of either IAA or some stimulus generated by IAA.

TABLE 3. Effect of TIBA on auxin-induced mobilization of ^{14}C-sucrose flowing either acropetally or basipetally in stem segments. Radioactivity (cpm) accumulated at ends treated with IAA.

Direction of ^{14}C flow	TIBA Treatment	
	Control	+ TIBA
Acropetal	640± 70*	110± 35
Basipetal	870±122	931±123

* Standard error of the mean. Ten replicates per treatment.

TABLE 4. Effect of applying IAA above and below the TIBA ring on ^{14}C-sucrose transport in stem segments. Radioactivity (cpm) accumulated at IAA-treated ends.

TIBA Treatment	Hormone Treatment	
	Lanolin	IAA
Control	786±82*	2019±248
TIBA	–	1797±368

* Standard error of the mean. Ten replicates per treatment.

The nature of this proposed IAA action awaits clarification but the above observations may suggest that IAA-induced transport depends on effects distant from the point of hormone application. Whilst the above work is at best tentative and open to other interpretations it is considered that the approach may prove profitable in further resolving the way in which hormones act on metabolite transport.

CONCLUDING COMMENTS

Currently, the central question in assessing the significance of hormonal control of metabolite transport is to distinguish whether hormones act directly and/or indirectly on phloem transport. Resolution of this problem will further define the sink's role in regulating metabolite distribution within plants. That is, does the sink regulate transport through its capacity to accumulate metabolites and/or, via hormones, which control the functional capacities of the transport processes? It is emphasized that the two possibilities are not necessarily mutually exclusive. For instance, given that transfer depends on the latter, then under these conditions, sink control may be exerted by rate-limiting metabolite efflux from the phloem. Indeed, the concept of rate-limitation should be considered when exploring the above possibilities.

Available evidence implicating hormonal action on phloem transport is tentative but sufficient to justify further examination of this possibility. Whilst such an exercise may not necessarily answer the problem, it is considered it will generate new approaches that may throw light on the mechanisms regulating metabolite distribution.

ACKNOWLEDGEMENTS

I am grateful to Professor P.F. Wareing, F.R.S. in introducing me to the problem and providing much of the stimulus for the work described. I am indebted to Mr R. McDonald for his industrious technical assistance. Much of the unpublished work was supported by funds from the Australian Research Grants Commission.

REFERENCES

Altman, A. and Wareing, P.F. (1975). The effect of IAA on sugar accumulation and basipetal transport of ^{14}C-labelled assimilates in relation to root formation in *Phaseolus vulgaris* cuttings. *Physiol. Plant 33*, 32-38.

Booth, A., Moorby, J., Davies, C.R., Jones, H. and Wareing, P.F. (1962). Effects of indolyl-3-acetic acid on the movement of nutrients within plants. *Nature 194*, 204-205.

Bowen, M.R. and Wareing, P.F. (1971). Further investigations into hormone-directed transport in stems. *Planta (Berl.) 99*, 120-132.

Davies, C.R. and Wareing, P.F. (1965). Auxin-induced transport of radio-phosphorus in stems. *Planta (Berl.) 65*, 139-156.

Eeuwens, G.J. and Schwabe, W.W. (1975). Seed and Pod wall development in *Pisum sativum* L. in relation to extracted and applied hormones. *J. Exp. Bot. 26*, 1-14.

Frydman, V.M., Gaskin, P. and MacMillan, J. (1974). Qualitative and quantitative analyses of gibberellins throughout seed maturation in *Pisum sativum* cv. Progress No.9. *Planta (Berl.) 118*, 123-132.

Glasziou, K.T. and Gayler, K. (1972). Storage of sugars in stalks of sugar cane. *Bot. Rev. 38*, 471-490.

Goldsmith, M.H.M. (1969). Transport of plant growth regulators

In: "Physiology of Plant Growth and Development" ed.
M.B. Wilkins. McGraw-Hill, England.

Hatch, A.H. and Powell, L.E. (1971). Hormone-directed transport of ^{32}P in *Malus sylvestris* seedlings. *J. Amer. Soc.
Hort. Sci. 96*, 230-234.

Hew, C.S., Nelson, C.D. and Krotkov, G. (1967). Hormonal
control of translocation of photosynthetically assimilated
^{14}C in young soybean plants. *Amer. J. Bot. 54*, 252-256.

Ilan, I., Gilad, T. and Reinhold, L. (1971). Specific effects
of kinetin on the uptake of monovalent cations by sunflower
cotyledons. *Physiol. Plant 24*, 337-341.

Lepp, N.W. and Peel, A.J. (1970). Some effects of IAA and
kinetin upon the movement of sugars in the phloem of
willow. *Planta (Berl.) 90*, 230-235.

Milthorpe, F.L. and Moorby, J. (1969). Vascular transport
and its significance in plant growth. *Ann. Rev. Plant
Physiol. 20*, 117-138.

Moorby, J. (1968). The effect of growth substances on transport in plants. In: "The Transport of Plant Hormones",
ed. Y. Vardar, Amsterdam: North Holland pp.192-206.

Morris, D.A., Kadir, G.O. and Barry, A.J. (1973). Auxin transport in intact pea seedlings (*Pisum sativum* L.): The inhibition of transport by 2, 3, 5-triiodobenzoic acid. *Planta
(Berl.) 110*, 173-182.

Mothes, K. and Engelbrecht, L. (1961). Kinetin-induced transport of substances in excised leaves in the dark. *Phyto
chem. 1*, 58-62.

Mullins, M.G. (1970). Hormone-directed transport of assimilates in decapitated internodes of *Phaseolus vulgaris* L.
Ann. Bot. 34, 897-909.

Nitsch, J.P. (1970). Hormonal factors in growth and development. In: "The Biochemistry of Fruits and their Products".
ed A.C. Hulme, Academic Press, London and New York.

Patrick, J.W. and Wareing, P.F. (1970). Experiments on the
mechanism of hormone-directed transport. In: "Plant Growth
Substances" ed. D.J. Carr, Springer-Verlag, Germany
pp.695-700.

Patrick, J.W. and Wareing, P.F. (1973). Auxin-promoted transport of metabolites in stems of *Phaseolus vulgaris* L.
Some characteristics of the experimental transport systems.
J. Exp. Bot. 24, 1158-1171.

Patrick, J.W. and Wareing, P.F. (1976). Auxin-promoted transport of metabolites in stems of *Phaseolus vulgaris* L.
Effects at the site of hormone application. *J. Exp. Bot.*
(In press).

Patrick, J.W. and Woolley, D.J. (1973). Auxin physiology of
decapitated stems of *Phaseolus vulgaris* L. treated with

445

indol-3yl-acetic acid. *J. Exp. Bot. 24*, 949-957.

Peel, A.J. (1974). "Transport of Nutrients in Plants". Butterworths, London.

Phillips, I.D.J. (1975). Apical dominance. *Ann. Rev. Plant Phsyiol. 26*, 341-367.

Sacher, J.A. (1973). Senscence and postharvest physiology. *Ann. Rev. Plant Physiol. 24*, 197-224.

Seth, A. and Wareing, P.F. (1967). Hormone directed transport of metabolites and its possible role in plant senescence. *J. Exp. Bot. 18*, 67-77.

Shindy, W. and Weaver, R.J. (1967). Plant regulators alter translocation of photosynthetic products. *Nature 24*, 1024-1025.

Shindy, W.W., Kliewer, W.M. and Weaver, R.J. (1973). Benzyl-adenine-induced movement of ^{14}C-labelled photosynthate into roots of *Vitis vinifera*. *Plant Physiol. 51*, 345-349.

Smith, D., Muscatine, L. and Lewis, D. (1969). Carbohydrate movement from autotrophs to heterotrophs in parasitic and mutualistic symbiosis. *Biol. Rev. 44*, 17-90.

Thrower, S.L. (1967). The pattern of translocation during leaf ageing. In: "Aspects of the Biology of Ageing" S.E.B. Symposia Vol. 21, Cambridge University Press.

Wardlaw, I.F. (1974). Phloem transport: Physical chemical or impossible. *Ann. Rev. Plant Physiol. 25*, 515-539.

Wareing, P.F. and Patrick, J.W. (1975). Source-sink relations and the partition of assimilates in the plant. In: "Photosynthesis and Productivity in Different Environments". ed. J.P. Cooper, Cambridge University Press.

Wareing, P.F. and Seth, A.K. (1967). Ageing and senescence in the whole plant. In: "Aspects of the Biology of Ageing" S.E.B. Symposia Vol.21. Cambridge University Press.

Weaver, R.J., Shindy, W. and Kliewer, W.M. (1969). Growth regulator induced movement of photosynthetic products into fruits of "Black Corinth" grapes. *Plant Physiol. 44*, 183-188.

Went, F.W. (1939). Some experiments on bud growth. *Amer. J. Bot. 26*, 109-117.

Wolswinkel, P. (1974). Enhanced rate of ^{14}C-solute release to the free space by the phloem of *Vicia faba* stems parasitised by *Cuscuta*. *Acta. Bot. Neerl. 23*, 177-188.

CHAPTER 38

Nutrient Mobilization and Cycling: Case Studies for Carbon and Nitrogen in Organs of a Legume

J.S. Pate

Department of Botany, University of Western Australia, Nedlands, W.A. Australia

INTRODUCTION

The interdependence of the autotrophic centres of its nodules and shoot for carbon and nitrogen, makes the legume a valuable experimental tool for investigating the co-ordination of nutrition and transport in vascular plants. *Lupinus albus*, the species selected here, has the added advantage that conducting elements of phloem and xylem can be tapped, - thereby allowing the construction of balance sheets for entry or exit of specific organic solutes through these channels. Transport activity can then be viewed alongside the overall functioning of the participating organs.

The paper selects three examples : (i) The developing fruit - importer of carbon and nitrogen through xylem and phloem, (ii) The nodulated root - importer and agent of cycling of shoot-derived carbon, exporter of fixed nitrogen in the transpiration stream, (iii) The leafy shoot - mobilizer to other plant parts through phloem of photosynthate and transpirationally-derived nitrogen.

TRANSPORT TO THE DEVELOPING FRUIT AND ITS ECONOMY IN TERMS OF CARBON, NITROGEN AND WATER

Fruits of legumes share with other fruits the properties of being heavily dependent for growth on capture of water and solutes through vascular linkages with the parent plant, and of having to establish the bulky enveloping structures of carpel and ovules before commencing to lay down reserves in their embryos. The carpel of most legumes is green and possesses stomata thus enabling a fruit to control its water loss, and to photosynthesize using carbon dioxide derived externally or collecting in the sealed gas space of the carpel from the ripening seeds (Bollard 1970; Flinn and Pate 1970; Crookston et al. 1974). These properties, combined with the high efficiencies with which legumes mobilize assimilates to their fruits and developing seeds, constitute a most intriguing situation for study.

The fruit tips of *Lupinus albus* bleed a phloem exudate when cut, and upper parts of shoots yield tracheal (xylem) sap when extracted under mild vacuum (Pate et al. 1974; 1975). So the proportions and likely concentrations of organic solutes delivered to fruits can be studied as fruit development proceeds. The terminally-borne fruits are easily enclosed for study of gas exchange and transpiration, measurements also essential to an appreciation of the fruit's physiology. The sections that follow assemble information derived from recent studies on fruit and seed development of the species (Atkins et al. 1975; J.S. Pate, P.J. Sharkey and C.A. Atkins, unpublished data). Information relates to the first fruit formed on the primary inflorescence.

1) The requirements For Carbon And Nitrogen For Dry Matter Production By The Fruit. - Data for dry weight gains and percentages of C and N in dry matter allow the information set out in Table 1 (Items 1 and 2) to be calculated. Demand for C and N rises rapidly over the first eight weeks after anthesis and then falls abruptly. A net loss of carbon is recorded for the interval weeks 10-12, suggesting that in late stages of ripening translocation fails to keep pace with respiratory losses. A very low input of nitrogen is recorded at this time.

2) The Exchange Of Carbon Dioxide Between Fruit And Surrounding Atmosphere. - Daily measurements of gas exchange through the fruit's life cycle show that during the day time it operates consistently above compensation point from week 2 until it commences to yellow and dry out at week 10. However, since night time losses are high, net carbon exchanges, expressed on a fortnightly basis, always show a negative balance for carbon. (Table 1, Item 3). Over three-quarters of the gaseous loss of carbon occurs over weeks 8-12, the period when developing seeds represent a large proportion of the fruit's respiration. Item 4 of Table 1 gives total requirement of the fruit for carbon and, dividing these figures by corresponding values for the fruit's nitrogen requirement (Item 2), the C : N "consumption ratio" of the fruit can be calculated (Item 5, Table 1). This ratio declines (16.9 to 10.7) as the fruit develops, a feature reflecting an early demand for carbohydrate in synthesis of cellular fabric of the pod *versus* a later requirement for increasing amounts of nitrogen as the seeds commence to synthesize protein.

During the 12 weeks of growth 883 mg C and 72.5 mg N are used giving an average C : N consumption ratio of 12.2 : 1.

448

TABLE 1. Carbon and Nitrogen Balance of Developing Fruits of *Lupinus albus* L.

| | Age of fruit (weeks) | | | | | | |
	0-2	2-4	4-6	6-8	8-10	10-12	0-12
1. Carbon increment as dry matter (mg C fruit^{-1})	65.5	86.7	257.2	309.5	102.3	-34.5	786.7
2. Nitrogen increment as dry matter (mg N fruit^{-1})	4.0	5.9	18.5	30.0	13.1	1.0	72.5
3. Carbon exchanged as CO_2 with atmosphere (mg C fruit^{-1})	-2.2	-0.7	-3.5	-14.0	-31.0	-45.2	-96.6
4. Total carbon requirement of fruit (1+3) (mg C fruit^{-1})	67.7	87.4	260.7	323.5	133.3	10.7	883.3
5. Ratio for fruits requirements of C and N (4÷2)	16.92	14.81	14.09	10.78	10.18	10.70	12.18

*3) Concentrations And Relative Composition of C and N
Solutes In Transport Channels Serving The Fruit.* - Information
relating to sugar, amides, and amino acids of phloem and amino
acids of xylem are set out in Fig. 1, these compounds carrying
over 97% of the C and virtually all of the N translocated to
the fruit. Phloem sap tends to become less concentrated in
sucrose but more concentrated in amino acids as the fruit ages.
(Fig. 1A).

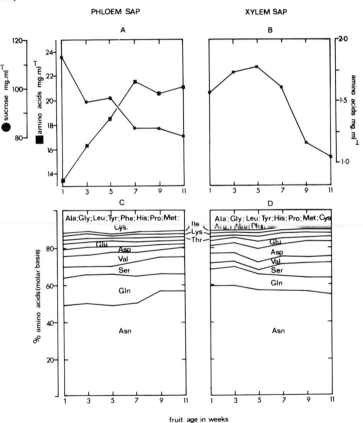

*Fig. 1. - Concentration and composition of organic sol-
utes in transport channels serving the lowest (first
formed) fruit of the primary inflorescence of* Lupinus
albus *L. A and C - phloem sap collected from tips of the
fruit. Fruit age is expressed in weeks after anthesis.
B and D - xylem (tracheal) sap from shoot segments immediately below the inflorescence. Abbreviations for amino
compounds are as listed in Biochem. J. 126, 773-780. (1972).
(Unpublished data, P.J. Sharkey and J.S. Pate).*

Xylem sap has no detectable sugar. The concentration of
amino compounds rises to a maximum at week 5 and then falls
(Fig. 1B, and similar results for *Pisum*, (Pate et al. 1965).

Despite the quite large changes in concentrations of amino
acids, the balance between individual constituents in either
transport channel remains remarkably constant during fruit
development (Fig. 1C and D). The close resemblance in composition between xylem and phloem fluids is discussed later.
Sucrose and asparagine are obviously the major sources of
nourishment to the fruit. Asparagine arrives greatly in excess
of its requirement for incorporation as such into protein or
soluble reserves of pod and seed, and studies using labelled
asparagine show that its amide nitrogen is used extensively as
a nitrogen donor for synthesis of protein amino acids, especially those not delivered in sufficient amount through the
phloem. The carbon of asparagine is contributed largely to
the non-amino compounds of the seed. An asparaginase is suspected to be responsible for cleavage of the amide (Atkins
et al. 1975).

   4) Transpiration And The Water Budget Of The Fruit. -
Measurements of water losses in transpiration and gains or
losses of water in tissues allow a water budget to be constructed for fruit development (Table 2, Items 1 - 3). The
negative values refer to the drying out of pod and maturing
seed. The total demand is 43.4 mls, by far the larger proportion of this lost in transpiration. On average, 22.5 mls
water are transpired per gram of dry matter produced, a transpiration ratio equivalent to that of the most frugal xerophyte
or succulent. Of course, the comparison is not fair as the
fruit receives elaborated assimilates, the carbon of which it
does not have to gain from the atmosphere. Indeed, transpiration of the complement of fruit on a mature plant of *L. albus*
represents only 2-5% of the plant's total water loss, showing
how heavily the fruit depends on water-demanding processes
conducted elsewhere in the shoot.

   *5) Estimation Of The Relative Importance Of Xylem And
Phloem In Delivery Of Materials To The Fruit.* - One approach
to this problem is to calculate the C : N ratios of fluids of
xylem and phloem and estimate from these values the proportions
of the two streams of solutes required to balance the measured
increments of carbon and nitrogen on the fruit (see Table 3).
Early in growth the ratio of C : N delivered in phloem is
significantly higher than the C : N requirement of the fruit,
implying that xylem-borne solutes of low C : N ratio supplement the nitrogen requirements of the fruit. As development
proceeds, however, the C : N ratios of phloem and growing

TABLE 2. Water Budget of Developing Fruits of *Lupinus albus* and Estimates of Delivery of Carbon, Nitrogen and Water by Xylem and Phloem

	Fruit Age (weeks)						
	0-2	2-4	4-6	6-8	8-10	10-12	0-12
1. Water loss in transpiration (mls fruit^{-1})	1.57	3.67	9.82	13.23	7.85	4.47	40.61
2. Increment/loss in tissue water (mls fruit^{-1})	2.05	4.20	1.55	0.35	-1.16	-4.22	2.77
3. Total water input (mls fruit^{-1})	3.62	7.87	11.37	13.58	6.69	0.25	43.38
4. Water intake through phloem (mls fruit^{-1})	1.13	1.69	4.87	6.92	2.98	0.24	17.51
5. Water intake through xylem (mls fruit^{-1})	2.49	6.18	6.50	6.66	3.71	0.01	25.87
6. Carbon intake through phloem (mg C fruit^{-1})	66.30	83.40	256.30	322.00	131.80	10.70	870.50
7. Carbon intake through xylem (mg C fruit^{-1})	1.37	3.96	4.36	3.53	1.52	0.00	14.70
8. Nitrogen intake through phloem (mg N fruit^{-1})	2.94	4.72	16.11	26.63	11.05	0.91	62.40
9. Nitrogen intake through xylem (mg N Fruit^{-1})	0.70	2.01	2.17	2.37	0.75	0.00	8.00

TABLE 3. C:N Ratios of Phloem and Xylem Fluids of *Lupinus albus* and Estimated Percentage Delivery of Carbon and Nitrogen to Fruits through these Transport Channels

	Fruit Age (weeks)					
	0-2	2-4	4-6	6-8	8-10	10-12
1. C:N ratio of fruit phloem sap*	19.2	15.59	14.01	10.75	10.64	10.48
2. C:N ratio of xylem (tracheal) sap*	1.96	2.00	2.03	1.96	2.05	2.06
3. C:N ratio for fruit's requirements for carbon and nitrogen (see also Table 1.)	16.2	14.81	14.09	10.78	10.18	10.70
4. Estimated % delivery of C + N through phloem**	84.4	94.3	100.0	100.0	94.6	100.0
5. Estimated % delivery of C + N through xylem**	15.5	5.7	-	-	5.4	-

* Based on sugar and amino acid levels shown in Fig. 1.

** Calculated from comparisons of C:N ratios of incoming phloem (Item 1) and xylem (Item 2) and the C:N requirements of the fruit (Item 3).

fruit become more closely matched, and, considerably less intake from xylem would therefore be anticipated. Indeed, at three of the later intervals of study, C : N ratios for fruit intake are slightly higher than in donor phloem (Table 3). If these differences were significant the fruit would have to rid itself of a slight excess of nitrogen – gaseous losses as ammonia or export from the fruit in the xylem would be possibilities. Several earlier workers (see Zimmermann 1969; Bollard 1970) have commented that xylem might drain excess water and solutes from developing fruits but concrete evidence on this appears to be lacking at least in legumes.

A second approach (Table 2) is based on the water budget of the fruit, and assumes that xylem and phloem operate as mass flow systems delivering assimilates in the concentrations and proportions found in tracheal and phloem sap (Fig. 1). Calculations can then be made to fit a situation in which the combined xylem and phloem input would fulfill the recorded carbon and water intake of the fruit and match as nearly as possible the fruit's increment of nitrogen. The results using this approach (Table 2) give a surprisingly good fit, for, in meeting the total water and carbon budget for the fruit's growth cycle, transport in xylem and phloem supply 97% of the requirements of the fruit for nitrogen (see Table 1 and 2). Xylem is estimated to deliver 60% of the fruit's water, 11% of its nitrogen and a mere 1% of its carbon, the corresponding values for phloem delivery being 40%, 86% and 96%. Unlike the calculations made earlier on the basis of C : N ratios, this form of calculation suggests that a net intake through the xylem occurs throughout fruit development, delivering, in fact, the equivalent of 64% of the water lost in transpiration. Xylem may therefore be of significance in providing the fruit with a continuous source of elements such as calcium which are sparingly mobile in the phloem (see data on mineral composition of transport fluids of *Lupinus albus* in another chapter of this volume).

A generalized equation for the total economy of the fruit of *L. albus* can now be drawn up.

Intake through xylem and phloem* : 43 mls H_2O + 1756 mg sucrose + 384 mg amides and amino acids.

Production of storage products in seeds of fruit : 418 mg protein + 132 mg oil + 110 mg perchloric acid soluble carbohydrate.

Left as residue in dried out pod : 700 mg**

Present as material additional to ergastic substances in seed : 440 mg

Respiratory loss of fruit : 230 mg**

* All values refer to the 12-week growth cycle of the lowest (first formed) fruit of the primary inflorescence.

** Expressed as equivalent weights of sucrose.

Thus, viewed as a processor of incoming organic solutes, the fruit converts some 31% by weight of its imports into 'useful' seed reserves, loses in respiration the carbohydrate equivalent of 11% of its total intake, and utilizes the remaining 58% in the manufacture of structural components, principally cellulose and lignin of pod and seed coats. The pod contributes most effectively as an organ of conservation of carbon respired by seeds, for, in operating above compensation point during the photoperiod of all but the last two weeks of its life it is likely to fix and cycle back to the seeds at least half of the total carbon respired during development. The *Pisum* fruit appears to be less efficient in this respect, losing 26% of its carbon intake in respiration (Flinn and Pate 1970; Pate 1974).

THE ECONOMY OF CARBON, WATER AND NITROGEN IN THE ROOT AND ITS NODULES

Whether nodulated or not, legume roots are important centres for assimilating inorganic forms of nitrogen from the rooting medium. In doing so they metabolize carbohydrate translocated from the shoot and return to the shoot *via* the xylem organic solutes of nitrogen (Pate 1962, 1973; Ogbnonorie and Pate 1972). Deprived of a source of combined nitrogen the legume is totally dependent for nitrogen on its nodules – organs which comprise only a few percent of the plant's mass yet engage in traffic of large fractions of the plant's organic solutes.

The approach followed and the techniques used are essentially those developed for *Pisum sativum* by Minchin and Pate (1973). Application to *Lupinus albus* (D. Herridge and J.S. Pate, unpublished) involved a period 21-56 days after sowing, from the commencement of nitrogen fixation in nodules until the start of flowering. A sequence of harvests from the plant population allowed carbon and nitrogen increments to be determined in dry matter of root, shoot and nodules. The respiratory output of enclosed nodulated roots of intact plants was measured using an *in situ* gas flow technique with Pettenkoffer assemblies to collect respired CO_2. The proportion of the root's respiration associated with nodules was assessed by gas exchange measurements of freshly detached nodules (see Minchin 1973). Uptake of water by plants during growth was determined and amounts built into plant tissues distinguished from those lost in transpiration. The types and concentrations of nitrogen exported from root and nodule were measured by analysis of sap bleeding from cut xylem of these organs (see Pate et al. 1969; Minchin 1973; Gunning et al. 1974).

Using the above information, flow diagrams for carbon, nitrogen and water can be assembled as shown in Fig. 2. The following assumptions are made: –

1) All nitrogen fixed by the root nodules leaves by the xylem and transpiration stream. Distal regions of the root and the growing apices of nodules do not have direct access to newly fixed nitrogen but acquire it indirectly after it has cycled through the shoot and returned to the root in the translocation stream. This is based on the results of ^{15}N feeding experiments on *Pisum* (Oghoghorie and Pate 1972; Pate and Flinn 1973). In *L. albus* the phloem stream moving to the root carries high levels of nitrogen as amino acids (M. Hill and J.S. Pate, unpublished).

2) Fixation products leave the nodule as a solution of amides and amino acids at 6.5 mg N ml^{-1}. The C : N ratio of these exports is 2.24 : 1 (data for *L. albus* from Minchin, 1973). Export of the 62 mg of nitrogen fixed during the period 21-50 days after sowing therefore requires 9.7 mls of water and carries 140 mg C back to the shoot in the xylem (Fig. 2).

The cycle for carbon shows that of every 100 units net gain of carbon by the shoot 29 are incorporated directly into shoot dry matter, and 71 translocated to the root system. The root respires 36 units of carbon and binds 19 into dry matter, whilst the nodules respire 4 units, use 3 in growth, and return 9 back to the shoot combined with the fixed nitrogen. The net result for the vegetative period of growth is that 38 units of carbon are incorporated into above ground portions, 22 into those below ground. Later in the life cycle when fruits are filling and roots have ceased to grow the picture is likely to change to one where the shoot is a much greater sink for both carbon and nitrogen. At this time also, relatively less carbon is likely to be donated to the root as translocate.

The data for carbon flow turn out to be very similar to those recorded for *Pisum sativum* (Minchin and Pate 1973) over a comparable period of growth. However a higher proportion of carbon is incorporated into the starchy tap root tissues of *Lupinus* than into the slender root system of *Pisum*, and as a result of this, the proportion of the root's net gain of carbon which cycles through nodules is higher in *Pisum* than in *Lupinus*.

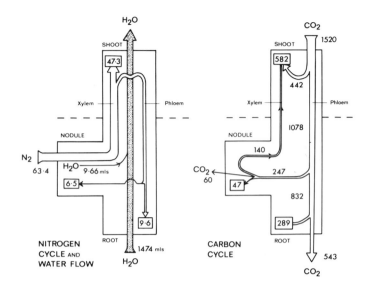

Fig. 2. - Flow patterns for carbon, nitrogen and water in Lupinus albus *plants relying solely on root nodules for nitrogen. The study period is 21-56 days after sowing, i.e. from the start of nitrogen fixation to the beginning of flowering. Quantities of carbon and nitrogen are given as milligrams per plant. (Unpublished data D. Herridge and J.S. Pate).*

The nitrogen budget (Fig. 2) shows a relatively higher retention of N by shoots than in the case of carbon, and a net export of nitrogen from the nodules of almost 90%. The requirement of water for export from nodules is insignificant relative to that passing through the whole root system.

Information from carbon and nitrogen flow cycles (Fig. 2) show that for every 7.2 units by weight of sucrose translocated to the root, 1 unit by weight of amino compounds, principally amide, returns to the shoot as products of nitrogen fixation. This respresents a 13.7% conversion of carbohydrate to amino compounds, a value much lower than the 22.8% conversion recorded by Minchin and Pate (1973) for *Pisum*. This again reflects the larger consumption of carbohydrate in the tap root of *Lupinus*. Viewed in terms of carbohydrate utilization by nodules the differences between the two legumes become insignificant, nodules of *Pisum* requiring 4.1 mg C as translocate to support fixation of 1 mg N (Minchin and Pate 1973), the comparable figure here for *Lupinus* being 3.9.

457

XYLEM TO PHLOEM TRANSFER OF AMINO ACIDS IN LEAFY SHOOTS: MAJOR PATHWAY OF NITROGEN TRANSPORT

A major function of the older parts of the plant shoot is to act as transpirational catchment for solutes ascending from roots in the xylem and to participate in the re-routing of these solutes in concentrated form to weakly-transpiring structures of high sink capacity. This activity is particularly well displayed for nitrogen in legumes, reaching a climax when fruits are filling and demands outstrip by far mobilization of resources already present in the plant. ^{15}N labelling of roots at this time shows that nitrogen assimilated in roots passes with speed and high efficiency to the fruits and seeds (Pate and Flinn 1973).

There is, however, a noticeable lag in transfer of xylem - administered label to seeds, and this is presumed to relate to the passage of the relevant compounds from xylem to phloem and their subsequent translocation to the fruit (Lewis and Pate 1973, Sharkey and Pate 1975). This cross transfer process can be studied relatively easily in *Lupinus albus* because the time course of arrival of xylem-fed label in the phloem sap of the fruit can be monitored, and phloem sap analysis shows to what extent the substrate has been metabolized in its passage to the fruit. A study of the fate of 12 different ^{14}C labelled amino acids in fruiting shoots of *L. albus* has been published (Sharkey and Pate 1975) and the following comments derive from this source.

Xylem amino acids vary considerably in the speed and efficiency with which their carbon is mobilized to seeds. Certain common amino compounds of the xylem stream (asparagine, glutamine, valine, threonine and serine) are transferred to phloem rapidly and intensely, transfer carbon to fruit and seed with high efficiency, and travel to the fruit largely attached to the same molecule as was supplied through the xylem. A typical 15-minute pulse labelling study shows a peak labelling of the phloem of the fruit at 45-60 minutes, after which there is a rapid decline in the rate of arrival of ^{14}C. 20%-40% of the ^{14}C fed has reached the fruit by six hours. This time course suggests bulk transfer to phloem by an effective route, probably largely by retrieval of the compound whilst still within the vascular tissue of stem and leaf. Vascular tissue of nodes and minor veins of leaf - situations where transfer cells abound - appears to be especially active in passing solutes from xylem to phloem and the quantitative implications of this are being examined by means of autoradiography and more detailed studies of organ labelling (D. McNeil, C.A. Atkins, J.S. Pate, unpublished).

A second group of amino acids - aspartic acid, glutamic acid, glycine and γ-amino butyric acid - behave somewhat differently. Though normally present in xylem, they are transferred to phloem very ineffectively. Transfer of their ^{14}C to seeds occurs only to the extent of 10-20% in a 6 hour period, and the ^{14}C which does appear in fruit phloem sap is attached mainly to compounds other than the molecule supplied. No sharp peak is observed in time course of labelling of phloem and there is a less marked decline in through-put of tracer during the 3-6 hour period after feeding than in the case of the other group of amino acids mentioned above. These findings suggest that a less direct route to the phloem might be employed, possibly *via* the mesophyll of the leaf. This suggestion is being tested.

The data from the labelling experiments enable a quantitative picture to be obtained of the flow of carbon from xylem amino acids to phloem of the fruit. Viewed alongside the

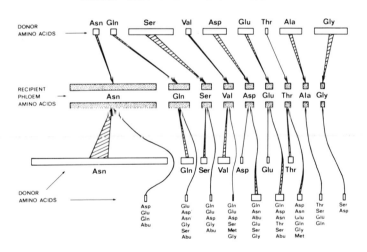

CARBON FLOW FROM PHOTOSYNTHESIS

CARBON FLOW FROM XYLEM

Fig. 3. - Flow of carbon from photosynthesis and from xylem amino acids to the amino acids of phloem serving fruits of Lupinus albus *L. The relative importance of the various pathways to the phloem is indicated by the width of the rectangles representing the donor amino compounds. Rectangles representing amino acids of phloem (central block of figure) are drawn in proportion to their abundance on a carbon basis. Abbreviations for amino compounds are as listed in Biochem. J. 126, 773-780 (1972). (data from Sharkey and Pate 1975).*

complementary flow of photosynthetically-fixed carbon to amino acids of the phloem the picture summarized in Figure 3 is obtained. The middle block of the figure represents the seven commonest amino compounds of the sap, rectangles for each constituent being drawn in proportion to its relative abundance on a carbon basis. The top line of rectangles of the figure shows the extent to which carbon of photosynthetically fixed $^{14}CO_2$ is donated to phloem amino acids. The lower two lines of blocks depict xylem amino acids which pass carbon directly or indirectly to the phloem amino acids. The amino acids are grouped according to the phloem compounds to which they donate carbon, and the relative importance of each cross transfer reaction from xylem is depicted in terms of the horizontal width of the rectangle which that compound occupies in the diagram. The relative intensity of carbon flow along a particular pathway from xylem to phloem is computed by taking into account (a) the relative abundance of the donor xylem compound on a carbon basis, (b) the overall efficiency with which it donates radiocarbon to the fruit, and (c) the relative amounts of radiocarbon of phloem recovered in the phloem amino acid(s) labelled directly or indirectly by the donor compound (see Sharkey and Pate 1975).

The flow diagram shows that asparagine of the phloem derives little carbon from photosynthesis or from the carbon of other xylem amino acids, but comes mainly from supplies of the amide currently arriving in the xylem. This represents a direct transport link from the nodule, where the compound is synthesized to the seed where it is used as a primary source of nitrogen for protein synthesis (see earlier sections). Since approximately half of the amino acid carbon and 60-70% of the nitrogen of phloem is represented by this compound, its role in transport and in nutrition of growing organs of *L. albus* cannot be overstressed. Glutamine, valine, threonine and serine are compounds which also pass largely in unchanged form from xylem to phloem, but the phloem pool of these compounds derives also from other amino acids. Serine, alanine, glycine, aspartic and glutamic acids are amino compounds which derive significant amounts of photosynthetically-fixed carbon.

The dicarboxylic amino acids aspartate and glutamate are exceptional in showing considerably less direct flow of carbon to phloem as unmetabolized amino acid than indirect flow channelled through other xylem compounds (Fig. 3). It is envisaged that these two amino acids might serve effectively as nitrogen sources for the older vegetative parts of the shoot, and thereby provide nitrogen essential for maintenance of photosynthetic tissues serving the fruit. Complications of this nature relating to the partitioning of specific compounds for different purposes within the plant body imply the existence of regulatory devices of much greater complexity and subtlety than had hitherto been suspected.

CONCLUSION

Organs of the legume differ to a surprising degree in the economy which they exhibit in respect of utilization of carbon, nitrogen and water. Once more is known of the budget arrangements of different parts of the plant and of the interactions of these parts in respect of solute and water transport a much clearer picture of plant functioning will be obtained.

ACKNOWLEDGEMENT

The author is greatly indebted to his colleagues for permission to use unpublished data. The instances are cited in the text.

REFERENCES

Atkins, C.A., Pate, J.S. and Sharkey, P.J. (1975). Asparagine in legume seed nutrition. *Plant Physiol. 56*, 807-812.

Bollard, E.G. (1970). The physiology and nutrition of developing fruits. In : The biochemistry of fruits and their products. (ed. A.C. Hulme). Vol. 1. 387-425. Acad. Press, London-New York.

Crookston, R.K., O'Toole, J. and Ozbun, J.L. (1974). Characterization of the bean pod as a photosynthetic organ. *Crop Sci. 14*, 708-712.

Flinn, A.M. and Pate, J.S. (1970). A quantitative study of carbon transfer from pod and subtending leaf to the ripening seeds of the field pea *(Pisum arvense* L.). *J. Exp. Bot. 21*, 71-82.

Gunning, B.E.S., Pate, J.S., Minchin, F.R. and Marks, I. (1974). Quantitative aspects of transfer cell structure in relation to vein loading in leaves and solute transport in legume nodules. *Symp. Soc. Exp. Biol. 28*, 87-124.

Lewis, O.A.M. and Pate, J.S. (1973). The significance of transpirationally derived nitrogen in protein synthesis in fruiting plants of pea *(Pisum sativum* L.). *J. Exp. Bot. 24*, 596-606.

Minchin, F.R. (1973). Physiological functioning of the plant-nodule symbiotic system of garden pea *(Pisum sativum* L. cv. Meteor). Doctoral Thesis, Queen's University of Belfast, N. Ireland.

Minchin, F.R. and Pate, J.S. (1973). The carbon balance of a legume and the functional economy of its root nodules. *J. Exp. Bot. 24*, 259-271.

Oghoghorie, C.G.O. and Pate, J.S. (1972). Exploration of the nitrogen transport system of a nodulated legume using ^{15}N. *Planta (Berl.) 104*, 35-49.

Pate, J.S. (1962). Root exudation studies on the exchange of ^{14}C labelled organic substances between the root and shoot of the nodulated legume. *Plant and Soil 17*, 333-356.

Pate, J.S. (1973). Uptake, assimilation and transport of nitrogen compounds by plants. *Soil Biol. and Biochem. 5*, 109-119.

Pate, J.S. (1974). Pea. In : Crop Physiology, some case histories. (ed. L.T. Evans) Cambridge University Press, Cambridge, England. 191-224.

Pate, J.S. and Flinn, A.M. (1973). Carbon and nitrogen transfer from vegetative organs to ripening seeds of field pea *(Pisum sativum* L.). *J. Exp. Bot. 24*, 1090-1099.

Pate, J.S., Gunning, B.E.S. and Briarty, L.G. (1969). Ultrastructure and functioning of the transport system of the leguminous root nodule. *Planta (Berl.). 85*, 11-34.

Pate, J.S., Sharkey, P.J. and Lewis, O.A.M. (1974). Phloem bleeding from legume fruits - a technique for study of fruit nutrition. *Planta (Berl.). 120*, 229-243.

Pate, J.S., Sharkey, P.J. and Lewis, O.A.M. (1975). Xylem to phloem transfer of solutes in fruiting shoots of legumes, studied by a phloem bleeding technique. *Planta (Berl.). 122*, 11-26.

Pate, J.S., Walker, J. and Wallace, W. (1965). Nitrogen containing compounds in the shoot system of *Pisum arvense* L. The significance of amino acids and amides released from nodulated roots. *Ann. Bot. 29*, 475-493.

Sharkey, P.J. and Pate, J.S. (1975). Selectivity in xylem to phloem transfer of amino acids in fruiting shoots of white lupin *(Lupinus albus* 1.). *Planta (Berl.). 127*, 251-262.

Zimmermann, M.H. (1969). Translocation of nutrients. In : The physiology of plant growth and development. (ed. M.B. Wilkins). McGraw Hill, Maidenhead, England.

CHAPTER 39

Remobilization of Nutrients and its Significance in Plant Nutrition

J.F. Loneragan[1], K. Snowball[2], and A.D. Robson[2]

[1]*School of Environmental and Life Sciences, Murdoch University, Murdoch, Western Australia 6153.*
[2]*Department of Soil Science and Plant Nutrition, Institute of Agriculture, University of Western Australia, Nedlands, Western Australia 6009.*

INTRODUCTION

The ability of plants to remobilise nutrients from leaves and transport them in the phloem to other organs is an extremely important property in plant nutrition. It is important in the relationship between nutrient supply and the development of nutrient deficiency and in the relationship between nutrient concentrations in plants and plant growth.

These relationships are reasonably well understood for those nutrients which move freely from leaves and for those which do not move at all. They are poorly understood for those nutrients whose behaviour is intermediate between these two extremes.

MOBILE AND IMMOBILE NUTRIENTS

A number of nutrients such as N, P, and K move readily to other plant organs from leaves in which they have accumulated. Indeed, their movement to young growing organs may continue to the detriment of the old leaves in which their concentrations may fall to deficient levels (e.g. Bouma 1975; Smith 1975).

Several consequences flow from the rapid mobility of these nutrients. Plants which have accumulated excess concentrations in their leaves may continue to grow unchecked even when they have no external supply of these nutrients. Moreover, deficiencies do not develop until the total amount of nutrient in the plant as a whole becomes inadequate. For these mobile nutrients the concentration of nutrient in the plant as a whole gives a reasonable indication of the nutrient status of the plant. The concentration of nutrient in old leaves also gives a good indication of nutrient status and an early indication of deficiency. By contrast, the concentration of a mobile nutrient in young growing organs generally gives a poor guide to nutrient status since it remains high even when the plant is deficient (Ulrich 1952). Once the young leaf is fully expanded, its concentration of nutrient may provide a useful guide to nutrient status of the plant as it did, for example, in tropical grasses in which the concentration of P in the youngest expanded leaf was as good an indicator of P status as that in the whole tops although still not as good as that in older leaves (Smith 1975).

These characteristics of nutrient deficiency for mobile nutrients stand in direct contrast with those for immobile nutrients such as Ca. Once deposited in plant leaves, Ca becomes virtually immobile. As a result, plant organs can only grow if they receive a continuous supply of Ca from the external medium or from the transpiration stream (Haynes and Robbins 1948; Harris 1949; Wiersum 1966). The development of Ca-deficiency is thus largely independent of the total amount of Ca in the plant since it develops as soon as the external supply of nutrient becomes inadequate regardless of how much excess nutrient may be stored in older leaves. For example, plants transferred from high-Ca to low-Ca solutions developed severe symptoms of Ca-deficiency even though they contained up to five times more Ca than healthy plants given a low but continual supply of Ca: the oldest leaves of the deficient plants retained an excess of Ca in their leaves while apices and young leaves on the same plant developed symptoms of severe Ca-deficiency (Loneragan and Snowball 1969).

Similar observations have been made for B (Brandenburg 1939), which is generally considered to be immobile in phloem. However, some evidence suggests that B may move from leaves to developing fruit in some species (Eaton 1944; Campbell et al. 1975). Clearly, the contrasting behaviour of mobile and immobile nutrients must be taken into account when devising procedures for deficiency diagnosis. Concentrations of immobile nutrients in old leaves and whole plants, unlike those of mobile nutrients, may be quite misleading in assessing the nutrient status of plants. On the other hand, their concentrations in young tissues give a good indication of nutrient status.

NUTRIENTS OF INTERMEDIATE OR VARIABLE MOBILITY

Studies with radioactively labelled nutrients applied to leaves led Bukovac and Wittwer (1957) to propose that some nutrients (Fe, Mn, Zn, Cu, Mo) have a mobility from leaves intermediate between the freely mobile and the highly immobile nutrients. Others (e.g. Epstein 1971) prefer to view the behaviour of these nutrients as indicating that they are mobile in the phloem but that the degree of mobility varies with plant species. We will review evidence that, even within a single species, the degree of mobility of some nutrients may be highly variable depending upon environmental conditions and upon the stage of plant growth: under some conditions they behave as if highly mobile and under other conditions they behave as if immobile.

In the case of three nutrients for which data are available, mobility varies strongly with the adequacy of supply of the nutrient itself: mobility is highest at luxury concentrations and lowest at deficient. Such behaviour may lead to some peculiar anomalies in the relationships among nutrient supply, nutrient concentration in plants, and yield. In addition, experiments on retranslocation in plants given a luxury supply of nutrient may be quite misleading if extrapolated to interpret the behaviour of plants during the development of deficiency of the same nutrient.

Zinc. - When applied with radioactive tracer to old leaves, a small proportion of Zn is rapidly mobilised to meristematic regions of the plant (Bukovac and Wittwer 1957). This evidence has caused some authors to regard Zn as a highly mobile nutrient (Crafts and Crisp 1971). Indeed, when given luxury supplies of Zn, several plant species have been shown to mobilise appreciable quantities of Zn from old leaves to developing inflorescence and grain (Wood and Sibly 1950; Williams and Moore 1952; Riceman and Jones 1958). But under conditions of Zn-deficiency the same species mobilise little if any Zn from old leaves even when they are senescing from Zn-deficiency. At the same time, young leaves of Zn-deficient plants retain high concentrations of Zn (Rosell and Ulrich 1964; Shedley and Robson, private communication).

Sulphur. - Biddulph (1956) and Bukovac and Wittwer (1957) regarded S as a highly mobile element since ^{35}S applied as sulphate to bean leaves moved rapidly throughout the plants. Indeed, much of the S present as sulphate in old leaves moved readily to new leaves when cotton plants were transferred from plus-S to minus-S solutions. By contrast, old leaves lost none of their protein-S or soluble organic-S during 35 days in the minus-S solutions. In this experiment, the old leaves remained dark green while leaves which developed after transfer became highly chlorotic (Ergle 1954).

Early chlorosis of young leaves has also been observed during the development of many other plant species (see Bouma 1975). These symptoms are characteristic of an immobile element and indicate, as Bouma (1975) has suggested, that analysis of young leaves should give a better indication of S-status in plants than analysis of older leaves or whole plant tops.

However, other factors may complicate the development of S-deficiency in plants. Nitrogen status appears particularly important. In the examples already discussed ample N was present. Where N is low or where plants are relying on symbiotic activity for their N supply, symptoms of S-deficiency may be indistinguishable from N-deficiency, developing most strongly in older leaves (Anderson and Spencer 1950; Eaton 1966). Apparently, under these conditions, protein hydrolysis is accelerated in older leaves leading to greater mobility of organic S from them. With such complications, it is not surprising that experiments on S-mobility sometimes produce conflicting results from various workers (Eaton 1966; Bouma 1975).

Clarification of the conditions affecting the mobility of S in the plant would assist in the development of better procedures for diagnosis of S-deficiency.

Copper. - Like Zn and S, Cu applied to old leaves moves rapidly to other parts of the plant (Bukovac and Wittwer 1957). In experiments which will be reported in detail elsewhere, we have also found that like Zn and S the extent to which Cu moves in wheat plants depends strongly on the level of Cu supply. Leaves of plants given a luxury supply of Cu lost more than 70% of their Cu during grain development. By contrast, leaves of Cu-deficient plants lost less than 20%.

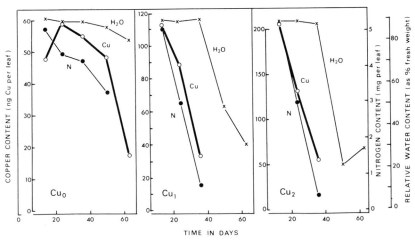

Fig. 1. - The change with time in contents of Cu, N and H₂O in the oldest leaf of wheat plants grown on a Cu-deficient soil to which nil (Cu₀), marginal (Cu₁), and luxury (Cu₂) supplies of Cu had been added (Lonergan, Snowball and Robson - unpublished).

Moreover, the rate of movement of Cu from old leaves of Cu-deficient plants was much slower than from leaves of Cu-sufficient plants (Fig. 1). Thus Cu had a low mobility in Cu-deficient plants and this may explain the characteristic development of Cu-deficiency symptoms in young tissues.

The reasons for the prolonged retention of Cu by the leaves of Cu-deficient plants are not understood. However, preliminary results suggest that the retention of Cu might be related to retention of N which was also prolonged in Cu-deficient relative to Cu-sufficient plants. In all treatments, movement of Cu from old leaves paralleled movement of N (Fig. 1). Perhaps protein hydrolysis is a prerequisite to the release of Cu in forms which can be incorporated into the phloem stream.

The low mobility of Cu in Cu-deficient plants combined with its high mobility in plants given luxury Cu also produced some anomalous relationships between Cu-concentrations in old leaves and yield. Thus 50 days after sowing, the concentration of Cu in the oldest leaf was higher in wheat plants suffering from such acute Cu-deficiency that they produced no grain at all than in plants given a marginal or a luxury supply of Cu (Fig. 2). Fig. 2 also shows how an understanding

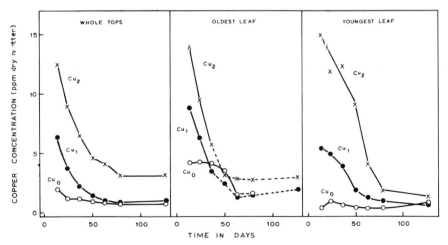

Fig. 2. - Effect of supply on the concentration of Cu in whole tops and in oldest and youngest leaves of wheat plants grown on a sand from Lancelin, Western Australia (Loneragan, Snowball and Robson - unpublished).

of the mobility of Cu in Cu-deficient plants can improve diagnosis for Cu-deficiency. Clearly the youngest leaf is more sensitive for diagnosing Cu-deficiency than either the oldest leaf or the whole plant.

CONCLUSIONS

Data reviewed in this paper show the importance to plant nutrition of the mobility of nutrients from old leaves. They also show that for many nutrients mobility is more complex than generally believed, varying considerably with experimental conditions and particularly with the level of supply of the nutrient under study. Mobility also varies greatly with stage of growth and with plant species. Clarification of the factors influencing nutrient mobility should assist greatly in improving our understanding of nutrient deficiencies in plants and in devising improved methods of defiency diagnosis.

REFERENCES

Anderson, A.J. and Spencer, D. (1950). - Sulphur in nitrogen metabolism of legumes and non-legumes. *Aust. J. Sci. Res. Ser. B.3*, 431-499.

Biddulph, S.F. (1956). - Visual indications of S^{35} and P^{32} translocation in the phloem. *Amer. J. Bot. 43*, 143-148.

Bouma, D. (1975). - The uptake and translocation of sulphur in plants. *In* "Sulphur in Australasian Agriculture" ed. by K.D. McLachlan, Sydney University Press, Sydney, pp. 79-86.

Brandenburg, E. (1939). - Über die Grundlagen der Boranwendung in der Landwirtschaft. *Phytopathol. Z. 12*, 1-112.

Bukovac, M.J. and Wittwer, S.H. (1957). - Absorption and mobility of foliar applied nutrients. *Plant Physiol. 32*, 428-435.

Campbell, L.C., Miller, M.H. and Loneragan, J.F. (1975). - Translocation of Boron to Plant Fruits. *Aust. J. Plant Physiol. 2*, 481-487.

Crafts, A.S. and Crisp, C.E. (1971). - Phloem Transport in Plants. W.H. Freeman and Company, San Francisco.

Eaton, F.M. (1944). - Deficiency, toxicity and accumulation of boron in plants. *J. Agr. Res. 69*, 237-279.

Eaton, F.M. (1966). - Sulfur. *In* "Diagnostic Criteria for Plants and Soils" ed. H.D. Chapman, Univ. Calif. Div. Agric. Sci. Berkeley pp. 444-475.

Epstein, E. (1971). - Mineral Nutrition of Plants: Principles and Perspectives. John Wiley and Sons, Inc., New York.

Ergle, D.R. (1954). - The utilization of storage sulphur by cotton and the effect on growth and chloroplast pigments. *Bot. Gaz. 115*, 225-234.

Harris, H.C. (1949). - The effect on the growth of peanuts of nutrient deficiencies in the root and the pegging zone. *Plant Physiol. 24*, 150-161.

Haynes, J.L. and Robbins, W.R. (1948). - Calcium and boron as essential factors in the root environment. *J. Amer. Soc. Agron, 40,* 795-803.

Loneragan, J.F. and Snowball, K. (1969). - Calcium requirements of plants. *Aust. J. Agric. Res. 20,* 479-490.

Riceman, D.S. and Jones, D.B. (1958). - Distribution of zinc and copper in subterranean clover (*Trifolium subterraneum* L.) grown in culture solutions supplied with graduated amounts of zinc. *Aust. J. Agric. Res. 9,* 73-122.

Rosell, R.A. and Ulrich, A. (1964). - Critical zinc concentrations and leaf minerals of sugar beet leaves. *Soil Sci. 97,* 152-167.

Smith, F.W. (1975). - Tissue testing for assessing the phosphorus status of green panic, buffel grass and setaria. *Aust. J. Exp. Agric. Anim. Husb. 15,* 383-390.

Ulrich, A. (1952). - Physiological bases for assessing the nutritional requirements of plants. *Ann. Rev. Plant Physiol. 3,* 207-228.

Wiersum, L.K. (1966). - Calcium content of fruits and storage tissues in relation to the mode of water supply. *Acta Bot. Neerl. 15,* 406-418.

Williams, C.H. and Moore, C.W.E. (1952). - The effect of stage of growth on the copper, zinc, manganese, and molybdenum contents of Algerian oats grown on thirteen soils. *Aust. J. Agric. Res. 3,* 343-361.

Wood, J.C. and Sibly, P.M. (1950). - The distribution of zinc in oat plants. *Aust. J. Sci. Res. Ser. B.3,* 14-27.

INTEGRATING SYSTEMS - SUMMARY AND DISCUSSION

Adapted from presentations given by the Session Chairmen :
T.F. Neales and L.G. Paleg

There was general agreement throughout the conference that, regardless of the process under examination, a much greater effort and more ingenuity will have to be directed to the exploration of "normal" conditions in intact plants.

A simple source-sink model can be used to cover the supply and partitioning of dry matter between the various parts of the plant. However a reminder is needed that the source-sink concept is in fact another expression of the components of growth analysis such as net assimilation rate, leaf area ratio, relative growth rates and harvest index. In modelling carbon flow the various compartments within the plant need to be considered. It is interesting that most agricultural plants place their assimilates more-or-less directly into economic yield, although in one plant, the artichoke, these initially move into stem storage tissue, and subsequently when the plant flowers are mobilized into the tubers. To define the relationship between source and sink and uncover the control mechanisms relating to the partitioning of assimilates, various experiments have been devised, such as those of Cook and Evans, where relative sink size and distance between source and sink have been manipulated. There is some evidence from this work that the larger sink will have a competitive advantage over a smaller, but otherwise identicial sink, in the utilization of current photosynthate. One thing apparent from this work is the difficulty in getting whole plant systems with a minimum number of uncontrolled, or unknown, variables. Although long term partitioning of dry matter was not considered at this Conference the question still needs to be asked to what extent the distribution of ^{14}C, following $^{14}CO_2$ labelling, mirrors this long term partitioning.

The control of a plant's carbon balance is an important aspect of productivity and there is evidence that the movement of carbon into the plant by photosynthesis can be regulated by a remote system dictated by demand. It is probably not warranted to assume that there is a simple association between product starch and sucrose build up in the leaf, and the rate of photosynthesis, and Kriedeman has discussed the role of abscisic acid and phaseic acid in this context.

There were several areas of agreement in connection with hormones, such as the nonvascular transport of IAA, but it

was equally evident that we still have much to learn about
where they are made, how they travel, where they work, and
their mode of action. The only growth substances that can
strictly be classified as hormones are probably IAA, which is
involved in the growth of coleoptiles and gibberellins, which
move from the embryo to the aleurone in seeds to control the
release of hydrolytic enzymes during germination. However
there is strong circumstantial evidence for others such as
the cytokinins, which are produced in the roots and may control leaf function and the floral stimulus, which moves from
the induced leaves to the shoot apex, but has yet to be characterized chemically.

The study of plant hormones is complicated by the need
to study their function in isolated plant tissues. It is
difficult to study apical dominance without using decapitated
seedlings, or polar movement without using sections. However
as pointed out earlier the interruption of vascular tissue
induces callose formation, surging of cell contents, P-protein plugging and a host of abnormal and largely undesirable
events. Realistically it seems likely that we may never es-
cape from the necessity to use tissue with vascular disrupt-
ions for some kinds of experiments, but the important aspect
is to recognize the limitations of the systems. Perhaps it
might be possible to take another idea from the animal physiologists and use perfusion techniques, or select an experimental system, such as the unpollinated ovary, where events
can be timed reliably, sink strength manipulated by regulating the number of ovules fertilized, or the production of
parthenocarpic fruit by hormones can initiate processes similar to those induced by pollination without any interference
with vascularization.

The physical concepts of water movement in plants are
well established, in contrast to the physical concepts associated with the partitioning of photosynthate. However, J.B.
Passioura has pointed to the fact that many water relations
studies deal largely with short term events and there is a
need to consider a much longer time base when looking at
water use by plants. Strategies for coping with drought usually involve long and medium term morphological changes,
such as adjustments to leaf area index and root:shoot ratio,
and the orientation of leaves. It is therefore important
that the study of the short term events be carried out in the
context of the longer term events and be directed, at least
in part, towards explaining how the longer term events come
about.

One important aspect of nutrition came out during the
papers and discussion period concerned with calcium ions. It

was pointed out that the development of calcium deficiency, in, for example, subterranean clover, could be largely independent of the total amount of calcium in the foliage, with a deficiency developing as soon as the external supply becomes inadequate. The whole situation is perhaps surprising when in many other ways the plant appears to have an excess capacity. It would be of interest to know if the reproductive development of the progenitors of our current varieties of crop plant were also balanced so finely, with respect to calcium, on transpirational supply and demand. This seems an unlikely trait to confer evolutionary advantage and human rather than natural selection may be the basic cause. The basis for calcium immobility is not fully understood, but it may be possible to find conditions which will enhance calcium remobilization and this is one area where a knowledge of the cell to cell transfer processes could be relevant to the distribution of an ion throughout the whole plant.

Finally brief mention should be made of the balance sheet approach made by J.S. Pate for the uptake and utilization of nitrogen. This perhaps best of all typifies the problems of transport, both short and long distance, throughout the plant and the interactions that occur between growth and nutrient uptake. There is a clear need to extend this type of approach to other nutrients and metabolites.

Index

A 6
B 7
C 8
D 9
E 0
F 1
G 2
H 3
I 4
J 5